The Biological Basis of Cancer

The text is designed to be used for an undergraduate course on cancer. It covers everything from the molecular to the clinical aspects of the subject, with a lengthy bibliography designed to assist newcomers with the cancer literature.

An introduction acquaints students with the biological principles of cancer and the human dimensions of the disease by considering genuine cases of cancer in fictionalized letters. Other chapters discuss cancer pathology, metastasis, carcinogenesis, genetics, oncogenes and tumor suppressors, epidemiology, and the biological basis of cancer treatment. A glossary of cancer-related terms is included.

Upper-division undergraduates with a background in freshman biology and chemistry, as well as beginning graduate students, will find this a most useful text.

The Authors

Robert G. McKinnell is a Morse/Alumni Distinguished Professor in the Department of Genetics and Cell Biology at the University of Minnesota and is a past president of the International Society for Differentiation. He was a Royal Society of London Guest Research Fellow at Oxford, and he has authored a number of books including *Cloning of Frogs, Mice and Other Animals* (University of Minnesota Press).

Ralph E. Parchment is professor in the Division of Hematology and Oncology at Wayne State University and was formerly at the Center for Drug Evaluation and Research of the U.S. Food and Drug Administration. His research interests lie in cancer treatment.

Alan O. Perantoni is a principal investigator and group leader for the Developmental Biology Working Group in the Laboratory of Comparative Carcinogenesis, National Cancer Institute. He has maintained a research interest in oncogene and suppressor gene involvement in experimental carcinogenesis. He is currently engaged in studies of genetic alterations in pediatric tumors.

G. Barry Pierce is Centennial Distinguished Research Professor of Pathology at the University of Colorado Health Sciences Center and world renowned for his recognition of the stem cell origin of and differentiation of cancer cells.

The Biological Basis of Cancer

Robert G. McKinnell
Ralph E. Parchment
Alan O. Perantoni
G. Barry Pierce

CAMBRIDGE
UNIVERSITY PRESS

PUBLISHED BY THE PRESS SYNDICATE OF THE UNIVERSITY OF CAMBRIDGE
The Pitt Building, Trumpington Street, Cambridge CB2 1RP, United Kingdom

CAMBRIDGE UNIVERSITY PRESS
The Edinburgh Building, Cambridge CB2 2RU, UK   http: //www.cup.cam.ac.uk
40 West 20th Street, New York, NY 10011-4211, USA   htt: //www.cup.org
10 Stamford Road, Oakleigh, Melbourne 3166, Australia

First published 1998

Printed in the United States of America

Typeset in Adobe Garamond, in QuarkXPress [GH]

*A catalog record for this book is available from
the British Library.*

Library of Congress Cataloging-in-Publication Data
The biological basis of cancer / Robert G. McKinnell . . . [et al.],
     p.   cm.
   Includes bibliographical references and index.
   ISBN 0-521-59298-4 (hb). – ISBN 0-521-59695-5 (pb)
   1.Cancer. 2. Carcinogenesis. 3. Cancer cells. I. McKinnell,
Robert Gilmore.
   [DNLM: 1. Neoplasms. 2. Molecular Biology.   QZ 200 B6147 1998]
   RC261.B6273   1998
   616.99′4 – dc21
DNLM/DLC for Library of Congress   97-42369
                                         CIP

ISBN 0-521-59298-4 hardback
ISBN 0-521-59695-5 paperback

We dedicate this book to our students

Contributors

Professor Robert G. McKinnell
Department of Genetics and Cell Biology
University of Minnesota
Saint Paul, Minnesota 55108-1095

Professor Ralph E. Parchment
Division of Hematology and Oncology
Wayne State University
Detroit, Michigan 48201

Dr. Alan O. Perantoni
Laboratory of Comparative Carcinogenesis
National Cancer Institute
Frederick, Maryland 21702

Professor G. Barry Pierce
Department of Pathology
University of Colorado Health Sciences Center
Denver, Colorado 80262

Contents

Preface

University of Minnesota undergraduate students were in the streets and at the barricades during the early 1970s. The war in Vietnam was of considerable concern to them and, of course, to many others. The streets were hazardous. Police helicopters hovered overhead, tear gas was in the air, and police truncheons came down on the young heads of demonstrators. The then president of the University of Minnesota, Malcolm Moos, entered into the midst of the fray and urged students to return to the classrooms. Moos simultaneously pleaded with professors to develop academic material that would entice demonstrating young people to return to the classroom. The Biology of Cancer was born of those skirmishes. It was reasoned that cancer was of more immediate interest to distressed undergraduates than the nephridia of earthworms or the aquatic larvae of sea urchins. That was indeed the case. A seminar room with eight seats was reserved for the first offering; 80 students showed up. The Biology of Cancer has been offered ever since at the University of Minnesota.

The course was (and still is) designed for undergraduates. The goal of this book is to provide an understandable text that relates directly to cancer. Some would say that if one understands modern cell biology (or molecular biology or genetic biology), then one understands cancer biology. That is not the belief of the authors. Cancer biology is *more* than simply modern cell biology. Certain areas of cancer research are not covered now, and never will be covered, in conventional biology classes. Pathology as it relates to cancer is an example. Who in a cell biology class explains why it is that the spread of cancer so often occurs prior to detection? This is discussed in Chapter 1. What biology class examines in detail contemporary therapy for cancer? The last two chapters are devoted to this subject.

We assume our students have an introduction to modern biology. Accordingly, we jump immediately into fundamental concepts of pathology as they relate to cancer. After all, this is why the students elect to take a course in cancer biology. Metastasis (Chapter 2), the most feared aspect of cancer, is why much of cancer treatment is palliative. Because it is so fundamental to an understanding of cancer, it is presented immediately after pathology. Metastasis occurs as a result of a number of steps, each of which has a counterpart in normal biology. We explain how these otherwise normal activities result in this profoundly malignant attribute of cancer. After the first two chapters define cancer, we next examine the causes of cancer (Chapter 3). Some cancers are hereditary. We discuss those known to be so and consider the nature of genetic change found in some tumors (Chapter 4). What we know about specific genes that regulate cancer cell expression is considered in Chapter 5. DNA-containing viruses may contribute to the cause of some cancers in humans – the oncogenic effect of DNA-containing viruses was discovered first in a frog cancer. The frog cancer and other animal cancers are described in Chapter 6. Cancer probably takes more years of life in the United States than any other disease (the death of a child with leukemia is not equivalent to a 90-year-old man dying of a stroke – in the former, perhaps seven or more decades of life are lost due to cancer; the old man in the latter instance, although dying of a common condition, has already outlived his life expectancy, and thus the number of life-years lost are nil. Nuns are more vulnerable to the ravages of breast cancer than random other women. Our knowledge of who is vulnerable to which cancer and how many people are involved is considered in epidemiology (Chapter 7). The design of drugs and treatment is not an arcane endeavor. The various modalities for cancer therapy are rational and based on biology. The biological basis of drug action is considered in Chapters 8 and 9. The appendix briefly describes some common cancers, and a glossary follows.

It is altogether too easy for a student working in a laboratory and taking courses to forget the reason for cancer research, the people afflicted with this most feared disease. Authentic cancer cases are described in the Introduction in the form of letters with comments concerning those cases. Although the study of cancer biology does not require a consideration of cancer victims, we have found that many students show compassion for these individuals and come to appreciate how cancer biology can enhance their understanding of what occurs in cancer patients. In some cases, this understanding augments their desire to learn more cancer biology.

Skeptics may argue that undergraduates do not have an adequate background to understand cancer properly. Of course they do not. Neither do they have an adequate background to understand ecology, plant physiology, parasitology, and evolution. That is why we have developed an undergraduate biology curriculum and a text to study the biology of cancer. Just as a student begins to comprehend other aspects of biology while taking introductory courses, so too does an understanding of cancer begin with taking a cancer biology course. Even in the late 1990s, courses in cancer biology are relatively rare. This book may encourage offerings of cancer biology in other colleges and universities.

The subject matter of cancer is fascinating to most people including nontraditional students. Attorneys, nurses, doctors, and other professional people occasionally attend our classes. We anticipate that libraries will make this book available to their patrons. Our book about cancer goes far beyond popular accounts, but it is not the turgid reading material found in a medical text. We intend that it be understandable to most educated adults. The text is not designed as a self-help manual to assist in diagnosis or treatment. Questions about one's medical care should be discussed with one's physician. Rather it is designed to provide understanding of the biological and medical basis of contemporary studies in cancer.

Compiling the pages of a text is labor. Our labors have been eased by Debra L. Carlson, Augustana College, Sioux Falls, South Dakota; Julie A. Thompson, Division of Epidemiology, School of Public Health, University of Minnesota; and Kristine S. Klos, Department of Genetics and Cell Biology, University of Minnesota. They read the text with a critical eye, surgically ablated sentences that were difficult to understand, rearranged prose to make better sense, caught errors in English, and applied their healing skills to phrases in need of treatment. We are in their debt. The authors are exceptionally appreciative of the cooperation and continued efforts of Françoise Bartlett and Christina Viera to enhance this book during its production.

Finally, we thank students who have and will take a course in the biology of cancer. In a conversation with a distinguished Scottish scientist, one of us commented that he wished he had more time for the laboratory. Perhaps, he thought, it would be better not to teach at all but rather devote full time to laboratory pursuits. The Scot, a fellow of the Royal Society of London, responded, "Appreciate your students and give thanks that you must teach – there will be days upon days when research does not go well – that is the way of research. But if you teach, you will leave the lab for a lecture, and your flagging spirits will be rejuvenated by the enthusiasm and youthful concerns of your students. Then, with vigor renewed, return to the lab." The authors of this book say, "Thank you, students."

Letters illustrating clinical aspects of cancer

G. Barry Pierce

*Whatever seemed ruthless, implacable, predatory could be analogized to cancer . . .
cancer was never viewed other than as a scourge; it was, metaphorically, the* barbarian
within [emphasis added].

Susan Sontag, Illness as Metaphor (1978)

The "barbarian within" occurs in the brain, throat, lungs, breasts, stomach and
colon, pancreas, liver, bones, muscles, joints, and many other sites. Indeed, it can
occur wherever cells divide, and as it develops, it ultimately deranges the organ,
causing a multiplicity of diseases relating to the organ in which it originated. This
characteristic has led to the erroneous conclusion that the barbarian is a multiplicity of diseases lacking commonalities. But commonalities there are, and they form
the basis for research into the cause, control, and therapy of cancer.

Some may read this book because they or a loved one has cancer, students of
science may read it because they are about to make career decisions, and students
of cancer may read it because of a desire to be acquainted with aspects of cancer
other than that with which they work. Although most students of cancer have
never seen a malignancy in a human, they are nevertheless expert in one or more
of the many fascinating and important aspects of the disease. They contribute to
the understanding of DNA and its replication, control of gene expression, receptors and growth factors, developmental aspects, immunology, prevention, and
treatment of cancer and a myriad of other important parts of the cancer problem.
We are all impressed by the recent compounding of knowledge, but even more so
by our lack of knowledge and understanding of how important facts generated in
one aspect of investigation bear on another. Students, unaware of the clinical
aspects of the disease, lack the information to make the correlations and see the
important problems that result. In addition, they may be driven in their endeavors

only by the intellectualism of their part of the problem. This book, and this introduction especially, are designed to provide insight and understanding into the
human aspects of this disease.

It is not our purpose here to present all-encompassing details of the clinical
behavior of patients with cancer that are necessary for diagnosis, medical practice,
and patient management. Neither is it our objective to detail clinical aspects that
could be misunderstood or inappropriately applied. Rather, we present a flavor of
the clinical aspects. The clinical problems are illustrated, and areas of ignorance
and points of attack identified.

Cancer is a dread disease, and the impact of death and suffering are often lost
when its research aspects are considered in isolation. Many believe research should
be driven only by the intellectualism of the problem and not by practical concerns. But recall that the control of diabetes with insulin was driven by an emo-
tionally charged clinical problem. It is often satisfying to be driven in research by
not only an intellectual charge but also a practical goal, providing perspective is
maintained.

Insight into the nature of the "barbarian within" is provided here in a series of
letters written by patients, their relatives, clinicians, and friends. The events
described in these letters are real, but of course the names of individuals and particular circumstances have been changed. They illustrate the diversity of clinical
conditions caused by cancer, the commonalities of the conditions, and the points
of research attack.

Colon cancer: Appendix, section A3; toxic effects of chemotherapy, Chapters 1 and 8

Dear Uncle Harvey,

I am writing to tell you about Dad. I know you would have come to see him if
you could have. I will bring you up to date about what has happened. You remember our feeling of fear and panic when his bowels became obstructed two and half
years ago. It did not seem reasonable—a man only 55 years old, a tough old
rancher. Yet, they said he would die within a week if the obstruction of his colon
was not "relieved" surgically. They removed part of his colon and anus, because the
tumor had spread into the anus. Then they did a colostomy on his belly (it is like
an artificial anus). The surgeon said the tumor had not spread to the liver or
lymph nodes, and Dad felt pretty good, even though he hated that new opening
with a passion.

About a year after surgery, he began to lose weight, but we didn't pay too much
attention to it, because he could stand to lose a few pounds anyway. Then, he
began to feel run-down and had some pain in his stomach, most on the right side
up high. They did some tests and the doctor told us the tumor had spread after all,
and the liver was enlarged with it. That was causing the pain. They gave him

chemotherapy and he was pretty miserable. Actually, it made him mad because what hair he had left all fell out. Why would a man's hair fall out? So then he wore his old baseball cap from high school. Anyway, he got much better, but then when they treated him with chemotherapy again, it didn't seem to do anything at all. They said his tumor was resistant. He was thin and sickly and had a good deal of pain. We couldn't get him to eat to get his strength up. The doctor was pretty good and gave him pain relievers. After that, it was just one thing after another – he got pneumonia that was cured with antibiotics, then Mom heard about a man, somewhere in Georgia, who had a new cancer cure. It was a secret cure, because he was afraid the American Medical Association would take it away from him. She wanted Dad to try the cure.

I went to the American Cancer Society and they said this man was a quack and all quacks have a secret. I explained this to Mom, but she said it was worth it, and besides, she had saved up $3,750 and they would use that. To make a long story short, the quack did not help, but Mom doesn't regret doing it even though she lost all her money.

Dad wanted to see all of my brothers and sisters. We talked and joked, but he was so frail and weak he could not spend much time with us. Would you believe he had lost 75 pounds? It broke our hearts. This was the guy who worked all day, partied all night, and wrestled the five of us to a standstill the next day.

Then he got really sick with a high fever and pneumonia, and they took him to the hospital. He died three days later. He had some kind of infection of the blood they couldn't treat. It was called septicemia. He had been in great pain, but there wasn't much we could do.

Mom is fine, but she feels guilty about Dad's death. I don't see what she could have done differently. I feel bad about some of his so-called "friends" who never came to see him, and I am angry about what happened to him. To see a strong, husky, independent man become bedridden and frail and have to be looked after hand and foot just didn't seem right.

Yours sincerely,

George

Author's note

This letter reveals much of what happens with people who have cancer of the colon. Changes in bowel habits, passage of blood in the stools, or some nonspecific and vague pains in the abdomen may occur. It is not uncommon for the individual to present ("present" is medical jargon for appearance before a clinician) with an obstruction, and it depends on the location whether the tumor can be removed and the bowel anastomosed (reconnected) end to end or whether a colostomy has to be made. As in this case, some patients find it extremely difficult to adapt to a colostomy, but fortunately organizations that can help patients deal with this or other cancer-related problems are available (in the United States, call the American

Cancer Society at 1 800 ACS 2345 or the National Cancer Institute's Cancer Information Service at 1 800 4 CANCER). At the time of surgery the tumor had spread, but it took time for the metastases to grow and become clinically evident (Chapter 2). Metastasis, wasting, and the death of the patient due to infection (in this case septicemia) are commonalities of the malignant phenotype. The loss of hair and diarrhea result from the killing of fast-growing normal cells of hair follicles and intestinal epithelium by the cytotoxic chemicals used in attempts to kill tumor cells. These chemicals lack specificity for cancer cells (Chapter 8).

Note the strength of will and determination of human beings and their loyalties and lack of loyalties. Friends are often embarrassed and afraid to visit people with cancer, and as a result the patients are lonely.

Breast cancer: Appendix, section A.1; genetics, Chapter 4; epidemiology, Chapter 7

Dear Ed,

Well, it is the big C again. We thought we had it licked, but it is back. I am writing to you for some of your free legal advice.

You will remember that Joyce had that "thickening" in the breast. Her doctor told her it was nothing and not to worry about it. She went for her annual checkup 10 months later and the doctor again said not to worry about it, that it was OK. Then, it began to enlarge, so she went to the doctor again, but he was away. His partner said she had a tumor and at her age it was probably cancer and should be operated on immediately.

When they removed the tumor, they also removed the fat from her armpit because this cancer usually spreads to the lymph nodes in the armpit first. They found it had spread to 12 of 15 lymph nodes. The doctors said this was a very bad sign. She had x-ray treatments and chemo. She didn't like the chemo treatments and was glad to be done with them. They nauseated her and gave her diarrhea.

She was well for 14 months, and then about 6 months ago she felt lousy and began to lose weight. We thought it was flu or something like that. Then she stepped off our back stoop and broke her hip. It turned out the cancer had spread and was quietly growing in her bones, of all places. It ate away the top of her leg bone, causing the fracture. Now she has pain in her back because the tumor has spread to her spine. She is losing weight and clearly cannot live much longer. I hate to say it, but this will be a blessing because of the suffering.

I am angry. I don't think that doctor knew what he was doing when he told her not to worry, that the lump was nothing. He let it grow and spread. I know that suing him won't help Joyce, but I don't think a man like that should be practicing medicine. If I sue him, I can say that publicly and maybe the warning will help other people. What should I do? Joyce doesn't want me to sue him, but if I don't I'll probably beat him up. I am desperate and angry, and I don't know what I am going to do. When you have been married to somebody for 35 years, it is difficult to reconcile what has happened. I will appreciate your advice.

Yours sincerely,
Ted

Author's note

As you will learn in this book, early tumors tend to behave less aggressively than later ones. The transition from less to more malignant is termed progression (see Chapter 1). Usually the earlier the tumor is diagnosed and treated, the better the prognosis (outlook) for the patient because the tumor has not reached the stage of invasion and metastasis (see other manifestations of progression, Chapter 2). Delay in therapy can result from patient fears or, as in this case, misjudgment or ignorance on the part of the physician. By the time definitive treatment was started in this patient, spread of the cancer to so many regional lymph nodes and distant organs had occurred that the patient's prognosis was hopeless. Note that some tumors often metastasize to particular sites. In this case, breast cancer has a predilection to spread to bone. Note the commonalities: growth leading to a rapidly enlarging mass, spread of cells (metastasis), and weight loss (cachexia).

Acute leukemia: Appendix, section A.12

Dear Aunt Molly,

I am sure you have heard that Jamie, our 3-year-old son, was diagnosed with leukemia. I didn't write at the time the diagnosis was made because we were overwhelmed by what happened subsequently. He was such a lovely child – so pleasant, so full of energy, and we miss him so. Jamie developed an acute fever, and overnight became seriously ill, with bleeding from his gums and nose. He also had large bruises, but we knew he had not been hurt. The doctor suspected leukemia and said the illness resulted from "displacement of normal white blood cells from the bone marrow by the malignant leukemia cells." As a result Jamie could not fight infection. The leukemia cells also displaced the cells that stop bleeding. The doctors said it was very serious, but most children with acute lymphatic leukemia are saved with new types of treatment. They took a sample of blood and bone marrow, but although he did have acute lymphatic leukemia, he was not one of the lucky ones. He was a little better after chemotherapy and x-ray, and then he developed septicemia, was bleeding, and was so lethargic. Last night Jamie died. It is hard to believe. He had been sick only 9 weeks but it seemed like ages. I hate to say it, but I feel relieved. He suffered so and he was so good. Now he doesn't have to suffer anymore. Please come and stay with me – I need your help.

Your loving niece,
Mary

Author's note

This letter speaks for itself. Malignant disease has a bimodal incidence and occurs most commonly in the young and the elderly. Acute lymphatic leukemia is a disease of the young, and ordinarily its treatment is considered one of the triumphs of modern chemotherapy. The disease, which before modern treatment caused death in a matter of weeks, now allows cure in 50% or more of patients. However, in this case, the malignancy failed to respond to treatment (Chapter 8). There is no way of knowing which patients will respond to chemotherapy. Note commonalities of rapid growth of malignant cells that invade and displace normal tissue and interfere with its function, in this case replacement of normal marrow with malignant cells. This loss of white and red blood cells and platelet-forming cells makes the patient prone to opportunistic infections, anemia, and bleeding, respectively.

Lung cancer: Appendix, section A7; epidemiology, Chapter 7

Dear Cousin Janet,

We have bad news to report. You will recall that my dad had that nasty chronic cough. Well, it got worse and he developed pneumonia. The doctor took x-rays and treated him with one of the new antibiotics. He got better, but then six weeks later he developed pneumonia again in exactly the same spot. It made the doctor suspect lung cancer, because Dad was 55 and had smoked two packs a day for the last 30 years. They made him cough up sputum and found malignant cells in it. These were "small cell lung cancer cells," which are the worst kind. They are so malignant that the doctors refused to operate because by the time these tumors are discovered, they have already spread all over. They gave Dad radiation treatments and chemotherapy, but they did not help much. He became disoriented and then he had some convulsions. The doctor said the tumor had spread to the brain. They suggested irradiating the brain, but my brother and I decided it was no use. We were not surprised that the doctors and nurses agreed with us. Dad died in his sleep just a month ago today. Mom is doing really quite well.

Sincerely,
Dorothy

Author's note

This individual was in the cancer age group with a smoking history (see tobacco and cancer, Chapter 7) that placed him at great risk for developing cancer of the lung. This case also illustrates the propensity of lung cancer to metastasize to the brain (Chapter 2). This occurs so commonly that if a person presents with signs and symptoms of a brain tumor, the clinician must always rule out the possibility of metastasis to the brain from a primary lung cancer.

Note an additional commonality: the cancer diagnosed as a small cell lung cancer lacks tissuelike, or "epithelial" or "glandular" differentiation as viewed in the light microscope. Lack of differentiation correlates with rapid growth and a poor prognosis.

Kidney cancer

Dear Fred,

I need to talk to someone, and since you are my oldest friend and a urologist to boot, you have been selected. I began to pass blood in my urine so I went to the doctor. He told me that because of my age, he had to rule out bladder cancer. I was cystoscoped and the bladder was OK, but then he injected a dye in my blood and took x-rays. The dye was passed from the kidney into the urine, and there in the kidney was a tumor filling this pelvis-thing (the part that drains the urine out of the kidney). I had to wait a week because there were other tests that had to be done. This was the worst week of my life because I had these horrible guilt feelings – l had never bought as much insurance as I should have because I wanted a nice home and things for the family. Now I am sure I have cancer and I am going to die. My wife is going to have trouble managing, and the kids won't be able to get an education. I really need some time, and I am angry besides. I didn't booze or smoke or play around, why should this happen to me? My doctor seems OK and I like him, but with something this important, I am not sure if I should let him do the surgery or go elsewhere. Please call and let me know.

Sincerely,

Charlie

Dear Charlie,

I was distressed to read your recent letter. I will be happy to come and help or do whatever is necessary. First, let me give you some advice. It sounds as though you may have a cancer of the pelvis of the kidney. This doesn't mean you will die tomorrow or the next day or even at all. But it does mean that unless this tumor is completely removed, it will in fact kill you someday. It may not, because about half of these tumors in the kidney pelvis, even though malignant, behave in a very benign manner, so possibly simple surgical removal of the kidney will cure you. In the event the tumor spreads, it will probably spread to the bladder. (It is a good sign that the bladder is not affected at this time.) That doesn't mean to say you are out of the woods, but it says the tumor cells have not become aggressive enough to spread through the urine to the bladder. If this spread occurs, very often it can be handled using the same type of procedure they used when you were cystoscoped. They can take out small tumors through a resectoscope inserted into the bladder through the penis. If the spread to the bladder recurs repeatedly and becomes extensive, you may lose your bladder, but this usually takes years. You'll have lots of living to do in the meantime, and it doesn't happen that often anyway. Finally, a few of these tumors spread via the lymphatics and bloodstream and go to distant organs. That is the worst possibility because the tumors do not respond to chemotherapy or irradiation and nothing can be done about them. I have my fingers crossed for you.

Yours sincerely,
Fred

Dear Charlie,

This get well card is sent with a great deal of relief. Gloria told me on the phone last night that your tumor was a grade 1, stage 1 noninvasive tumor (Chapter 1), and we are all optimistic you have been cured by surgery. Since you and I have always leveled with each other, I feel I should tell you that there is still a chance the tumor may have spread, but it is remote and with any kind of luck you're home free. I'm going to be in the Rockies this September. Let's go trout fishing!

Fred

Author's note

Fred's second letter to Charlie says it all. The degree of differentiation of tumors is graded 1 to 4 with 1 the most differentiated and least aggressive. Grade of tumor plus stage (in this case noninvasive) sets the prognosis as excellent. The combination of stage and grade together gives a more accurate prognosis than either alone (tumor grades and stages are discussed in Chapter 1).

Squamous cell cancer: Appendix, section A.4

Dear Eileen,

I am writing to tell you about Grandpa, who is having a terrible time. Ever since Gram died four years ago he has not been quite right, if you know what I mean. He had this sore on the inside of his gum and refused to have it looked after. He claimed it was caused by his dentures. So he whittled and sanded his dentures with his knife, but the sore did not go away. We coaxed him to go to the doctor, but he refused. After about a year of tinkering with his dentures, they finally broke. The dentist told him he had cancer of the gum – a squamous cell cancer. It's strange, but that's a skin cancer. Well, he'd figured out all along he probably had cancer, but he didn't want to know. He claimed Gram was OK until they told her she had cancer, and he was damned if he would let them do it to him.

The cancer has spread to lymph glands in his neck and to the floor of his mouth and jaw. They told Grandpa they were going to do radical surgery to take out part of his tongue, jawbone on the right side, and all of the lymph glands in his neck. "The hell you say," said Grandpa. "You can't cure me, and you aren't going to cut on me."

The doctors have really been working on me to get him to submit to surgery. They say, in the first place, if he had gone and had this looked after when the sore first developed, he could have been cured with very little effort. Now they say if they do radical surgery, they can spare him a lot of discomfort and even if they take out his jaw, they can rebuild another one.

He smells pretty bad because of the infection in the tumor. Some of it is decaying, and it would appear we are in for a bad time. Grandpa just looks me in the eye and says the medical profession is after his money and he would rather give it to his grandkids than to those lousy doctors. He says he knows it is going to get bad for him, and when it gets too bad he will just cash in his chips. I don't know what to do with him, but I think it might be a good idea if you came to visit while he is still able to do the things he enjoys.

Yours sincerely,

Nelly

Author's note

In this situation the patient delayed seeing a physician because he was afraid he might have cancer. Thus he will die of a disease that, under ordinary circumstances, is curable with modern therapy. Note his strength of will, characteristic of many elderly people who have been through the school of hard knocks. It would be interesting to know the events at the terminus. Did he lose his resolve and accept therapy? Many people say they will refuse therapy, but when they are faced with dying, they often opt for treatment even though they know it offers little chance for cure. Note the commonalities: growth, invasion and destruction of tissues, and distant metastasis (Chapter 2). Weight loss will follow, then infection and death (Chapter 1).

Testicular cancer: Appendix, section A5

Dear Dennis,

This letter is to thank you for your advice and support during our trials and tribulations with my son's testicular teratocarcinoma. Do you remember in medical school how the professors teased us when they lectured about testicular cancer and how they made light of such a grim disease? Never did I think I would come face to face with the realities in my own family. Bill is 26 years old. We were at the lake and he told me he had a large testicle, but it didn't hurt, so he had not done anything about it. It proved to be a teratocarcinoma that had metastasized. We did a retroperitoneal lymph node dissection and then treated him intensively with chemotherapy. He has been disease-free for two years, which means he is almost surely cured.

He had some side effects of the chemotherapy: he lost his hair – about which he was embarrassed – and although he can have sexual relations, he has no ejaculate. Luckily, he and Marge had the baby before he got sick.

I look back over this nightmare and realize in a sense how fortunate we are. He was cured with modern chemotherapy, which did not exist ten years ago. I guess temporarily losing your hair and having "dry" ejaculations is not too heavy a price to pay for life.

Your help and counsel were much appreciated,

Sam

Author's note

This individual was in the typical age group for a germ cell tumor of the testis. He was probably embarrassed to seek medical advice, which caused delay and the development of a large tumor. Because these tumors have a tendency to metastasize (Chapter 2) via the lymphatics along the aorta, the surgeons removed all of these lymph nodes and any tumor that might have spread into them. Then the patient was given massive doses of chemotherapy with the attendant side effects, which occur because the toxic chemicals lack specificity. Even so, such cures are among the marvels of modern chemotherapy (Chapter 9), and over 70% of such patients are cured. Unfortunately, it is not known why chemotherapy cures one kind of cancer but not another. The cause(s) of the differential response of tumors to chemotherapy has not been a high priority of the medical establishment.

Stomach cancer

Dear Mom,

I am enjoying my first year of residency very much, and after all the stainless steel and scientific medicine we learned it is fun to see some of the old-timers practice. An old Scot came into clinic the other day not feeling very well with some vague upper abdominal pains that sounded like dyspepsia. I was going to give him some medicine to tide him over the weekend when the attending physician came in. The doctor is an equally dour old Scot, and he learned that after eating porridge every morning for fifty years, this patient no longer had a taste for oatmeal. He winked at me and said we better work him up. Well, we worked him up exhaustively and found he had a tiny adenocarcinoma of his stomach. People with this disease may lose their taste for a favorite food, and this old doctor knew about it. Well, we operated on him and the tumor had not spread, and we think he is going to be one of those lucky people who beats stomach cancer. When you consider that only about 10% of patients with stomach cancer survive, you see how lucky he really is. Old Doc MacAllister just winks and says it's all in a day's work. He got the tip-off about the oatmeal because he spends a lot of time with his patients, talks to them, teases them. I don't see how he can make a living, but he sure is a good doc.

With love,
Jennifer

Author's note

This is a clear example of good luck in life. The patient came under the care of an old-time physician who practiced the art of medicine and spent adequate amounts of time with his patients to know and understand them. The physician picked up an apparently trivial point in the clinical history which raised the possibility that the patient had stomach cancer, a disease with a cure rate of less than 10%. He followed through with a vigorous and thorough workup and discovered a small curable cancer of the stomach. It had not reached a stage where it had spread. This is clinical medicine at its best.

Melanoma: Appendix, section A.8; epidemiology, Chapter 7

Dear Jill,

I have the strangest story to tell you, and at the beginning I must stress how lucky I am. I went to the doctor three months ago for a Pap smear. Because I am a redhead and live in sunny California, she always checks my moles. I have about four, the largest about the size of your thumbnail. One is on my back just above my bikini top and the others are on my legs. Anyway, the one on my thigh had changed color and it itched a little. It had been a light brown and now it had a dark and light area. The doctor was worried it was becoming malignant.

She took it off, and the report was superficial melanoma. It had not spread at all and if you have to have a melanoma, this is the best kind. Apparently, melanomas spread widely, but mine had no evidence of invasion and the odds are better than 3 to 1 that I am cured. I prefer odds of 10 to 1 or 100 or 1, but with this kind of disease you take what you can get.

Sincerely,
Amy

Author's note

Most people have a dozen or more pigmented spots on their bodies. Because a significant number of melanomas develop from such spots, the problem is to know which ones to treat and when. Redheads typically lack the pigment that protects skin cells from ultraviolet light, a carcinogen, and are thus at risk for developing melanoma (see section on ultraviolet radiation and skin cancer, Chapter 7). Unfortunately, too few risk factors are known for malignant disease, and more and better approaches to cancer epidemiology (Chapter 7) are needed. Changes in existing moles may be early manifestations of malignant change and warrant prompt treatment. Had this tumor infiltrated even 1.5 mm, instead of a 3 to 1 chance for cure, the odds would have been 3 to 1 in favor of death. Clearly, we need more understanding of why and how malignant cells invade and metastasize (Chapter 2). The usual noteworthy changes in pigmented lesions are itching, pain, and increase in size, all of which usually indicate the lesion is invading. The trick is to diagnose melanoma early.

Neuroblastoma: Appendix, section A.9

Dear Andrea and Bob,

You did not get a Christmas card from us last year because we were overwhelmed with Gilbert's problems. But there is good news after the earlier bad news.

A year ago November, Gilbert, our bouncy 9-month-old, had a lump in his stomach. I felt it while I was bathing him. It turned out he had a malignant tumor of his adrenal gland, called a neuroblastoma. Well, it was removed early in December last year and although it seemed to be confined to the adrenal gland, the doctors, just to be sure, gave him chemotherapy. He was awfully sick for a while, but now he is healthy and his CAT scans and everything are OK. We have been assured he is one of the lucky 40% or so who are cured with modern treatment.

I was worried about the treatments and what they would do to him, but he seems OK.

Sincerely,

Sandra

Author's note

Neuroblastoma is the most common tumor of children in the United States, and although it can present with fever and pain, it is more commonly found by the parent while bathing the child. These tumors are rapidly growing, invasive (Chapter 2), and can involve the adjacent organs, such as kidney and liver, by direct invasion, and they can metastasize via the bloodstream. For reasons unknown, tumors found in very young children have a better prognosis than those that develop in older ones. Prior to chemotherapy (Chapter 9) the outlook was poor, but with modern chemotherapy about 40% of the children can be cured.

Spontaneous regression (Chapter 1) has been reported in neuroblastoma, but its occurrence is so rare that it offers no hope for the individual.

Summary

These few letters illustrate some of the reactions of human beings to the multiplicity of diseases known as cancer. Cancer can occur in almost any tissue of the body, and because different signs and symptoms are produced by different tumors, the tendency is to view cancer not as a single entity but as a series of diseases. It is no wonder that people are terrified by the diagnosis of cancer. Early diagnosis and extirpation of the tumor before it has invaded and spread offer the best hope for cure. Once cancer has spread, with about only a few possible exceptions (e.g., acute lymphatic leukemia, choriocarcinoma, testicular cancer, Hodgkin's disease), chemotherapy and radiation therapy are not curable, merely palliative. Their initial positive effects, which provide freedom from pain and provide useful life, are all too soon lost as the tumor becomes resistant.

Note the commonalities of malignant tumors in various locales: a mass that grows, becomes progressively more malignant, and alterations in differentiation of the cells of the mass eventuate in rapid growth, invasion, and metastasis of the undifferentiated cells with destruction of normal cells and tissues. Weight loss (cachexia) and the inability to mount good immune and anti-inflammatory responses result in death usually by infection.

These commonalities point to the deficiencies in our knowledge of cancer. We must understand growth regulation of cells to be able to control the growth of the mass. Understanding the mechanism of progression could lead to means of preventing tumors from becoming more and more malignant with time. Understanding the mechanisms of invasion and metastasis could prevent spread. Understanding cachexia could provide useful life. Development of specific therapies could lead to cure. Because society cannot afford to treat all patients with cancer, however, there is an imperative need to identify causes of malignancy and thus minimize or prevent this dread disease. These are among the issues discussed in this book.

1

The pathology of cancer

G. Barry Pierce

1.1 Introduction

Our first task is to provide you with a working knowledge of the pathologic terms and concepts used throughout the text. This chapter defines terminology, compares and contrasts malignant and benign tumors, considers characteristics and behavior of malignant cells, and discusses how invading malignancies kill an individual. Tumors with time undergo changes that lead to autonomy. This progression of events is also examined.

An appreciation of embryology leads to a consideration of the origin of stem cells and the concepts of determination and differentiation. Both are important to understanding the origin of cancer cells, and such comprehension may lead to new modalities for treating cancers. Most textbooks of cancer biology begin with a discussion of cells. But we start with an examination of what cancer is to help you get a better grasp of the material that follows. Metastasis is difficult to understand without a prior foundation in the concepts of pathology. Similarly, carcinogenesis or chemotherapy is incomprehensible without knowledge of what a malignant cell is and how it behaves.

Much of our knowledge about tumors dates from antiquity. The streaks of hard gray tissue that extend from a tumor into the normal tissues reminded the Greeks of a crab, so they named the condition cancer (from the Latin word meaning crab). The term "tumor" denotes a mass, whether neoplastic, inflammatory, pathologic, or even physiologic. Today, tumor is used generically to describe any neoplasm, irrespective of its origin or biologic behavior. The term "cancer" is generic for any malignant tumor.

Willis (1967) defined a neoplasm as a mass, the growth of which is incoordinate with the surrounding normal tissues and that persists in the absence of the inciting stimulus. It is worthwhile considering this definition in detail. First, the mass, like any other tissue, is composed of parenchymal cells and stroma, which are the essential parts of an organ. The parenchymal cells of the mass may be well differentiated, organized as normal tissues, and proliferate slowly, or they may be poorly differentiated, rapidly proliferating, and have little or no organization. In either situation, the host is induced to supply a stroma for it. This host response is mediated by angiogenic factors (Folkman 1985, 1993), which are synthesized by the parenchyma of the tumor and stimulate proliferation of all stromal cells, including fibroblasts and vascular cells.

Accrual of mass could be the result of a decrease in cell cycle time, the period of time required for a cell to make the arrangements for cell division and to divide. More rapid cell cycles (i.e., decreased cell cycle time) would generate more cells. However, most neoplastic cells *do not* cycle faster than their normal counterparts. For example, 40 kg of gastrointestinal cells and 10 kg of white blood cells are produced annually by a human of average size (Donald Coffey, personal communication). This is a prodigious effort in cell replication when we consider that the fetus in utero requires 9 months to achieve 4 kg of weight. In contrast, a patient may die harboring a tumor of less than 5 kg that took several years to develop.

Thus, tumors increase in size not because the tumor cells cycle faster than normal cells but because so many tumor cells are cycling. This fact has implications for cytotoxic chemotherapy, which is often designed to kill dividing cells. Cytotoxins do not distinguish between normal and malignant cells. Therapies dependent on interference with DNA synthesis also kill rapidly proliferating normal cells. Destruction of normal white blood cells and gastrointestinal cells may result in infection, bleeding, and diarrhea. Fortunately, normal cells recover faster from the poisoning than cancer cells, so the clinician administers the cytotoxins in cyclic fashion, hoping to rescue the normal cells while achieving a cumulative toxic effect on the tumor (Goldin and Schabel 1981).

The definition of neoplasm by Willis (1967) also states that the mass persists in the absence of the inciting stimulus. This is an important consideration and distinguishes the neoplasm from the modulations of growth that also result in changes in mass. These modulations are considered to be normal cellular responses to environmental stimuli, and they persist only as long as the environmental stimulus is present.

Some tissues are capable of renewing themselves: as normal cells become senescent and die, the cells responsible for renewal, known as stem cells, proliferate in a controlled manner to replace the precise number of cells lost. In the presence of certain environmental stimuli, these tissues can become hyperplastic. By definition, the hyperplastic organ is larger than normal because of an increased number of normal cells, and it remains hyperplastic as long as the inciting stimulus is present. If the inciting stimulus is removed, the organ returns to normal size. As an example, the cells of the prostate gland respond to administration of testosterone by dividing, and the gland becomes enlarged (hyperplastic). Upon withdrawal of the hormone, the gland atrophies and returns to normal size. The breast undergoes hyperplasia

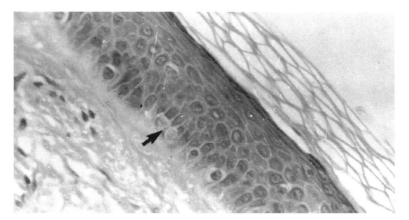

Figure 1–1. Photomicrograph of normal squamous epithelium from skin. The stem cells are located in the basal layer (arrow). Above the arrow, keratinocytes differentiate from their cuboidal form and mature progressively in an orderly manner to form flat surface keratinocytes lacking nuclei, which then form keratin. The basal layer of cells rests on the basement membrane, which separates epithelium and stroma, but is invisible unless special stains are employed. Aside from the layers of keratin on the surface, this is identical to squamous metaplasia.

during pregnancy, and after delivery the gland returns to normal size (Figure A–2b). Atrophy, the converse of hyperplasia, is an acquired reversible decrease in size of an organ as a result of a diminution of size and/or number of cells in the organ. In addition to withdrawal of a hormone, atrophy can be caused by a variety of factors including reduction of blood supply.

Other important modulations in cells to consider are not necessarily associated with changes in mass, but rather with changes in differentiation. Metaplasia is a case in point: when an epithelium, for example in the respiratory tract, is chronically injured (as with use of tobacco), it may change from respiratory epithelium to squamous epithelium (Fig. 1–1). The respiratory epithelium has mucous cells and ciliary cells (Figure 1–2). The latter sweep mucus containing bacteria and garbage to cough-sensitive areas in large bronchi where it is expelled by coughing. This is the housecleaning mechanism of the airway. The metaplastic squamous epithelium, although normal in all respects but its location, is not ciliated and cannot provide this service. As a result, mucus laden with dust and bacteria accumulates behind the patches of squamous epithelium and leads to infection. The squamous metaplasia that occurs at the bifurcations of bronchi in cigarette smokers can undergo complete reversion to normal respiratory epithelium over a period of time after cessation of smoking.

Thus, metaplasia is a reversible change in phenotype in response to environmental stimuli. Interestingly, the metaplastic squamous cells are believed to be the target in carcinogenesis in the bronchi; bronchi normally contain no squamous cells, yet squamous cell carcinoma of the bronchus is common, particularly in longtime cigarette smokers. Similarly, squamous metaplasia can occur in the

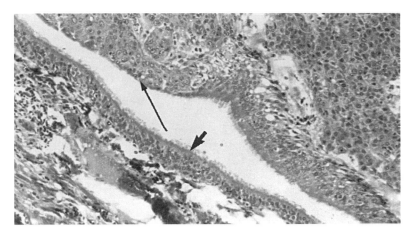

Figure 1–2. Photomicrograph of respiratory epithelium. Note ciliated cells that line the airway (short arrow). A large squamous cell carcinoma is growing, invading, and destroying the mucosa (long thin arrow). Thus the housekeeping mechanism of the airway is destroyed, predisposing to infection.

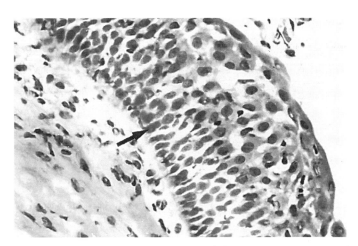

Figure 1–3. Photomicrograph of a frozen section of dysplastic squamous epithelium. The arrow points to the basal layer. Whereas in the normal situation there is orderly differentiation of basal cells to keratin (see Figure 1–2), in dysplasia the order is mixed up. Poorly differentiated cells are present near the keratin layer.

epithelium of the pelvis of the kidney, ureters, or bladder when these epithelia are chronically injured as by exposure to pus-laden urine from abscesses in the kidney (pyelonephritis). In both of these situations the metaplastic squamous epithelium is normal in every sense except its location (Kumar, Cotran, and Robbins 1992).

Epithelium may be dysplastic. In this situation, undifferentiated or partially differentiated cells may be mixed with the differentiated ones (Figure 1–3), changing the orderly pattern of organization of the tissue. Dysplasia can be a precursor to

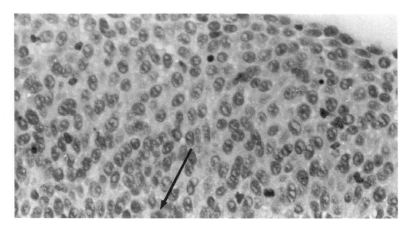

Figure 1–4. Photomicrograph of carcinoma in situ. Unlike the dysplastic specimen (Figure 1–3), no differentiation from basal (arrow) to superficial layer is evident. There is no evidence of invasion into the peribronchial tissue. In other words, the basement membrane has not been penetrated by the anaplastic cells.

carcinoma, but dysplasia is also a modulation because it disappears with removal of the inciting stimulus. This distinguishes dysplasia from carcinoma in situ (Figure 1–4), which is an irreversible state and a stage in progression to invasive carcinoma.

A neoplasm is distinguished from the reversible cellular modulations by persisting after the inciting stimulus has disappeared, which makes it difficult to establish cause and relationship effects in carcinogenesis. It also means that, unlike most inflammatory or infectious diseases in which the etiologic agent can be ascertained and treated, in neoplasia, the inciting agent is long gone, so the resultant mass of cells is the target of treatment. This is an important point.

1.2   Benign versus malignant tumors

1.2a   Benign tumors

Benign tumors ordinarily do not cause death, but there are rare exceptions to this rule. Death may result from synthesis of toxic levels of hormones by the benign tumor cells or by the position of the benign tumors in situations essential for life.

Benign tumors are slow growing. Rarely is a mitosis present in a microscopic field. The cells are well organized and well differentiated (the cells closely resemble those of the normal tissue). They may synthesize the gene products specific to the differentiated state (often termed luxury molecules because they are not essential for the survival of cells, but are essential for the well-being of the organism). Luxury molecules may be demonstrated in the tumor cells by special histochemical procedures. Normally, benign tumor cells synthesize fewer of these molecules per cell than those of the normal tissues, but when the tumor becomes large, toxic

Figure 1–5. Photograph of a pheochromocytoma of the adrenal medulla. The adrenal cortex (medium arrow), which synthesizes cortisone among other endocrines, overlies the medulla (long arrow), which secretes epinephrine. Note the tumor, which lies in the medulla. It has grown, compressed, and thinned the cortex, the cells of which have progressively atrophied, died, and their stroma has formed a capsule (short arrow). This tumor secreted epinephrine in toxic amounts, causing acute episodic hypertension.

amounts of the molecules can be synthesized. For example, a benign tumor of the cells of Islets of Langerhans may secrete enough insulin to cause insulin overdose, resulting in hypoglycemia, and death. Similarly, a benign tumor of the adrenal medulla may produce the hormone epinephrine. The extremely high blood pressure induced by the hormone can result in death. A benign tumor of the pituitary gland of a child may synthesize enough growth hormone to cause the individual to become a giant. Under normal conditions, production of molecules for specialized function is carefully controlled to meet the needs of the host, but a benign tumor produces these molecules with no regard for the needs or safety of the host.

As a benign tumor grows, it transgresses on adjacent normal tissues. This is an expansive type of growth that pushes the normal tissue ahead of it and compressing the thin-walled capillaries of the normal parenchyma. With insufficient blood to nourish the normal parenchymal cells, atrophy results. As the normal cells atrophy and eventually die, only the connective tissue stroma of the normal tissue remains. The stroma is compressed and forms a capsule around the tumor (Figures 1–5 and 1–6a). The pressure can also cause destruction of nearby vital centers, resulting in death. For example, a benign and slow-growing tumor in the coverings of the brain may eventually cause death of the host by pressure atrophy and destruction of vital centers.

Benign tumors lack the intrinsic features discussed next that give the lethal potential to malignant tumors. The vast majority of benign tumors are just that – benign – and cause the host little or no damage. A few, however, by virtue of their positions near a vital center or by their abilities to synthesize large amounts of biologically active molecules, may destroy their hosts.

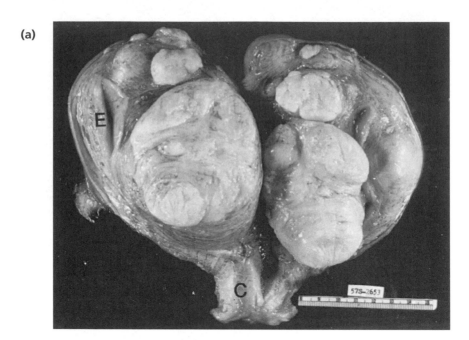

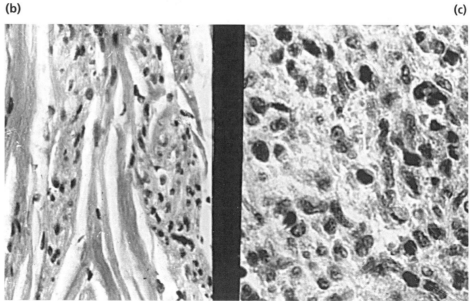

Figure 1–6. (a) Photograph of uterus opened to display multiple leiomyomas. Note compression of surrounding muscle and displacement of endometrial cavity by the noninfiltrative growth characteristic of benign tumors. Such tumors may cause problems during pregnancy, excessive bleeding at or between menses, but they are not life threatening to the host. Note cervix at C and endometrial cavity at E. (b) Photomicrograph of a leiomyoma (benign tumor of smooth muscle) of the uterus. The smooth muscle cells, normal in appearance, are arranged in organized muscle bundles. The features of differentiation are contained in the abundant cytoplasm. (c) Photomicrograph of a leiomyosarcoma (malignant tumor of smooth muscle) of the uterus taken at the same magnification as the specimen illustrated in (b). Note the large and variable size of the nuclei, lack of differentiation of cell cytoplasm, and lack of organization of cells into muscle bundles. This specimen illustrates many of the features of anaplasia.

1.2b   Malignant tumors

The cells of malignant tumors have the intrinsic ability to kill the host unless they are removed or killed. In contrast to benign tumors (Figure 1–6b), numerous normal and abnormal mitotic figures are present in malignant tumors (Figure 1–6c). These cells are programmed for proliferation and have large vesicular nuclei with large nucleoli (where ribosomes are made). The cytoplasm contains many polysomes that are unattached to membranes and synthesize structural proteins required for cell division. The cells are pleomorphic in size, shape, and staining reactions and have increased nucleocytoplasmic ratios.

Malignant cells are less well differentiated than their benign counterparts, some so poorly that they defy histopathologic identification. Such tumors are said to be anaplastic ("anaplasia" literally means a condition without form, but to the pathologist it represents the summation of all of the microscopic attributes by which malignant tissue is diagnosed; compare Figures 1–6b and 1–6c). The anaplastic cells invade and destroy the normal architecture of the organs, and replace it with masses of disorganized malignant cells (Figures 1–7, 1–8, and 1–9). For example, a well-differentiated adenocarcinoma may produce glandular epithelium, but the epithelium lacks the normal relationship with stroma (Figure 1–9). A less differentiated adenocarcinoma might make solid plug-like masses of tumor cells without glandular lumens. An even less differentiated one may have columns of single anaplastic epithelial cells forming rows of cells that penetrate between the normal stromal cells. Finally, an adenocarcinoma may make no such organized structures, and the individual anaplastic epithelial cells would be recognized as adenocarcinoma cells

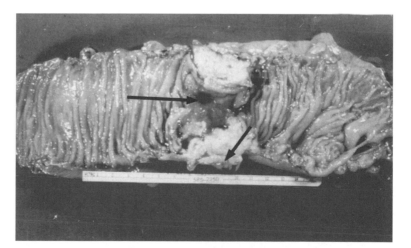

Figure 1–7. Photograph of an adenocarcinoma of the colon. The bowel has been opened along its long axis. Note the napkin ring appearance imparted by the adenocarcinoma that has invaded the mucosa forming a malignant ulcer at the long arrow and through the bowel wall and into the fat (short arrow). This stage of malignancy carries a poor prognosis.

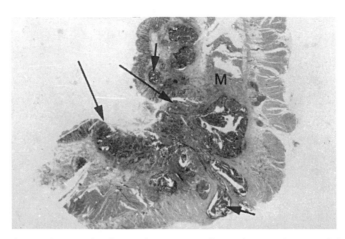

Figure 1–8. Photomicrograph taken at low power of an adenocarcinoma of the colon. Normal mucosa at the top is disrupted by a "malignant" ulcer (between long arrows). The base of the ulcer is composed of anaplastic cells in a fibrous stroma. Clumps of these cells (short arrows) have invaded beneath the epithelium at the margin of the ulcer, deep into and through the smooth muscle (M) and into the adjacent fat of the serosa. This stage of the disease has a poor prognosis.

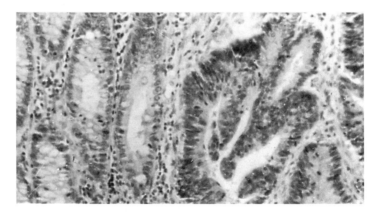

Figure 1–9. Photomicrograph taken through the margin of a low-grade adenocarcinoma of the colon. Normal glands are on the left and illustrate the orderly arrangement of the cells with basally placed nuclei and well-differentiated cytoplasm forming regular-shaped glands. Contrast this with the appearance of glands to the right where multiple nuclei are piled up, the cytoplasm is darker staining, and the glands are irregular in shape. These are evidences of anaplasia, the hallmark of malignancy.

only because some of them contain mucin, a marker for glandular epithelium (Figure 1–10) (see section on tumor markers).

Contrast this appearance with that of an adenomatous polyp of the colon (Figure 1–11). These benign masses grow into the lumen of the colon, are composed of normal-appearing cells, and the patient is cured by simple removal. Malignant change can occur in these lesions, but if the abnormal cells have not invaded the

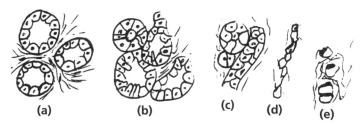

Figure 1–10. Degrees of anaplasia commonly found in adenocarcinoma cells. (a) Normal acini with regularly shaped cells and connective tissue stroma between acini. (b) Adenocarcinoma, low grade, with sightly irregular acini with some variability in cell shape, size, and number. Note glands are back to back with no stroma between them. (c) Adenocarcinoma high grade. Note the cells are unable to arrange themselves in glandular acini and form plugs of anaplastic cells. (d) A further degree of anaplasia; the cells cannot form plugs, only columns of single cells. (e) The most anaplastic tumor; the cells cannot form single-cell columns faintly reminiscent of their epithelial origin. They make mucin, however, a marker for epithelium.

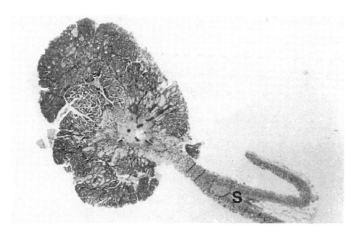

Figure 1–11. Low-magnification photomicrograph of adenomatous polyp illustrating the entire lesion. No invasion into the stalk means that cure has been achieved by local removal.

stalk of the polyp, again the patient is cured by removal of the polyp, its stalk, and a small cuff of normal mucosa. If not treated, the malignant cells will invade not only the stalk, but the bowel itself. A malignant ulcer results. Further invasion leads to obstruction of the bowel (Figure 1–7).

The suffix *-ous* means "-like." Thus, adenomatous means "like an adenoma." Malignant change rarely if ever occurs in adenomas. Adenomatous polyps are in reality focal areas of hyperplasia that become polypoid by the passing fecal stream. They contain numerous stem cell targets for carcinogens and predispose the patient with familial polyposis to cancer and death.

Large areas of necrosis are apparent in rapidly growing tumors. Cell death in such tumors was attributed to rapid tumor growth with inadequate blood supply. However, dead cells were often found adjacent to blood vessels. It is now known that much of the cell death is due to apoptosis (see section 1.16).

1.3   The diagnosis of benign and malignant tumors

No absolute rules apply in diagnosing benign and malignant tumors. Success is based solely on the knowledge, skill, and experience of the observer. This may be a frightening thought, but even in our scientifically sophisticated society no better method has been developed, despite decades of experimentation. The pathologist, given a specimen, is expected to render a diagnosis and define a prognosis for the patient. In addressing these tasks, every bit of information concerning the patient that can be elicited is used. The history of the patient, including age and sex, the presenting signs and symptoms, the clinical diagnosis, the organs involved, and so on: all are considered. Often a biopsy is taken while surgery is being performed because the surgeon wishes to know if the lesion is benign or malignant. The pathologist examines the gross specimen, freezes selected portions of it in a cryostat, sections and stains it, and examines the tissue with a microscope. Evidence of disorganization of the normal histologic architecture by neoplastic cells is sought (Figures 1–9 and 1–10). Invasion into capillaries or lymphatics suggests distant metastasis may have occurred. The diagnosis, if it is possible to make one, is given to the surgeon within 5 to 10 minutes of receipt of the specimen. The surgeon then proceeds in a manner consistent with the pathologist's diagnosis.

After surgery the resected tissue is preserved and examined grossly. The pathologist notes whether the tumor is confined to the organ in which it originated or whether it has penetrated the organ and invaded surrounding tissues and organs (Figure 1–7). Regional lymph node involvement is also noted. These findings are confirmed by the microscopic examination of numerous specimens taken during the gross examination. The microscopic examination allows the tumor to be graded and confirms the stage of spread of the tumor.

1.4   Tumor grading and staging

Tumor grading is an indication of the degree of malignancy and correlates to a degree with prognosis. Tumor staging describes the size of the primary lesion, the degree of invasion, the presence of lymph node involvement as well as metastases. Staging also plays a role in establishing prognosis. Grading, or the degree of anaplasia, is assessed from 1 to 4, with 4 the most malignant. Cytologic grading should correlate with the prognosis, but, unfortunately, this is not always the case. Sometimes a highly malignant tumor (as judged by grade) can be completely extirpated early, curing the patient. Tumors with low degrees of malignancy (as judged by grade) may be discovered late, after they have invaded and metastasized, and

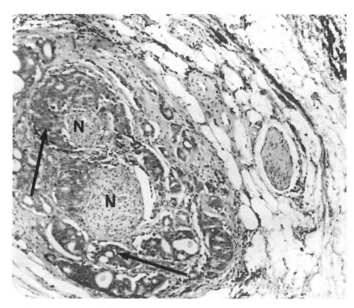

Figure 1–12. Photomicrograph illustrating invasion of cells of an adenocarcinoma of the prostate (arrows) along a nerve tract (N). This is a poor prognostic sign, especially when it occurs in the capsule of the gland.

may be incurable. Thus, the stage, or degree of spread of the tumor, must be considered when grading a tumor to obtain accuracy in prognosis. The gross examination of the specimen aids in assessing the stage, but microscopic examination is also important because the extent of invasion can be assessed more accurately and the early involvement of small blood vessels, lymphatics, and lymph nodes can only be seen with the microscope (Figure 1–12). A system of staging has been adopted worldwide (Rubin and Cooper 1993).

Carcinoma in situ is a neoplastic stage wherein the tumor cells are confined to the epithelium of origin and have not transgressed the basement membrane to invade the capillaries, venules, and lymphatics of the stroma. This is the cancerous stage with the best prognosis. These lesions are cured by extirpation. The patient's prognosis deteriorates progressively if the tumor has invaded the stroma of an organ but has not penetrated it. The prognosis is worse if the wall is penetrated (Figures 1–7 and 1–8) and even worse than that if spread to lymph nodes has occurred. It is dismal if there are distant metastases (Figures 1–14 and 1–15).

1.5   Classification and nomenclature

Tumors are named according to the tissue or organ in which they originate, and then a suffix is added to denote whether the tumor is benign or malignant (Figure 1–13). The suffix *-oma* literally means "a tumor of" and is construed to mean a benign tumor. Accordingly, a fibroma is a benign tumor of fibrous connective tissue.

	Benign	Malignant
Epithelial Endothelial } Origin	Adenoma of liver, salivary gland, colon, kidney, etc.	Adenocarcinoma of liver, salivary gland, colon, kidney, etc.
Mesenchymal connective tissue } Origin	Lipoma Fibroma Chondroma etc.	Liposarcoma Fibrosarcoma Chondrosarcoma etc.
		Neuroblastoma Retinoblastoma Nephroblastoma
Germinal tumors	Teratoma	Teratocarcinoma Embryonal carcinoma Seminoma
Other		Hepatoma synovioma Melanoma Leukemia

Figure 1–13. Chart of nomenclature.

A lipoma, chondroma, and neuroma are benign tumors of fat, cartilage, and neural tissue, respectively. Adenoma is a benign tumor of glands, irrespective of the glandular organ in which the tumor develops. For example, an adenoma of the breast is a benign glandular epithelial tumor of the breast (Figure A–3).

Three suffixes identify malignant tumors: *-carcinoma, -sarcoma,* and *-blastoma.* The suffix *-carcinoma* implies a malignant tumor of epithelial origin; for example, carcinoma of the lung signifies a malignant tumor derived from epithelium of the lung. Adenocarcinoma of the breast signifies a malignant glandular epithelial tumor of the breast. Squamous cell carcinoma of the skin is a malignant tumor of squamous epithelium of the skin. Adenocarcinoma of the colon is a malignant tumor of the glandular epithelium of the colon, and so on. *Sarcoma* is the designation given to malignant tumors of connective tissue. For example, osteosarcoma is a malignant tumor of bone-forming cells, and chondro-, lipo-, and lymphosarcoma are respectively malignant tumors of cartilage, fat, and lymphocytes. Finally, a group of highly malignant childhood tumors is denoted by the suffix *-blastoma.* These include neuroblastoma, originating in the neuroblasts of the adrenal medulla; retinoblastoma, originating in the retina of the eye; and nephroblastoma, originating in embryonic cells of the kidney (Figures A–16 and A–17a).

The blastomas are so undifferentiated that they reminded our scientific ancestors of the inner cell mass of the blastocyst, hence the suffix *-blastoma.* Carcinomas were distinguished from sarcomas because the former were believed to metastasize first via the lymphatics, whereas sarcomas seldom do. The terminology is archaic, but nothing can be done about it.

Some tumors have names that must be remembered as exceptions because they do not conform to the rules of nomenclature. The following are all malignant tumors, although their names imply they are benign: melanoma, pigmented skin cancer; hepatoma, liver carcinoma; synovioma, sarcoma of the cells lining the joints; and leukemia, which means white blood, refers to a sarcoma of leukocytes with so many malignant cells in the blood that it may be creamy in color. If numerous malignant lymphocytes are circulating in the blood, the condition is referred to as a lymphatic leukemia, either acute or chronic. Lymphoma is a generic term for any sarcoma of reticuloendothelial cells, with the exception of leukemia.

1.6 Metastasis

Malignant cells have a capacity to grow along tissue spaces, nerves, and vessels, and finally penetrate lymphatics, venules, and capillaries as well as the body cavities (Figure 1–12). Single or small clumps of tumor cells then break away from the original mass and are carried to distant organs, where they implant. This process of dissemination of malignant cells is called metastasis. The original tumor is called the primary tumor, and the process of metastasis establishes secondary tumors in distant organs where they incite the development of a stroma and develop into new tumors, which in turn invade and metastasize (Figures 1–14 and 1–15). Metastasis has been defined by Willis (1967) as the discontinuous growth of tumor cells. Chapter 2 presents an extended description of metastasis.

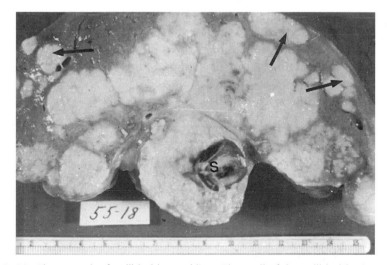

Figure 1–14. Photograph of gallbladder and liver. The wall of the gallbladder is normally thin and soft like a piece of green velvet. In this specimen, adenocarcinoma cells of the gallbladder have grown and invaded the organ converting the wall into a thick, cheeselike mass surrounding a gallstone (S). The tumor has invaded the liver and metastasized within the liver (arrows).

Figure 1–15. Photograph of the surface of a lung of a woman to illustrate metastases of leiomyosarcoma. The pleural surface is studded with nodules of metastatic sarcoma, each capable of invading and metastasizing. Remember: when tumors metastasize, the metastases are multiple.

Cells of the embryo may invade, and migrate. For example, primordial germ cells of the bird invade embryonic blood vessels in the yolk sac, where they are formed and migrate (metastasize?) to the primitive gonad. How do normal migration and metastasis differ? One difference is the nature of the cells involved. The germ cells are normal and under homeostatic control, whereas the malignant ones are not and form tumors where they implant. It is conceivable that the mechanisms are the same. One might ask if there is a gene for migration that is mutated, enabling malignant cells to metastasize, or if there is expression of a gene for migration in malignant cells that had been repressed since embryonic life. This issue is also discussed in Chapter 2 on metastasis.

The most important rule to remember about metastasis is that secondary tumors are almost always multiple and therefore not amenable to surgery or irradiation as curative procedures. Currently, the only hope for prolonging useful life or obtaining cure for the person with metastasis is chemotherapy (see Chapter 8).

1.7 Tumor markers

Tumor markers are molecules synthesized by tumors that are characteristic of the tumor and properly employed lead to the identification of it. As discussed, malignant tumor cells produce far fewer luxury molecules than normal cells, but so many tumor cells are generated that enough luxury molecules may be produced to cause clinical manifestations. For example, choriocarcinoma, the malignant tumor of trophoblast, can produce large amounts of human chorionic gonadotropin (HCG), which causes gynecomastia in men (breast hyperplasia). The presence of the hormone in the blood of a person with a testicular tumor indicates the tumor will be or contains choriocarcinoma (Bagshawe 1992).

In addition to the synthesis of histiotypic (tissue-specific) molecules by malignant cells, some cancers make molecules that are biologically active but bear no apparent relationship to the histiotype of the tumor. If enough are produced for a clinical response, the syndrome is referred to as a paraneoplastic syndrome (Gerwitz and Yallow 1974; Bunn and Ridgway 1993). For example, adrenocorticotrophic hormone (ACTH) normally is produced by the pituitary gland. However, a squamous cancer of the lung may synthesize enough ACTH to stimulate the adrenal gland to produce cortisone, causing Cushing's syndrome. When the primary tumor is removed, Cushing's syndrome disappears. After surgery, elevated levels of ACTH in the blood of such patients indicates that metastases have occurred, even though they may not have grown to appreciable size.

HCG and ACTH are good examples of tumor markers that are helpful clinically (Sell 1980; Schwartz 1993), but there are others such as alpha fetoprotein and carcinoembryonic antigen that are also useful in making diagnoses and following the course of patients. The latter are markers for malignant tumors of endodermal origin (colon, liver, stomach, etc.). Markers are sometimes elevated in non-neoplastic conditions, so care must be exercised in their use in diagnosis.

It is important to consider what tumor markers represent biologically (Sell 1980). Braun (1956) showed that a plant teratoma (see Appendix A) synthesized two essential growth factors that were not synthesized by adult plant cells but were synthesized by embryonic cells. Thus, we infer that synthesis of markers by tumors of mammals reflects synthesis of the same molecules by the normal cell lineage during its development in the embryo. Why these particular genes should be activated in carcinogenesis is not known. If a tumor makes a nonhistiotypic molecule, it seems likely that the normal embryonic stem cells should also make it. Because these are functional molecules, they probably play a regulatory role in the development of the corresponding tissue in the embryo.

1.8   How cancer kills

1.8a   Organ failure

Tremendous reserves have been built into most of the organ systems: for example, an otherwise healthy adult can survive with one-half of a healthy kidney. After removal of one kidney, the other undergoes hypertrophy and/or hyperplasia, which compensates to a degree for the loss of tissue. A person can survive easily with a single functional lung. In the case of the liver, an animal can survive after removal of two-thirds of its liver, and, over the course of a week, regenerate the lost tissue (Bucher and Malt 1971; Fausto and Webber 1993). Thus, there is an enormous reserve of normal tissue plus a built-in mechanism to regenerate more functional tissue as required. One can see why, during development of a tumor, signs or symptoms relating to organ failure are seldom seen before the tumor is far advanced. Although tumors can cause symptoms indicating an organ is affected (e.g., heaviness, and if the capsules of organs are stretched there may be pain, plus all of the nonspecific side effects of tumors such as fevers, cachexia, and weakness), there may be little or no evidence of organ failure. This explains why tumors of solid organs are silent for so long, why diagnosis is often delayed, why metastases are often present at the time of diagnosis, and why people seldom die of organ failure. Leukemia is the notable exception to this rule. Leukemic cells overgrow the marrow, causing bleeding, infection, and anemia as a result of destruction of megakaryocytes, normal leukocytes, and erythroid precursors.

1.8b   Obstruction of the gastrointestinal tract, ducts, and hollow organs

The situation is somewhat different if the cancer involves the hollow organs such as the gastrointestinal tract or ducts carrying secretions from one organ to another. Carcinoma developing in such hollow organs has a propensity to invade the mucosa, ulcerate it, cause bleeding, and thereby the possibility of anemia and weakness. The tumor also invades and grows circumferentially around the wall of the organ or duct in a shape resembling a napkin ring. This ring thickens with time, and ultimately the lumen is obstructed (Figure 1–7). This is best seen in the descending colon where the fecal stream is solid and contributes to intestinal obstruction, a common presenting sign of cancer in that locale. Such a patient will die in about a week unless the obstruction is relieved surgically. Thus, the surgeon is often faced with a life-threatening emergency that must be relieved immediately. Even if the tumor has spread beyond the colon and the patient is incurable, relief of the obstruction affords palliation and offers useful life to the patient.

Squamous cell cancer of the esophagus results in an ulcerative lesion and causes difficulty in swallowing, which exacerbates the cachexia associated with malignancy. In addition, the tumor may penetrate the wall of the esophagus and grow into and through the wall of the adjacent trachea or bronchus causing a fistula (abnormal passageway). Aspiration of food through such passages into the bronchus and lung may result in pneumonia. Distant metastases occur, and the patient becomes cachectic and dies of infection.

Some of the cancers in the walls of hollow organs are fungating (mushroomlike in appearance) and grow into the lumen. They often occur in the right side of the colon where the fecal stream is fluid. These cause obstruction less frequently than the infiltrating napkin ring–like tumors that occur commonly in the descending colon. These fungating tumors often cause severe anemia because of massive ulceration with bleeding. This exacerbates the cachexia that occurs as the tumor grows and spreads to lymph nodes and distant organs. Again, death usually results from cachexia and infection.

Death or serious illness seldom results from the compression of ducts. For example, the ureters may be displaced by massive metastases from teratocarcinoma of the testis, but the outflow of urine is seldom obstructed sufficiently to cause signs and symptoms. A major exception to this rule is the involvement of ducts and ureters by the matted masses of lymph nodes of Hodgkin's disease. There is often a fibroblastic response (excessive growth of connective tissue) in Hodgkin's disease. The resulting scar tissue may surround and obstruct ducts. The resultant backup of secretions can give rise to symptoms and serious disease. Similarly, obstruction of the urethra by adenocarcinoma of the prostate (a portion of the urethra passes through the prostate) can cause acute and chronic retention of urine, which predisposes the individual to cystitis. A patient with invasive squamous cell carcinoma of the cervix also usually dies from the complications of urinary obstruction. In this case, the cancer cells invade around the ureters where they enter the bladder, obstruct them, and cause uremia. Invasion from an adenocarcinoma of the head of the pancreas frequently results in infiltration of and obstruction of the bile ducts, which causes jaundice.

The central nervous system is also a hollow organ with walls of various thickness. A developing tumor, whether benign or malignant, can cause disease by destroying vital centers in the brain (the thick-walled portion of the hollow nervous system). In addition, the central nervous system is contained within rigid walls, and the increase in mass of a tumor can increase intracranial pressure, causing serious effects on the brain. Moreover, tumors can grow into the canalicular system of the brain, obstructing the flow of cerebrospinal fluid with resulting damage.

If an obstruction can be removed surgically and the duct anastomosed (reconnected) successfully, then the patient is relieved of the immediate effects of the tumor. If an adenocarcinoma of the colon is close to or has involved the anus, removal of the tumor requires removal of the anus and the sphincters, and the individual is incapable of controlling expulsion of feces. An artificial anus, a colostomy, is made on the anterior abdominal wall (see letter to Uncle Harvey in Introduction). Similarly, individuals with carcinoma of the esophagus may have a feeding tube placed directly into the stomach to bypass an obstruction. Drainage tubes may be placed in the biliary tree of the liver to relieve jaundice and to compensate for loss of bile ducts due to destruction by the cancer or the surgical procedures used. A person whose bladder has been destroyed by cancer may have an artificial bladder constructed, or the ureters, which drain the liquid waste from the kidneys, may be brought through the abdominal wall so they can drain into a bag. Similarly, shunts may be made between the canals carrying cerebrospinal fluid, draining dammed up fluid through tubes running down the neck into the thorax or abdominal cavity. Shunts relieve the increased intracranial pressure due to the backup of fluid, and the procedure eliminates stroke and other complications of the condition. These procedures are all palliative and designed to provide additional useful life.

Because of the essential role of the heart, it might be supposed that involvement of the heart would be a primary cause of death in cancer. Interestingly, primary tumors of the heart are extremely rare, and metastatic tumors involving the heart are also rare. (Carcinoma of the lung by direct extension may involve the pericardium, but seldom involves the musculature or seriously interferes with the heart's function.) Melanoma may metastasize to the heart, but this is an exceedingly rare cause of death.

Obstruction of ducts can lead to death, but the duct obstruction is often a presenting sign or symptom that can be relieved surgically, as described earlier. If recurrence of cancer causes reobstruction, or if obstruction occurs as a late event in the history of the illness, the surgical trauma may be too great for the debilitated patient to bear, and other palliative measures may be undertaken. In this regard, it is possible to eradicate focal tumors by irradiation, thereby relieving pain or pressure. Duct obstruction is usually not an immediate cause of death.

1.8c   Cachexia and infection

One of the syndromes most commonly associated with the late stages of malignancy is termed "cachexia" (Calman 1992). The term refers to the wasting of terminal cancer patients. Cachexia results from the starvation and debilitation of the patient by the cancer (Figure A–16). The person is wasted, weak, and incapable of mounting adequate anti-inflammatory responses. It is this wasting and the resulting incapacity that terrifies people as much as or more than the pain associated with cancer and metastasis. The immediate cause of death of such debilitated patients is infection, even with massive antibiotic therapy. The saprophytic organisms (organisms that grow on dead organic matter) that normally live in the mouth or nose, for example, invade the patient and cause pneumonia, septicemia (blood infection), abscesses, and death. The tumor represents the underlying cause, but infection is the immediate cause of death in most terminal cancer patients. Because of the efficiency of antibacterial therapy, fungi and yeast are often the agents responsible for the terminal events, and it is clear that if the cachexia could be controlled, prolongation of useful life could be achieved. Interestingly, just as the malignant neoplasm can literally starve the host, the fetus in utero also appears to have priority for the metabolic resources of the maternal organism and can lead to the death of a debilitated mother.

Cachexia is possibly a paraneoplastic syndrome in its own right (Beutler and Cerani 1986). In this regard, Japanese scientists postulated the presence of a hormone secreted by the malignant tumor that resulted in poisoning of the host. Although such a molecule has not been identified, a fragment of a protein related to an immunoglobulin molecule, called tumor necrosis factor (TNF), was found to produce many of the features of cachexia when administered to rats. TNF is synthesized by a variety of inflammatory cells that are somehow activated by malignant tumor cells (Beutler and Cerani 1986). It is likely that TNF contributes to cachexia.

As described here (and also in the letter to Uncle Harvey; see Introduction), cachexia with superimposed infection, usually by saprophytes, is the typical cause of death in patients with cancer. The cancer is the predisposing cause of death; infection is the immediate cause.

Because certain tumors grow rapidly and are vascular, hemorrhage either from the tumor into body cavities or into the tumor itself can be a terminal event. This is rare but does occur. For example, hepatomas are observed in some young women with long histories of taking birth control pills. These very rare tumors are also extremely vascular and soft because of their sinusoidal makeup. They may rupture resulting in fatal hemorrhage, but this is an ultrarare event. Choriocarcinoma, the malignant tumor of trophoblast, may occur in women after abortions or pregnancies and in males from germ cells in the testis. It is also vascular and prone to hemorrhage. Like hepatoma, choriocarcinoma is extremely rare in North America. As a result, hemorrhage is not a common cause of death in cancer patients in the United States.

It bears repeating that we are finally left with infection as the most common immediate cause of death in the terminal cancer patient.

1.9   Spontaneous regression

Does spontaneous regression (the complete spontaneous cure of a patient) offer hope for the cancer patient with metastasis? The answer is *no!* Although spontaneous regression is a fact of tumor biology, it is so rare that little scientific attention can be or has been directed toward it (Lewison 1976).

The classic paper in spontaneous regression involved a baby with neuroblastoma treated by Cushing and Wolback (1927). The baby was sent home to die because of the stage of the disease. Imagine the surprise when the individual, now a teenager, presented with an inflamed appendix, which was removed. While the surgeons were exploring the abdomen, they looked for evidence of the neuroblastoma and found multiple small tumors in the peritoneal cavity. These proved to be ganglioneuromas. In the embryo, neuroblasts (the normal equivalent of neuroblastoma cells) differentiate into ganglion cells. Apparently, in this particular case, all of the neuroblastoma cells (malignant) had differentiated into ganglioneuroma cells (benign). Thus, the patient had a spontaneous remission of a proved malignancy as evidenced by the differentiation of all the malignant cells.

Cases of spontaneous regression are documented in which a highly malignant incurable tumor diagnosed by pathologic methods completely disappears (Lewison 1976). The mechanisms are unknown. Tumor immunologists believe they are rejected by an immune or host defensive mechanism. With the burgeoning knowledge of growth factors, it is conceivable that such tumors are responsive to negative growth factors which caused their disappearance (chalones?). Spontaneous regression probably will be explained someday on the basis of regulation of the malignant process by nonimmulogic factors. Thus, we are left with a rare phenomenon with a paucity of experimental approaches to it, and yet with a population desperately in need of relief from the problems caused by cancers. Individuals who believe that relief will come by spontaneous regression, and therefore ignore treatment, are doomed. Their tumors will, with extraordinarily rare exception, kill them.

1.10 Dormancy

Do we have inactive malignant cells lurking in our bodies? Dormancy is not uncommon clinically. Two histories illustrate the phenomenon. A 60-year-old man, with only one eye, presented with an enormously enlarged liver and aspects of tumor cachexia. The history revealed that the eye was removed for a melanoma 30 years previously. Biopsy demonstrated that the liver contained melanoma of the same type removed from the eye. The accepted explanation is that the melanoma had metastasized to the liver prior to removal of the eye, and melanoma cells that had metastasized had lain dormant in the liver for many years only to be reactivated by unknown agents.

A more common example is that of a 45-year-old woman with an adenocarcinoma of the breast which was removed by radical mastectomy. In this operation the breast plus the underlying muscles and contents of the armpit are all removed. The patient was well for 10 years, and then a pea-sized nodule (about 4 mm in diameter) was discovered in the mastectomy scar at an annual examination. It proved to be the same type of cancer that was present originally, and in a few months the patient succumbed with widespread metastases. Apparently, a few cells were left in the incision at the time of mastectomy and remained dormant for the 10-year period, at which time the malignant phenotype was reactivated.

The behavior of cells, normal or neoplastic, their interactions with each other and the environment, and reactions to growth factors is slowly becoming understood. It is in such mechanisms that the ultimate explanation of dormancy probably will be found. An experimental model for dormancy was developed by Fisher and Fisher (1967) in which injection of large numbers of tumor cells into the portal vein of rats caused massive liver metastases and death within a short period of time. When only 50 cells were injected in the portal tracts, the animals survived apparently tumor free, and no signs of tumor were observed when these animals were examined surgically. The animals were closed up and observed. They died within a few weeks postsurgery from the effects of massive overgrowth of tumor cells in the liver (Fisher and Fisher 1967). Apparently, a few tumor cells lay dormant in the liver and were activated by the growth hormones of the repair process. This experiment is important because it demonstrates that, as in normal development (Grobstein and Zwilling 1953), a critical number of cells is required for the expression of a phenotype, in this case the malignant one, and that environmental circumstances, in this case surgical trauma with release of growth factors, can stimulate growth of the dormant cells.

In conclusion, cancer cells are not necessarily unresponsive, autonomous cells as we have classically been led to believe. They do respond to some environmental perturbations. Intensive effort must be made to identify the positive and negative regulators of neoplastic growth.

1.11 Initiation, latency, promotion, and progression

The development of cancer is a multistep process. The stages of carcinogenesis, initiation, latency, promotion and progression interact sequentially in the formation of a malignancy. Initiation is the first step in which a carcinogen interacts with DNA (Freidwald and Rous 1950). It is a rapid process that permanently alters cells. Initiated cells do not develop into tumors in the absence of a suitable environment and are termed latent. If the environment contains promoting agents, proliferation of cells will be preferentially stimulated. The effects of promoting agents are reversible, and if the promoter is removed, disappearance of the expanding clones of cells will result. They will reappear if the promoter is reapplied. The second step, progression, is characterized by events that result in the autonomous state; at this time the effects are not reversible. The mechanism of progression is addressed later.

1.12 Latency

Latency is the period of time between the application of a carcinogen (Chapter 3) and the appearance of a tumor. During latency a small number of cells initiated by the carcinogen are genetically programmed for the malignant phenotype. Initiation changes the determination of stem cells from normal to neoplastic, but initiated cells do not appear different than their fellow stem cells. (Note that initiation does not alter the histiotypic determination of the stem cell, it only superimposes the malignant phenotype on it.) In this sense malignant stem cells are remarkably like determined, undifferentiated normal stem cells in the embryo, but they have acquired a change in their potential to proliferate and differentiate. As in the normal situation, latent cancer cells express the malignant phenotype only in the appropriate environments. Multiple factors are required to express embryonic phenotypes as well as malignant ones. Over a period of time, and especially with the application of promoters (Chapter 3), initiated cells begin to express the malignant phenotype, and tumors develop. To reiterate, the expression of the malignant phenotype, just like the expression of the embryonic phenotype, depends on environmental conditions. Is it possible to prevent development of tumors by gaining an understanding of the social relationships of cells in dormancy and latency?

1.13   Progression to the autonomous state

Foulds (1969) defined progression as "the gain or loss of unit characters leading to the autonomous state." A unit character could be growth rate, ability to metastasize, inability to respond to a hormone, a differentiated function, or a morphologic feature. All of the events in the experiments described here illustrate the concept of progression as outlined by Foulds on the basis of studies of breast cancer in mice and man. The gains or losses of unit characters were uncoordinated and, once lost, were never regained by the tumor. Thus, tumors are dynamic and continually become more and more malignant.

Greene (1957) was interested in the early development of breast and uterine adenocarcinoma of rabbits. He transplanted rabbit tumors into the anterior chamber of the eyes of guinea pigs, an experimental form of metastasis. When the tumors were small and had not metastasized in the rabbit, they would not grow in the guinea pig's eye (an immunologically protected area). Transplants of the tumors, which had grown large and had metastasized in the rabbit, grew in the eye of the guinea pig. Greene concluded that tumors undergo changes with time. In this case, early tumors could not metastasize, whereas older ones could. They had progressed, expressed the ultimate in malignancy in the host rabbit (i.e., metastasized), and were autonomous.

Charles Huggins performed a series of brilliant experiments that characterized important features of tumor development (Huggins and Hodges 1941). He found that the prostatic hyperplasia which occurs in elderly dogs was dependent upon androgens. Antiandrogen therapy reduced the hyperplasia. As a consequence, Huggins, a urologist, decided to treat human patients afflicted with adenocarcinoma of the prostate by castration to remove testosterone. Dramatic results were obtained. Bedridden men with painful metastases that were destroying their spines were relieved of pain, their spines healed, and they were able to return to work and lead productive lives. Interestingly, these individuals were never cured. The tumors eventually reappeared, were now insensitive to hormones, and the individuals died. The tumors had progressed and lost their dependence on androgens (Hanks et al. 1993).

Jacob Furth (1953) performed experiments leading to the understanding of the concept of dependency. He interfered with the regulation of thyroid function by the pituitary gland. The pituitary gland produces thyroid stimulating hormone (TSH), which stimulates the thyroid gland to produce thyroid hormone. When enough thyroid hormone is produced, it in turn causes a feedback control of TSH production by the pituitary gland, thereby maintaining status quo. Furth blocked the production of thyroid hormone in mice, which in turn caused the pituitary glands to synthesize TSH. The thyroid glands responded to the TSH by becoming hyperplastic (producing more cells to alleviate the apparent need for thyroid hormone). Eventually, metastasizing adenocarcinomas of the thyroid developed in some of them. Transplants of these thyroid metastases did not grow when transplanted in appropriate strains of animals, but if the recipients had been thyroidec-tomized (resulting in high levels of TSH) the tumors grew. Thus, the growth of these metastasizing adenocarcinomas of the thyroid gland depended on high levels of TSH. After repeated transplantation to thyroidectomized animals, this depen-dence on TSH was lost, and the tumors grew in animals with a thyroid in the presence of normal levels of TSH. In current parlance, it would be said the tumors had progressed and lost their dependence on TSH.

The development of mammary cancer in mice has also given us important insights into the events in carcinogenesis. A baby mouse may receive the mam-mary tumor virus when it first suckles, but tumors do not appear on the average until about 40 weeks of age (latent period) (Bittner 1936; Held et al. 1994). The first tumor usually appears during pregnancy, grows rapidly, but surprisingly regresses and may even disappear between pregnancies. The tumor reappears dur-ing the next pregnancy, and may again regress after delivery. Eventually, either dur-ing or between pregnancies, the tumor begins to grow in the unrelenting progres-sive manner typical of adenocarcinomas of the breast (Foulds 1969). In this case, these tumors were initially dependent on the hormones that normally control mammary development during pregnancy and lactation, but the tumors lost their dependence on those hormones as the tumors progressed to the autonomous state. Are there lessons in these facts that can be used clinically?

1.14   Selection and cellular heterogeneity

Tumors are not comprised of homogeneous populations of cells; if they were, all cells of the tumor would be expected to behave identically. For example, all cells of a tumor would have the competence to metastasize or no cells of that tumor would have that capability. Similarly, all cells of a particular tumor would be expected to respond to a single chemotherapeutic agent in a like manner. Because we know that cells within a tumor behave differently, it logically follows that tumors are composed of multiple populations of cells with different abilities to respond to environmental stimuli. Some of these cells have growth advantages over other cells and can be selected for. Then the phenotypic traits of these cells are expressed and the behavior of the tumor changes.

Selection has long been recognized as a mechanism of progression. Selection can occur only if there is a heterogeneous population of cells. Thus, in Huggins's adenocarcinoma of the prostate and in Furth's adenocarcinoma of the thyroid, many malignant cells were responsive to androgens and to TSH, respectively. Apparently, there were other stem cells that were independent and eventually over-grew the responsive ones as evidenced by the eventual loss of dependence on the hormones. Cells are selected for or selected against by the environmental condi-tions. The net effect of progression is increased malignancy.

Selective pressures may also be generated by the tumor itself. Stevens (1967) developed a strain of teratocarcinoma called OTT6050. The teratocarcinoma cells, developed from cells of an early mouse embryo, were transplanted into the testes of adult mice. The teratocarcinomas that resulted were distributed to investigators throughout the world. Interesting cell sublines were isolated from them. One of these sublines, called F9, is an undifferentiated embryonal carcinoma, which, when appropriately treated with retinoic acid, differentiates primarily into endoderm (Strickland et al. 1980). Rizzino and Crowley (1980) demonstrated that large numbers of F9 cells grew successfully in defined media, failed to differentiate, and could be maintained indefinitely. But if small numbers of F9 cells were placed in the defined medium, they failed to grow. Rizzino and Crowley concluded that F9 cells each secreted an autocrine growth factor that reached critical concentration and stimulated cell growth, at appropriate cell density. Another subline from OTT6050, called C44, makes a toxic substance to which ECa 247 (a line of pretrophectodermal embryonal carcinoma derived from OTT6050) is especially sensitive (Gramzinski, Parchment, and Pierce 1990).

Thus, F9 and C44, components of the parent strain of OTT6050, made a growth factor or a toxic factor that affected themselves or other sublines of OTT6050. The conclusion is inescapable that autoselection by clonal lines within the tumor must play a major role in the selection that is an essential part of progression (Pierce and Parchment 1991).

If tumors are heterogeneous in their cellular composition – which indeed many, if not all, are – it logically follows that they are mosaics in exactly the same sense the multicolored skin of the cow is a mosaic. This would mean that each subline of heterogeneous populations of cells would have a patch size. It is known from studies of early development that a threshold number of cells is required for the expression of a phenotype and its function (Grobstein and Zwilling 1953). Thus, one could argue that in the parent OTT6050 tumor, the C44 cells would produce enough toxic material to kill ECa 247 cells only if the C44 cells had attained an adequate patch size (Pierce and Parchment 1991). Reconsider the experiments of Greene (1957) on breast and uterine adenocarcinoma of rabbits transplanted into guinea pig eye: conceivably the patch of cells capable of metastasis was not large enough to express the phenotype in the small nonmetastasizing tumors, but with time, patches of metastasizing cells became large enough to express their phenotype. Similarly, in Furth's experiments with thyroid metastases and TSH dependence, cells not dependent on TSH would eventually form a large enough patch in the metastases of TSH-dependent thyroid carcinoma cells to express their phenotype, overgrow the rest of the TSH-dependent patches, and ultimately dominate (Furth 1953). Molecular explanations for these phenomena are not known.

Heppner (1982) made fine contributions to the understanding of progression and the importance of heterogeneity of tumors. She developed subpopulations of cells from a viral-induced adenocarcinoma of the mammary gland in mice that are named 68H, 4.1, and 168. These cell lines varied in growth rates and ability to metastasize when grown subcutaneously. In addition, the growth of 68H was stimulated by media conditioned by 168, and growth of 168 was slowed by 4.1 tumor cells. Intratumor selection must play a role in progression.

What remains now is an explanation for the mechanism of producing cellular heterogeneity in tumors. The early tumor is composed of a mixture of clones representing the heterogeneity of normal cell types responding to the carcinogenic event, and, as shown earlier, selective pressures tend to reduce this heterogeneity so the tumor becomes monoclonal, homogeneous in composition, and autonomous (Pierce and Parchment 1991). Nowell (1976) has postulated the occurrence of mutations to explain heterogeneity in monoclonal tumors. Some mutations might impart a growth advantage on the mutated cell. Ultimately, a clonal patch would develop, of sufficient size, to express whatever characteristics were specified by the mutated cells. This phenomenon is called clonal expansion. It is also conceivable that chromosomal translocation could place promoters for one gene in control of another, cause a growth advantage, and/or other effects which could again lead to cellular heterogeneity and the changes in karyotype that are common in progressed cancer.

Whatever the mechanism of heterogeneity, it is clear that during progression selective events tend to make the tumor both monoclonal and more malignant. These events are made in the face of other pressures such as genetic instability that cause diversity of cell types. The result is a dynamic interaction leading to the development of tumors that are ever more anaplastic and autonomous with time. Tumors are continually changing in their biologic behavior.

1.15   A developmental concept of cancer

Cancer has been described variously as aberrations of growth, differentiation, or organization of cells. These are the basic processes of development that in a coordinated and controlled manner evolve the adult organism. Prior to puberty, they are regulated for accrual of mass, but when adult stature has been achieved reregulation occurs to maintain status quo of tissues, a process known as tissue renewal. Accurate knowledge of tissue renewal dates from the use of isotopes (Doniach and Pelc 1950). Prior to that, tissues of the adult were viewed as well differentiated, and little thought was given to their maintenance, much less to the presence in them of determined but undifferentiated cells from which the functional differentiated cells could be renewed or could serve as the target in carcinogenesis. In the absence of this information, an explanation was required for the mechanism by which undifferentiated malignant cells developed in mature differentiated tissue.

It was decided erroneously that dedifferentiation was the process by which differentiated cells became anaplastic during carcinogenesis. The rationale was based on the well-known fact that dedifferentiation occurs normally in amphibians. For example, if the lens of the eye of an amphibian is removed, the cells of the iris dedifferentiate. They extrude their pigment granules and become undifferentiated in appearance, and during the recovery period, they redifferentiate not only iris but a new lens as well (Yamada and McDevitt 1974). It is the ability of dedifferentiated cells to regain potential that was repressed in the process of determination (in the case cited, the iris cells reacquired the potential to make lens) that distinguishes dedifferentiated cells from undifferentiated ones which do not regain potential. A malignant tumor has never regained embryonic potential.

As we show subsequently, the variable degrees of differentiation in carcinomas are not a reflection of varying degrees of dedifferentiation, but represent the converse, varying degrees of differentiation of malignant stem cells. Malignant stem cells are, in turn, derived from undifferentiated determined normal stem cells of the tissues responsible for tissue renewal (Wylie, Nakane, and Pierce 1973). In carcinogenesis they give rise to equally undifferentiated malignant stem cells. The normal stem cells are derived by differentiation in the embryo. In the adult, if they become neoplastic, they form a caricature (a gross misrepresentation) of tissue renewal, and, in the case of teratocarcinomas, of embryogenesis (Pierce and Speers 1988). To understand the developmental concept of cancer better requires a short excursion into embryology.

Embryos begin as a single cell with the genetic potential to create the adult organism given the appropriate environment. Each of the cells at the 8-cell stage in the mouse is totipotent (Graham and Wareing 1976; Slack 1983), but speciation begins at the 8-cell stage and the potential of the involved cells is reduced by the 16-cell stage (Tarkowski and Wroblewska 1967). For example, the ball of 16 cells is composed of an outer layer surrounding one or two inner cells. The outer will form extra embryonic tissues including the trophectoderm, and its descendants will attach the free-floating embryo to the mother and form the placenta. The embryo is cystic at the 64-cell stage and known as the blastocyst. The cyst wall is made up of 52 trophectodermal cells, and the inner cells, now 12 in number, referred to as the inner cell mass (ICM), are attached to the inner surface of the blastocyst (Figure 1–16). They will form the embryo.

The first differentiation in cells destined to become the embryo occur in the embryo at the late blastula stage. The surface layer of cells on the ICM become different from the rest and form the primitive endoderm, which will form the digestive tract of the adult. The balance of the ICM cells become embryonic epithelium. Mesoderm develops after collapse of the blastocyst and implantation of the embryo into the wall of the uterus. It occurs where primitive endoderm lies over embryonic epithelium (Slack 1983), and there is speculation that the endoderm induces the embryonic epithelium to form the mesodermal cells. Mesoderm has the potential for production of connective tissue, fat, bone, cartilage, and so on. The embryonic epithelium has the potential to make brain or skin.

Ectoderm, mesoderm, and endoderm are referred to as the three germ layers (Figure 1–17), and each of these new tissues is said to be determined for a particular dif-

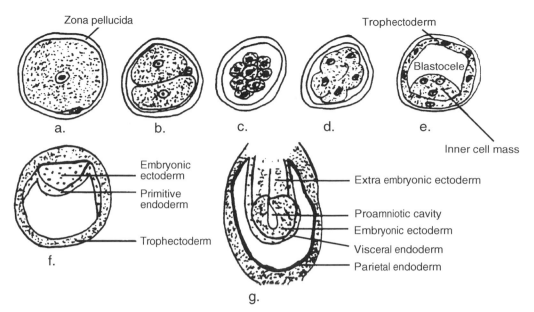

Figure 1–16. Chart of early development of the mouse. (a) Fertilized egg contained in the zona pellucida. (b) and (c) Two- and eight-cell stages. (d) Late eight-cell stage after compaction. This creates a layer of outer cells which encase one or two inner cells that will form the embryo. (e) Early blastocyst composed of 32 cells; the inner cell mass will form the embryo proper. (f) This is the late blastocyst, which has hatched from the zona pellucida. Primitive endoderm has differentiated from the surface of the inner cell mass and begun to line the trophectoderm. (g) Postimplantation mouse embryo. The embryonic epithelium and overlying proximal endoderm will form the embryo. Embryoid bodies resemble this part of the embryo. All of the other tissues present will form the placenta and extraembryonic membranes.

Embryonic origin	Adult derivative
Ectoderm	Skin
	Brain
	Breast
	Sweat glands
	etc.
Mesoderm	Fibrous tissue (connective)
	Cartilage
	Bone
	Muscle
	etc.
Endoderm	Gut
	Liver
	Lung
	Pancreas
	etc.

Figure 1–17. Three germ layers and the tissues derived from them.

ferentiation (Gehring 1968). Each has had its potential reduced in relationship to its precursor, cannot revert to the precursor, and can only express the differentiated phenotype in the appropriate environment. Note that these primitive cells are undifferentiated and do not express the gene products of the lineage for which they are determined. For example, determined skin cells have the potential to become keratinocytes and express keratin, but in the early embryo the environment allows them to proliferate only and cover the embryo. In the changed environment of the adult, these same cells express keratin.

Proof for the idea that the expression of the phenotype by determined cells depends on the appropriateness of the environment and that determination of cells is permanent stemmed from studies of drosophila (Gehring 1968). Clumps of embryonic cells named imaginal discs develop in drosophila embryos at the time organs form. An imaginal disc will develop into abdomen, another into wing, another into thorax, and so on. Imaginal discs determined for thorax were transplanted into the abdomens of adult flies. This foreign environment allowed for proliferation, but the cells did not express the differentiation. They were serially transplanted for several generations, and when returned to the appropriate place in the embryo, they differentiated into thorax (Gehring 1968). Two conclusions are drawn from this study: the determined state is stable and heritable, and its expression depends on environmental conditions. This leads to the interesting conclusion that the offspring of determined stem cells of the embryo which do not express the phenotype are the same as the ones in the adult that do so. The latter are regulated not for accrual of mass, as in the embryo, but for maintenance of status quo. When a cell is lost, a stem cell divides and sends another along the maturational pathway.

It was postulated that the undifferentiated determined stem cells of adult tissues respond to carcinogenesis and become stem cells of tumors (Pierce 1974). The malignant phenotype is superimposed on the histotypic determination, and the resultant tumor is a caricature of the process of tissue renewal.

As mentioned previously, after puberty when growth is complete, the rate of cell division is reregulated to maintain tissue. In a variety of tissues as exemplified by colon, intestine, skin, testis, bronchus, or bladder, tissue renewal is mediated by a population of precisely controlled stem cells. As stated, these are undifferentiated cells determined for their particular differentiation. When they divide, one cell remains behind as a stem cell; the other enters the maturational pathway, differentiates, functions, and in the intestine after about four days becomes senescent, dies, and is cast from the body (Figure 1–18). In the intestine, the stem cells are at the base of the villi in crypts (Snell 1978). They do not migrate along the villi; they know their place and respond to stringent homeostatic control because cell production equals cell loss. They retain the ability to respond to environmental perturbations by undergoing hyperplasia, atrophy, and so on. Stem cells are the reactive cells of a tissue; their differentiated progeny can only differentiate or die. The regulation of these processes is rapidly becoming understood with the identification of numerous regulatory molecules.

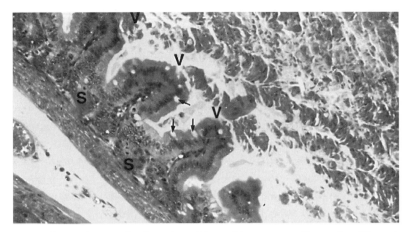

Figure 1–18. Photomicrograph of section of bowel in teratoma illustrating life cycle of interstitial epithelium. Stem cells are in the crypts (S), one remains behind as a stem cell, the other migrates along the wall of the villus expressing the determined phenotype as it goes (mucous cell at arrows). After four to five days the cells become senescent and are sloughed from the tips of the villi (V) into the fecal stream.

In this regard the existence of negative regulation of stem cells by chalones has been postulated. Chalones are believed to be tissue-specific, reversible inhibitors of stem cell proliferation synthesized by the differentiated cells of a tissue (Bullough 1962). If, for example, there were too many differentiated cells, enough chalone would presumably be produced by them to inhibit replication of only the stem cells of that differentiated tissue, and over time the differentiated cells would become senescent and die, restoring status quo. Conversely, if there were too few differentiated cells there would be correspondingly less chalone, and the stem cells would replicate in response to positive stimuli and restore the tissue. The need for such regulation is widely accepted, but the chalone field fell into disrepute when, despite intensive effort, molecules with attributes of chalones could not be identified (Langen 1985).

In the past it was difficult to identify the cells initially targeted in carcinogenesis because of the long latent period. Embryonal carcinoma cells, the stem cells of teratocarcinomas, develop only if primordial germ cells (the stems cell of the species) are present in the fetal testis at the time of carcinogenesis (Stevens 1967). This established the origin of this cancer from normal stem cells that had undergone transformation. In this situation, the normal and malignant stem cells were equally undifferentiated by ultrastructural criteria. Irrespective of the mechanism of carcinogenesis, there is no need to consider dedifferentation of the primordial germ cells as an explanation for the generation of undifferentiated cells of the embryonal carcinoma, or indeed of other malignant tumors. The determination of the normal stem cells was altered in the process of carcinogenesis, however, and given the appropriate environment, the initiated cells will express the malignant phenotype that is superimposed on the normal phenotype. The malignant stem cells derived from the normal stem cells are undifferentiated from the outset and bear close resemblance to their normal counterparts (Pierce, Stevens, and Nakane 1967). The histiotypic determination of the normal stem cell is not altered during carcinogenesis because malignant stem cells, if they can differentiate, always express the appropriate determined histiotype. What appears to be altered in the malignant phenotype is the regulation of cell proliferation, differentiation, and invasiveness of the neoplastic cells. The malignant cells are no more or less differentiated than their normal precursors, have no increase in potential, and thus cannot be dedifferentiated.

Not all tissue types in the adult have stem cells. What, then, accounts for the development of tumors in these tissues? Mesenchymal tissues such as fibrous tissue, cartilage, bone, or fat do not appear to have a stem cell compartment. Under certain conditions, such as wound healing, the fibrocyte nucleus becomes vesiculated and enlarged in becoming a fibroblast. The fibroblast synthesizes DNA and divides. There is no contradiction of facts or logic in applying the concept that carcinogenesis involves undifferentiated cells capable of dividing, whether from tissues with or without a stem cell component.

But what about brain tumors? There is no renewal of nerve cells in the adult. If nerve cells do not divide, are brain tumors an exception to the rule that tumors occur only in cells capable of division? Tumors of nerve cells do not occur in the adult. Neuroblastoma (the name indicates an origin in embryonic nerve cells) occurs congenitally or in very young children. The other brain tumors are derived from glial cells that are capable of cell division. Glial cells are the equivalent in the brain of connective tissue in other organs, and give rise to astrocytoma, glioblastoma multiforme, and oligodendroglioma.

Mature liver cells, given the appropriate stimulus, can similarly divide (Bucher and Malt 1971). Differentiated functional liver cells retain the capacity to divide and serve as the target in carcinogenesis. They give rise to carcinomas, and the tumor cells closely resemble normal liver cells morphologically and functionally. In addition, Sell and Leffert (1982) have reported stem cells in the liver, and it is a small leap of imagination to suppose that they too are targets of carcinogenesis.

In this regard, the stem cells of leukemias closely resemble their corresponding hematopoietic (blood-forming) stem cells (Greaves et al. 1983). Even though it has not yet been shown directly that the normal stem cell gives rise to the leukemic stem cell, in view of the experience with embryonal carcinoma, and in view of the similarity of appearance of normal and malignant stem cells, it can be stated that normal hematopoietic stem cells are the target in leukemogenesis. In a similar vein, adenocarcinoma cells of the colon resemble normal colonic stem cells and transformed fibroblasts in vitro closely resemble normal fibroblasts. Thus, it can be concluded that undifferentiated neoplastic cells originate from normal determined, undifferentiated stem cells, and the anaplastic appearance of tumors is the result of the preponderance of these cells with their inability to organize appropriately, in relationship to the number that differentiate.

We have seen in the previous discussion that malignant stem cells are no more or less differentiated than their normal stem cells, but have the malignant phenotype superimposed on the determined histiotype. What is the significance of this for cancer biology? There is no need to consider dedifferentiation as a mechanism of carcinogenesis, and the neoplastic cells given the appropriate environment should have capacity to differentiate. But is there evidence for this?

Numerous experiments have demonstrated differentiation of malignant stem cells into benign cells (Pierce and Speers 1988; see also Sachs 1993; and discussion in Chapter 6). The experiments have involved embryonal carcinoma cells, the stem cells of teratocarcinomas (see appendix), squamous cell carcinomas cells (see appendix), and leukemic cells (Gootwine et al. 1982). The malignant stem cells in each of these tumors were shown to differentiate into mature functional apparently normal cells of the appropriate lineage. It is now accepted unequivocally that some cancer cells can differentiate into terminally differentiated benign, if not normal cells, whereas others display abortive differentiation, and some display so little differentiation that it is almost impossible to identify the tissue involved and make a diagnosis. Because of the lack of differentiation in some tumors, many oncologists view cancer cells as having a block in differentiation.

Some embryonal carcinoma cells display no evidence of differentiation but can be induced to differentiate with retinoic acid or other agents (Strickland, Smith, and Marotti 1980; McBurney 1993). This indicates that the anaplastic cells of cancer have the potential for differentiation, although they may not express it, and reinforces the idea that carcinogenesis does not alter the original histiotypic determination of the cells, it superimposes the malignant phenotype on it.

Retinoic acid is capable of inducing differentiation in a variety of other tumors. Is this the prototype experiment for a new type of therapy, differentiation therapy, for cancer with metastases?

Like the determined imaginal disc cells of the fly, cancer cells only express the normal (nonmalignant) phenotype when placed in the appropriate environment. Embryonal carcinoma cells express the malignant phenotype. But if they are placed in the blastocyst, they give rise to cells that differentiate normally. In the blastocyst they behave as normal embryonic cells in the sense that they respond to signals produced by the embryo, differentiate appropriately, and in the adult their offspring respond to homeostatic controls.

How universal is the fact that embryos can regulate their closely related cancers? It has been shown that leukemic stem cells injected into 10-day-old mouse embryos may be normalized to the point that the animals are chimeric in their leukocytic tissues (Gootwine et al. 1982). Neuroblastoma cells, injected into the neural crest migratory route, do not express tumors in expected numbers, but the mechanism of suppression of tumor formation is not known (Podesta et al. 1984).

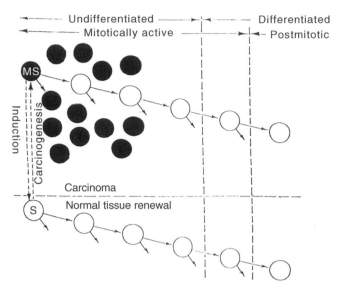

Figure 1–19. Chart of carcinoma as a caricature of the process of tissue renewal and embryogenesis. The normal process of renewal is illustrated at the bottom. Stevens demonstrated development of carcinoma from stem cells, Pierce demonstrated differentiation of malignant stem cells, Brinster demonstrated regulation of embryonal carcinoma by the blastocyst. The caricature, the gross misrepresentation, is the overproduction of malignant cells in relationship to the number that differentiate.

Melanoma cells injected into the embryonic skin on the day normal melanoblasts arrive in the skin do not produce melanomas in expected numbers. In this situation, destruction of the melanoma cells by a mechanism responsible for programmed cell death apparently plays an important role (Parchment and Pierce 1989).

The concept that carcinoma is a caricature of the process of tissue renewal uses the term "caricature" only in the sense of a gross exaggeration of the normal (Pierce 1983). The caricature is illustrated in Figure 1–19. The malignant stem cells (MS) generate a myriad of anaplastic carcinoma cells for every cell that enters the differentiation pathway. This is the gross exaggeration.

The normal process followed by stem cells is illustrated in the lower part of Figure 1–19. The normal stem cell proliferates only in response to an unknown environmental signal that says "a senescent cell has been sloughed, and a new one is required." The malignant stem cell has the capacity for proliferation, producing many copies of itself for each one that differentiates. The mechanism for carcinogenesis is not known, but current dogma suggests mutation occurs in genes either for growth factors or receptors, or stable changes in regulatory circuits that precludes normal responses (Chapter 4). We are in the unhappy situation of knowing what cells can be made to do, but we do not know what they normally do! What is important is that an appropriate environment, such as the embryonic environment in the formation of chimeric mice, can correct the aberration. Essential to differentiation therapy is understanding differentiation mechanisms such as how the blastocyst corrects the embryonal carcinoma cell and incorporates it into the inner cell mass as a functional and apparently normal cell.

1.16   Apoptosis

Nearly a century ago it was discovered that certain cells in both adult and embryonic tissues of many organisms, from *C. elegans* to mammals, die independently of any exposure to toxic xenobiotics (Kerr, and Currie 1972; Wyllie, Kerr, and Currie 1980; Bowen and Bowen 1990; Hengartner and Horvitz 1994). This naturally occurring type of cell death in healthy tissues acquired the name *apoptosis*, which described the appearance of the dying cell as subcellular structures fall away from the residual structure as "leaves fall from a tree in autumn." Apoptosis plays many important biologic roles, specifically eradicating redundant or spent cell types during development to prevent the formation of ectopic tissues (Pierce et al. 1990; Parchment 1993); removing embryonic tissues to generate the morphology of the adult (e.g., between the prospective digits in the developing hand and foot, or the tail of anura), balancing mitosis in renewing tissues to regulate tissue mass and selecting clones during lymphocyte maturation (Goel 1983; Sandford et al. 1984; McCarthy et al. 1992), and removing genetically damaged cells to prevent oncogenesis (Lennon, Martin, and Cotter 1991).

Apoptosis causes the appearance of several distinctive cytologic features. The cytoplasm and chromatin condense (*pyknosis*), and the plasma membrane contracts, causing the dying cell to withdraw from its neighbors except at points of adhesive plaques; yet most, if not all, of the intracellular organelles, including mitochondria and lysosomes, remain active (Bowen and Bowen 1990). Then the nucleus fragments into vesicles called apoptotic bodies that contain condensed chromatin fragments surrounded by condensed cytoplasm. Near the end of the process, the apoptotic cells and bodies are recognized and phagocytosed by adjacent cells, to be destroyed subsequently in lysosomes. Although molecular markers in some model systems can be used to monitor apoptosis, there appears to be no universal molecular marker at this time. Degradation of genomic DNA into nucleosome-sized fragments, activation of a Ca^{2+}, Mg^{2+}-dependent endonuclease, flip-flop of phosphatidyl serine from the inner to the outer leaflet of the plasma membrane, and up-regulation of cell surface receptors for vitronectin and cell surface transglutaminases have all been proposed as markers of apoptosis (Kyprianou et al. 1991).

These cytologic features provide a straightforward distinction between apoptosis and a second process of cell death called *necrosis*, in which the dying cell shows swollen cytoplasm and mitochondria, clumped chromatin, and nuclear membrane and plasma membrane rupture, although late-stage apoptosis in the absence of phagocytosis and lysosomal degradation can resemble necrosis (Kerr et al. 1972). Perhaps the most important distinction between apoptosis and necrosis is a biologic one: apoptosis is a controlled process that keeps intracellular substances contained within membranes at all times and therefore does not culminate with an inflammatory reaction; necrosis results in cytolysis and an inflammatory reaction which can often cause serious tissue damage (Kerr et al. 1972). Especially in the renewing epithelial tissues of the adult, apoptotic cells can be found completely surrounded by differentiated or proliferating cells.

This precise control of apoptosis, that is, its regularity, predictability, and localized nature, led to the notion of an intracellular, genetically programmed pathway that kills a cell from within and protects surrounding cells from the presumed lethal hit inside the cell, and hence the term "programmed cell death." The presence of isolated dead cells in the midst of a milieu of viable cells did not seem explainable unless there was a mechanism for such cellular suicide, and suicide is a reasonable explanation for the lack of adverse effects on neighboring cells. However, extracellular substances, such as tumor necrosis factor and monoclonal antibodies against epitopes on T- and B-cells, trigger apoptosis, and in the murine blastocyst, extracellular H_2O_2 triggers apoptosis of a few cells, expressing a developmental program that is no longer needed but spares the majority (Pierce, Llewellyn, and Parchment 1989; Parchment 1993). These data prove that death from within ("suicide") is not required to obtain an isolated apoptotic cell in the midst of living cells. Highly selective apoptosis can also be caused by extracellular factors to which all cells in the vicinity are exposed but to which only a handful respond (Parchment 1993). This more recent data indicate that apoptosis can be "murder" as well as suicide.

Recently, oncogenes and suppressor genes have been implicated in the regulation of apoptosis and the role of cell cycle progression in the commitment to the cell death program, both transcription factors and signal transduction mechanisms. Many of the functions of proto-oncogenes are required for apoptosis, and dysfunction of oncogenic forms can inhibit apoptosis and lead to an accumulation of cells (hyperplasia) that can contribute to malignancy (see Chapter 5).

1.17 Summary

This is the background on which the medical profession currently works: once cancer cells have progressed and metastasized, they cannot be extirpated or killed with x-ray; the only hope is treatment with cytotoxic agents. These lack specificity and destroy rapidly growing normal cells as a complication of treatment. The oncologist has little range for error in using these extremely toxic poisons, but this is the best hope for palliation and rarely a cure for these patients.

The need for an understanding of the pathology of cancer is underscored by the fact that carcinoma of the pancreas, for example, has a cure rate of about 5%. Doesn't that tell us that current treatments are inadequate? Should we persist in administering cytotoxic therapy with its deleterious side effects in such tumors hoping that a particular patient will be one of the lucky few? Should we admit we really don't have a therapy for such malignancies and expedite the search for new cytotoxins, or even better, for alternatives to cytotoxic treatment for such carcinomas? The key to this approach may lie in the observation that embryonic fields in every situation tested are able to reregulate malignant cells. In order to develop new and effective modalities for treating cancer, we must first understand the mechanisms of determination (Plasterk 1992) and the controls of proliferation and differentiation in the embryo.

2

Metastasis

Robert G. McKinnell

Metastasis appears to be a highly selective process that is regulated by a number of imperfectly understood mechanisms. Surprisingly, this view may be more optimistic than one that postulates that cancer metastasis is a random event. A random event cannot be characterized or manipulated.

I. J. Fidler and I. R. Hart 1982

The detection of metastases constitutes decisive evidence for categorizing a proliferating primary lesion, previously of uncertain potential, as neoplastic and "malignant," and the phenomenon is a topic of unparalleled importance in cancer medicine and biology.

D. Tarin 1992

2.1 Introduction

Cancer is infrequently a localized disease. This statement refers to the propensity of malignant neoplasms to disseminate and grow as secondary tumors in the body of the host, frequently before the primary tumor is discovered. Cancer cells detach from the primary tumor, translocate to distant sites, and grow as secondary colonies at the new anatomic locations. The establishment of secondary tumors, no longer contiguous to the primary tumor, is known as metastasis (Figure 2–1). Normal cells are strictly confined to defined anatomic domains; malignant cells' proclivity for growth at sites remote from their origin is a cardinal attribute of malignancy.

Although cancer has been known at least since the time of Hippocrates, who named this group of diseases (Ewing 1928) (Chapter 1), it was not until Recamier (1829) described a brain tumor that had developed from a primary carcinoma of the breast that such secondary tumors were named. He coined the term "metastasis"

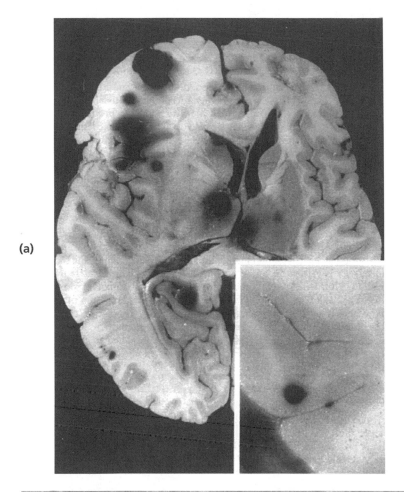

(a)

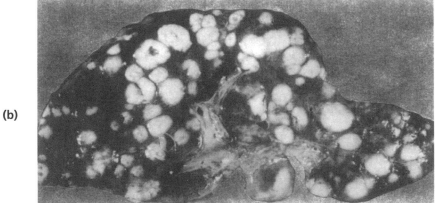

(b)

Figure 2–1. The devastation of metastatic cancer to human tissue: (a) a gastric adenocarcinoma metastatic to the brain (inset shows a metastatic lesion in the cortex-subcortical white matter interface). (From Mørk, S., Laerum, O.D., and deRidder, L. 1984. Invasiveness of tumours of the central nervous system. In *Invasion, experimental and clinical implications,* ed. M.M. Mareel and K.C. Calman, pp. 79–125. Oxford: Oxford University Press. By permission of Oxford University Press.) (b) multiple metastatic lesions to the liver. (From Warren, J.R., Scarpelli, D.G., Reddy, J.K., and Kanwar, Y.S. *Essentials of General Pathology.* New York: Macmillan Publishing Company. Copyright © 1987 The McGraw-Hill Companies. Reproduced with the permisssion of the McGraw-Hill Companies.)

to describe the local infiltration, invasion, and growth of secondary tumors (Wilder 1956). Later, Paget observed that many metastatic tumors are not capriciously disposed in the body; rather, their distribution follows a pattern specific for each cancer. He likened disseminated cancer cells to seeds that grow best in hospitable soil (Paget 1889).

James Ewing, in the first issue of the *Journal of Cancer Research,* compiled characteristics of cancer cells; he listed hyperplasia,[1] loss of polarity, promotion of connective tissue growth, anaplasia, infiltrative growth,[2] and metastasis. Of these six characteristics of cancer, *metastasis* was asserted to be the "most convincing evidence of lawless growth" (Ewing 1916). Although metastasis is still convincing evidence of cancer, it is clearly not lawless. However, the dissemination of malignant cells from a primary tumor with subsequent growth as multiple secondary colonies is known to follow a number of predictable steps in most patients with advanced metastatic disease. But the process is inadequately understood to provide significant hope to those unfortunate individuals with metastatic cancer.

Lawful (nonrandom) aspects of metastasis include the stepwise process of cell activities ("the metastatic cascade," Viadana, Bross, and Pickren 1978a) that leads to successful dissemination of malignant cells. Components of the metastatic cascade are cellular activities shared with normal cells. It has been postulated that knowledge of the mechanisms of normal cell translocation in a healthy body may lead to enhanced understanding of the comparable phenomena in neoplastic cells. The predictability, in general, of metastatic tumor sites is another aspect of the lawful behavior of translocated cancer cells. Clearly, metastasis is not the result of fortuitous or haphazard events – were it so, there would be no comprehension of this cardinal aspect of malignant cell behavior.

It is perhaps wise to consider the significance of therapeutic approaches to cancer in the context of the present chapter. Cancer treatment (Chapter 8) may include surgery, radiation, and chemotherapy. If a tumor is localized, then surgery is an adequate treatment for the obvious reason that a localized cancer ceases to exist with surgical obliteration. Radiation therapy is also directed at a limited area of the patient afflicted with cancer, and, with few exceptions, is palliative. Chemotherapy, however, is designed to eliminate disseminated cancer cells at *all* sites in the body, which is of course not possible with either surgery or radiation. Thus, chemotherapy is, in fact, therapy of metastasis, and it, too, is essentially palliative.

In the context of the preceding paragraph, note that cancers are not comprised of homogeneous populations of cells (Chapter 1). For example, a parent tumor

---

[1] Ewing's article was written in 1916 and his use of the term "hyperplasia" is now antiquated. Contemporary pathologists use the term "increased cellularity" or "cellular hyperproliferation" (Kern and Vogelstein 1991) and retain hyperplasia only for benign conditions of increased mitotic activity.

[2] "Invasion" is a more generally recognized term.

cell line was shown to contain coexisting cells with differing metastatic competence (Fidler and Kripke 1977). Similarly, aggressive neoplasms may have rapidly expanding subpopulations (clones) of cells with different chromosome constitutions coexisting within the primary tumor (Nowell 1992). Not surprisingly, cancer cells exhibit heterogeneity with respect to cell surface receptors, enzymes, growth, morphology, and also competence to invade and form metastases (Fidler 1978, 1990; Nicolson 1982; Tarin and Price 1979). Subpopulations of neoplastic cells differ with respect to their sensitivity to antineoplastic agents (Heppner et al. 1978; Heppner 1993). Accordingly, although cytotoxic chemotherapy has the potential to kill many kinds of mitotically active cells, tumors are "societies" of diverse cell types, not all of which have the competence to establish metastases and, thus, not all of which need be the target of exceptionally toxic therapy. The minority of cells that can form metastases must have the competence to perform a series of steps known as the metastatic cascade.

2.2 The metastatic cascade

A cascade may be defined as a series of interrelated stepwise waterfalls. Interruption of any of the component falls disrupts water flowing at lower parts of the cascade and at the mouth of the cascade. So it is with the metastatic cascade (Viadana, Bross, and Pickren 1978a, 1978b; Weiss 1985; Fidler 1997). A series of interrelated events, a "multistep process" (Mareel, Bracke, and Storme 1985; Mareel, Van Roy, and Bracke 1993), leads to metastasis, and disruption of any of the events in the cascade, at least in theory, restrains the establishment of disseminated cancer (Bastida 1988; Stracke and Liotta 1992).

The interrelated events of the metastatic cascade are listed in Table 2–1 and illustrated diagrammatically in Figure 2–2. They are considered individually in the text that follows.

Table 2–1. *Events in the metastatic cascade*

•Disruption of the basement membrane

•Cell detachment (separation)

•Cell motility

•Invasion

•Penetration of the vascular system

•Circulating cancer cells

•Arrest (stasis)

•Extravasation and proliferation

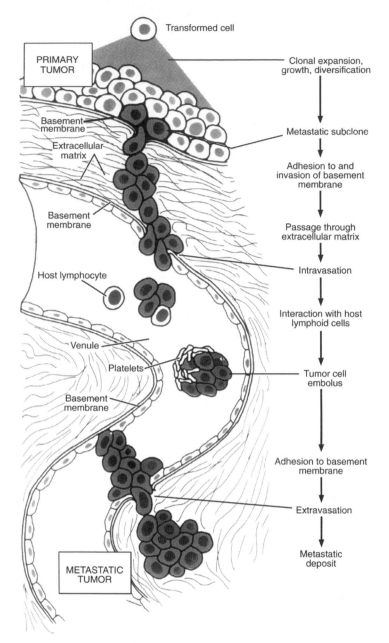

Figure 2–2. Cascade of events that lead to metastasis. Reading from the top of the diagram to the bottom, a transformed cell gives rise to mitotic progeny that form a primary tumor. The tumor is a mosaic of cells, some of which have the competence to give rise to metastatic colonies. The metastatic subclones (indicated with darkened cytoplasms and nuclei) bind to and disrupt the basement membrane on which they rest. They invade the extracellular matrix. Subsequently, they bind to and disrupt the basement membrane of a vessel as a prelude to intravasation. Thus, invasion requires, among other traits, cell motility and ultimately cell detachment from the primary tumor. The cells must also have the competence to penetrate the vascular or lymphatic system (intravasation) and circulate (either alone or as a component of an embolus) until they arrest by adhesion to the basement membrane of a vessel at a distant site. The nascent metastatic cells then exit the vessel (extravasation) at a site where they form a metastasis. (From Cotran, R.S., Kumar, V., and Robbins, S.L. 1994. *Pathological Basis of Cancer,* 5th ed. Philadelphia: W.B. Saunders Company.)

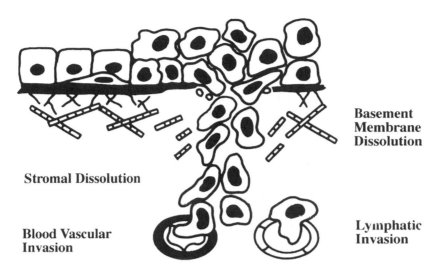

Figure 2–3. Destruction of the basement membrane and invasion with vascular intravasation. Carcinoma cells, like other epithelial cells, rest on a basement membrane (shown as a black line here). A critical stage in tumor metastasis is the transgression of the basement membrane after membrane disruption. This permits carcinoma cells to invade the connective tissue stroma and gain access to capillaries and lymphatic vessels. Blood- or lymphborne carcinoma cells are then transported to distant sites in the body. (From Liotta, I A., Rao, C.N., and Barsky, S.H. 1983. Tumor invasion and the extracellular matrix. *Lab Invest* 49:636–649.)

2.2a   Disruption of the basement membrane

Most malignancies are carcinomas, which are tumors of epithelial cells (Chapter 1). Epithelial cells are supported by basement membranes, thin sheets of extracellular material. Basement membranes separate epithelial cells from the vascularized connective tissues (basement membranes also are found surrounding nerves, muscles, and blood and lymph vessels). Blood capillaries and lymphatic vessels are *not* found in epithelia, and epithelial cells must derive their sustenance and oxygen from underlying vascularized connective tissue. Thus, basement membranes must be breached if carcinoma cells are to gain access to vessels for subsequent dissemination and formation of metastases (Figure 2–3). Enzymes have been found that digest components of the basement membrane and facilitate transgression of that boundary.

Basement membranes contain type IV collagen as a major component (Sinha et al. 1991), as well as a number of other constituents, including laminin, fibronectin, heparan sulfate proteoglycan, and entactin. Destruction of basement membrane is commonly a feature of malignancy; it is not surprising, therefore, that type IV collagenase (an enzyme that degrades type IV collagen) activity was described in the Lucké renal adenocarcinoma of frogs (Shields et al. 1984) and metastatic mouse cell lines (Liotta and Rao 1985; Turpenniemik-Hujanen et al.

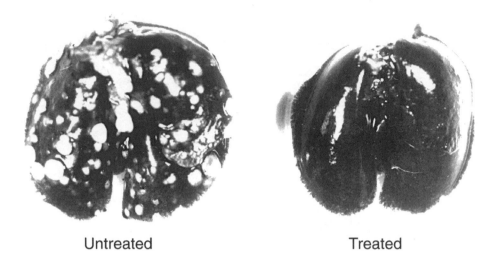

Untreated   Treated

Figure 2–4. Lungs of animals injected with a highly metastatic rat embryo cell line. The lungs on the left were of untreated hosts that received tumor cells plus sterile buffer intraperitoneally. The lungs on the right were of an animal that received tumor cells and was treated with intraperitoneal injections of a recombinant human inhibitor of metalloproteinases (rTIMP). Lung tumor colonies were inhibited by 83% in the rTIMP-treated animals. (From Alvarez, O.A., Carmichael, D.F., and Declerck, Y.A. 1990. Inhibition of collagenolytic activity and metastasis by a recombinant human tissue inhibitor of metalloproteinases. *J Natl Cancer Inst* 82:589–595.)

1985). A melanoma cell line that elaborates significant quantities of type IV collagenase readily invades basement membranes and is metastatic (Welch et al. 1991). Spontaneously occurring mammary cancers of mice (Shields et al. 1984) and human tumors (Shields et al. 1984; Fidler 1997) similarly produce type IV collagenase. Certain nonmetastatic mouse cells can be transfected with the *ras* oncogene (see Chapter 5), which induces the metastatic phenotype (Muschel et al. 1985). It is interesting, in the context of this section, that the *ras*-transfected cells, which exhibit metastatic competence, also produce elevated levels of type IV collagenase (Garbisa et al. 1987).

If collagenolytic activity is important in invasion of the basement membrane by cancer cells, one might prognosticate that inhibiting the degradation of collagen would impede invasion and metastasis. This has been reported (Reich et al. 1988; Albini et al. 1991). A reduction in the potential for tumor cell colonization of the lung (Figure 2–4) by a rat cell line resulted from the use of an inhibitor of metalloproteinases (a metalloproteinase is an enzyme that requires a metal ion for its activity; collagenases are metalloproteinases) (Alvarez, Carmichael, and Declerck 1990; see also Wojtowicz-Praga, Dickson, and Hawkins 1997). It has been suggested that carcinoma in situ (Chapter 1) treated with agents which maintain the integrity of the basement membrane may remain just that, carcinoma in situ, and fail to disseminate (Kohn and Liotta 1995).

Laminin, fibronectin, collagen, proteoglycan, and other basement membrane components are digested by the proteolytic enzyme cathepsin B. A role in metastasis was suggested for cathepsin B (Sloane and Honn 1984; Sinha et al. 1989), which was found localized in invasive cells and capillaries of human prostate cancer (Sinha et al. 1993, 1995). The enzyme may serve as a marker for metastasis. Serum and urinary cathepsin B levels were found to be higher in gastrointestinal cancer patients than in healthy controls. Further, and perhaps more importantly as it relates to this chapter, patients with liver or lung metastasis had even higher levels of the enzyme than cancer patients without metastasis. Serum cathepsin B levels returned to normal after resection of the malignancies (Hirano, Manabe, and Takeuchi 1993).

Urokinase plasminogen activator regulation may also be involved in invasion. Receptor-bound urokinase plasminogen activator on the surface of human colon cancer cells activates plasminogen to form the proteolytic enzyme plasmin, which is involved in basement membrane and connective tissue digestion (Wang et al. 1995). The digestive competence of cancer cells as it relates to invasion and metastasis is discussed further in Fidler (1997).

Enzymatic digestion of the basement membrane allows carcinoma cell access to connective tissue, and therefore, it is no surprise that aggressiveness of tumor cell behavior and levels of all classes of proteolytic enzymes is positively correlated (Kohn and Liotta 1995). However, although transgression of the basement membrane is an important early step for carcinoma cells, those cells must be competent to disseminate if metastasis is to occur. In other words, cells must detach and move away from the primary tumor.

2.2b   Cell detachment

As soon as it was recognized that metastasis results from the translocation of cancer cells, it became obvious that a fundamental property of those cells which detach must be reduced cohesion. Ordinarily, normal cells do not separate from their cohorts in a tissue; they stay in place. Cancer cells, however, exhibit a deficiency in cohesiveness, and this loss is essential in the metastatic cascade (McCutcheon, Coman, and Moore 1948). A number of years ago, Coman (1944) studied both normal and malignant lip and cervical cells. He reported that it required more force to separate normal cells from each other than it did to separate their malignant counterparts. The cell pairs were impaled on microneedles, and the force necessary to separate them was calculated by the bend of the needles as the cells were pulled apart. Normal pairs clung to each other tenaciously resulting in considerable cell distortion during separation, in marked contrast to the cancer cells, which separated easily (Figure 2–5) (Coman 1944, 1953). An alternate way to study cohesiveness is to shake tissue contained in a culture medium and count the cells that detach. In each case, more cells were detached from cancers of the stomach, colon, rectum, thyroid, and breast than from their normal tissue controls. In another

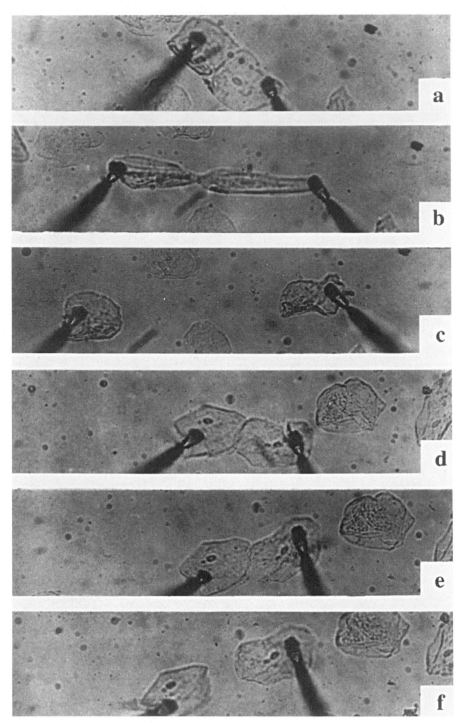

Figure 2–5. Separation of living squamous epithelium cells from a normal lip and their malignant counterparts from a carcinoma of the lip. Cell pairs are impaled on microneedles attached to a micromanipulator. One of the needles is moved while the other remains stationary. Note that the normal cells (a) are tenaciously attached to each other, and the microneedles stretch and distort the cells (b) prior to their being separated (c). In contrast, cancer cells (d) readily detach with minimal distortion and little movement of the microneedles (e and f). (From Coman, D.R. 1944. Decreased mutual adhesiveness, a property of cells from squamous cell carcinomas. *Cancer Res* 4:625–629.)

study, direct observations were made of normal and malignant cells in transparent chambers placed in rabbit ears. Tumor cells were able to "wander" in contrast to their nonliberated normal cell counterparts (Wood 1958; Wood, Baker, and Marzocchi 1968).

Complex factors relate to why tumor cells are less cohesive than their normal counterparts. In vitro analysis provides an opportunity to study factors affecting tumor cell detachment. Normal frog kidney cells do not separate from each other when agitated in a buffered calcium- and magnesium-free electrolyte solution at any physiologic temperature. However, with gentle agitation Lucké renal adenocarcinoma cells (malignant kidney cells of the frog) readily detach from a tumor fragment at elevated physiologic temperatures (18° to 28° C), but not at temperatures cooler than 18° C (Seppanen et al. 1984).

In studies of the Lucké tumor, conditions that permit decreased cell cohesion are similar to conditions that permit the elaboration of type I collagenase (see section 2.2d on invasion). As already noted, tumor cell dissociation readily occurs at elevated temperatures, and temperature-dependent elaboration of type I collagenase was shown to occur at 18° C and warmer but not at cooler temperatures. Next, it was observed that ethylene diamine tetraacetate (EDTA) *inhibited* tumor cell dissociation. (This is in contrast to what was expected; EDTA has been used for years in embryologic studies to *enhance* dissociation.) EDTA also inhibits collagenase. Similarly, excess cysteine inhibits both cell detachment and the enzymatic activity of type I collagenase (Seppanen et al. 1984). These observations are consistent with the hypothesis that cell detachment in the Lucké renal adenocarcinoma is mediated, at least in part, by the activity of an enzyme that is, or is similar to, a collagenase.

The Lucké renal adenocarcinoma cultured under noninvasive conditions is highly differentiated, whereas the tumor appears anaplastic when maintained under conditions that are permissive for invasion (McKinnell et al. 1986). In other words, poorly differentiated frog tumors metastasize more frequently than highly differentiated ones. The relationship of differentiation state and invasion competence of the frog malignancy holds for other tumors as well. The expression of the Ca^{2+}-dependent cell-cell adhesion molecule E-cadherin, a glycoprotein also known as uvomorulin, is related to the retention of the epithelial state and normal morphology with a noninvasive phenotype. Carcinoma cells that do not express the cell-cell adhesion molecule E-cadherin lose their characteristic morphology and become invasive (Frixen et al. 1991; Birchmeier, Hülsken, and Behrens 1995). Down-regulation of E-cadherin expression thus seems to be related to metastasis (Takeichi 1991; Mareel et al. 1993; Graff et al. 1995) and may, for example, be used as a clinical cancer cell marker with prognostic capabilities in thyroid cancer (Scheumman et al. 1995). As with almost all phenomena in cancer, cell detachment as it relates to E-cadherin is complex. E-cadherin is connected to the cytoskeleton by means of catenins. Down-regulation of α-catenin was reported to be associated with increased invasion in human colon cancer cells (Vermeulen et al. 1995). As stated before, interruption of any step in the metastatic cascade should abrogate the malignant process, and this applies to the cell-cell adhesion molecule E-cadherin (Berx et al. 1997; Fujimoto et al. 1997). Transfection of invasive cells with E-cadherin cDNA, with the resumed production of this molecule important in cell adhesion, blocks invasion. Clearly, the deceptively simple phenomenon of detachment of cancer cells from the primary tumor mass is of critical importance for metastasis.

Detached cells depend on motility for passage through the extracellular matrix and for ultimate contact with capillaries or lymph vessels. Factors related to cell motility are described next.

2.2c   Cell motility

For years pathologists have reported the occurrence of single cells or small clumps of cells that appeared to have become detached from a primary tumor. The observation of these detached cells was interpreted to mean the cells were released from the restraints of the tumor mass and had locomotive competence (Lambert 1916; Strauli and Haemmerli 1984). Over a century ago, the pioneer German pathologist Rudolf Virchow reported morphologic observations consistent with the concept that cancer cells possess "amoeboid" motility (Virchow 1863). Confirmation of such intrinsic cell motility by direct visual observation of wandering cells (Figure 2–6) was made by Carrel and Burrows (1911), Lucké (1939), Enterline and Coman (1950), and others. Motility is still being studied, especially as it relates to the capacity of cancer cells to invade contiguous normal tissue and to gain access to and exit from blood and lymph vessels.

Obviously, cancer cells have the capacity to become motile under the proper conditions. Given this capacity, are there ways to quantitate motility? The answer to that question is yes. One in vitro assay for motility involves counting cells that squeeze through microscopic pores in a membrane separating the cells from a substance which elicits a migratory response. This method allows quantitation of the response of cancer cells to a diversity of conditions that may affect locomotion. A cell line, PNKT4B, derived from a frog tumor, will migrate across eight micrometer pores in a membrane in response to the chemoattractant fibronectin. The migratory response is not unique to the tumor cell line and is shared with cells from primary frog tumors and normal embryos (Hunter, Tweedell, and McKinnell 1990). Curiously, it was noted that PNKT4B cells retain substantial migratory activity at temperatures that preclude invasion in vitro (McKinnell et al. 1988). Hence, a temperature-dependent factor other than cell migration probably inhibits invasion of PNKT4B cells at reduced temperature.

A potentially practical rationale for the quantitation of cell motility relates to the metastatic potential of cancer cells. If malignant cells with a high potential for metastasis differ in cell motility from malignant cells which are less aggressive, then that difference may become useful in identifying patients for intensive early treatment. Moreover, malignancies with high cell motility and high metastatic

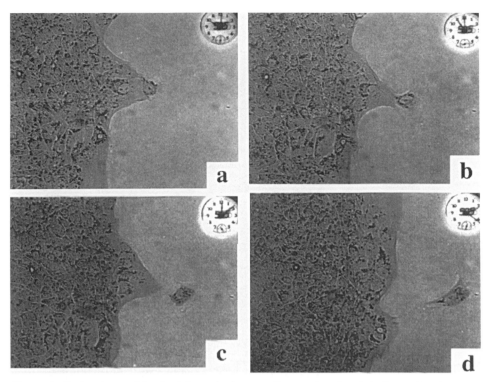

Figure 2–6. A culture of the Lucké renal adenocarcinoma showing detachment and out-migration of a tumor cell. The initially attached cell (a) begins its detachment (b) and wanders away from the tissue mass (c and d). Motility of this cell was recorded during a period of 4.5 hours. (From Lucké, B. 1939. Characteristics of frog carcinoma in tissue culture. *J Exp Med* 70:269–276. By permission of The Rockefeller University Press.)

competence suggest a need to focus drug research on agents that affect motility directly. It has been shown that cell lines obtained from a rat prostate cancer that exhibit high cell motility also have high metastatic competence. The expression of glyceraldehyde-3-phosphate dehydrogenase, a glycolytic (anaerobic) enzyme, correlates with cell motility and metastatic potential in the rat cells (Epner et al. 1993). The enzyme is a marker for metastasis. If human prostatic cancer cells behave similarly to the rat prostate cell lines, then it may become possible to predict which malignancies have a high potential for metastasis on the basis of the expression of the enzyme glyceraldehyde-3-phosphate dehydrogenase.

Motility is most likely a characteristic of all cancer cells that are free of a restraining tissue organization. Invasion into adjacent tissue depends on that motility. Chemoattractants that are important in motility have been identified (Lam et al. 1981; McCarthy et al. 1985), but what initiates locomotion? Recently, a novel cytokine that is liberated by and stimulates the motility of human melanoma and breast carcinoma cell lines has been partially characterized. The cytokine is known as autocrine motility factor (AMF) and stimulates both random (chemokinetic)

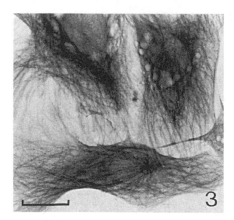

Figure 2–7. Light microscope photograph of frog tumor cells cultured on glass and stained with an antiserum against tubulin. Microtubules are revealed as fine lines in the cytoplasm of the three cells in this photograph. (From McKinnell, R.G., DeBruyne, G.K., Mareel, M.M., Tarin, D., and Tweedell, K.S. 1984. Cytoplasmic microtubules of normal and tumor cells of the leopard frog. Temperature effects. *Differentiation* 26:231–234.)

motility as well as directional (chemotactic) motility. AMF is believed to induce pseudopod protrusion and augmented levels of laminin and fibronectin receptors. AMF may also be involved in the adhesion and release of pseudopodia during locomotion, as well as production and liberation of enzymes essential for motility (Schiffmann 1990; see also Fidler 1997).

The cytoplasmic microtubule complex (CMTC) (Figure 2–7) is structured of tubulin and implicated in directional migration. The chemotherapeutic vinca alkaloids vinblastine, vincristine, vindesine, and the synthetic compound nocodazole are agents that depolymerize microtubules and also inhibit invasion. Chemotherapeutic agents, including 5-fluorouracil and mitomycin C, which do not depolymerize microtubules, do not inhibit invasion in vitro (Mareel et al. 1982; Mareel, DeBaetselier, and Van Roy 1991). Adriamycin, another type of chemotherapeutic agent, inhibits tumor cell motility and invasion in vitro at a concentration that does not have an effect on tumor cell growth (Repesh et al. 1993). Most would agree, considering these experiments, that motility plays a significant role in tumor cell spread.

2.2d Invasion

Carcinoma cells that have detached from the primary tumor mass, penetrated the basement membrane, and exhibited motility may now come in contact with connective tissue (Figures 2–2 and 2–3). Capillaries in the connective tissue are a biologically significant destination for motile tumor cells. In a sense, the problem of gaining access to capillaries is similar to that of breaching the basement membrane; that is, the cells need to digest a pathway to nearby small vessels. The most abundant protein in the body is type I collagen, a principal component of the intercellular matrix of connective tissue. Some tumor cells secrete an enzyme, type I collagenase, which can digest type I collagen.

The Lucké renal adenocarcinoma of the frog is aggressively invasive in vitro and metastatic in vivo at temperatures of 18° to 28° C, but not at lower temperatures (Lucké and Schlumberger 1949; McKinnell et al. 1986). One can culture frog tumor fragments at metastasis-restrictive and -permissive temperatures to identify cellular properties that correlate with, and may be important to, metastatic behavior (McKinnell and Tarin 1984; Tarin 1992). Relevant to the present issue is the observation that frog renal carcinoma maintained at 30° C elaborated high levels of type I collagenase, whereas little enzyme was detected at reduced, nonmetastatic temperatures. Normal kidney released negligible levels of collagenase at both high and low temperatures. The type I collagenase data could lead to the notion that normal frog kidney is metabolically less active (or moribund) at 30° C compared with the Lucké renal adenocarcinoma. However, a study of glucose uptake by normal and malignant kidney revealed that the normal renal tissue was alive and functioning at both the initial low and subsequent elevated temperature (Figure 2–8). What is demonstrated here is a metastasis-permissive enzyme response of the neoplastic tissue (collagenase release) that does not occur in viable kidney held under comparable conditions (Ogilvie, McKinnell, and Tarin 1984). Hence, the capacity to elaborate this proteolytic enzyme correlates with, and may be important for, invasion by the frog renal adenocarcinoma.

Spontaneously occurring mouse mammary carcinomas were characterized by their capacity to form secondary tumors after intravascular injection into normal, nontumorous hosts. Some tumors formed numerous colonies in the lungs of injected hosts; others formed only a few or no colonies. The mammary tumors that had a high colonization potential elaborated high levels of type I collagenase. The mouse tumors with low colonization potential produced little type I collagenase (Tarin and Price 1979; Tarin, Hoyt, and Evans 1982).

Human breast neoplasms, as well as other human malignancies, produce variable quantities of collagenase. Unfortunately, at the present time, no relationship seems to exist between quantity of enzyme produced by human breast neoplasms and subsequent clinical behavior (Tarin 1992).

Invasion assays of various kinds have been developed. The assays are designed to discriminate between normal noninvasive cells and their malignant counterparts. A good invasion assay will permit the identification of factors that may preclude malignant invasion. Obviously, carcinoma cells, which rest on a basement membrane, must transgress that membrane to gain access to connective tissue and capillaries. Various basement membranes, both real and artificial, have been used in invasion studies. Recently, sea urchin embryo basement membranes have been used for that purpose. The sea urchin basement membrane is biochemically similar to its mammalian counterpart; for example, it contains laminin, fibronectin, collagen type IV, and heparan and chondroitin sulfate proteoglycans. It is of interest to

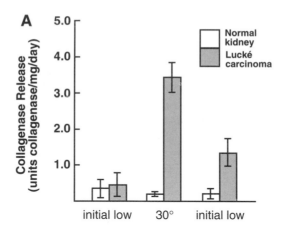

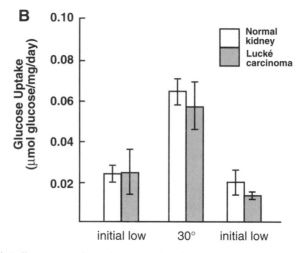

Figure 2–8. (a) Collagenase release by normal and neoplastic renal tissue at two temperatures. The initial low temperature, 7° C, is metastasis inhibiting; the second temperature, 30° C, is metastasis permissive. The Lucké renal adenocarcinoma releases significantly greater quantities of type I collagenase than does normal renal tissue cultured under identical conditions. (b) Glucose uptake by normal renal tissue at both temperatures attests to the viability and functioning of the normal tissue. Thus, the type I collagenase response illustrated in (a) is tumor specific and probably related to invasion. (From Ogilvie, D., McKinnell, R.G., and Tarin, D. 1984. Temperature-dependent elaboration of collagenase by the renal adenocarcinoma of the leopard frog, *Rana pipiens. Cancer Res* 44:3438–3441.)

note that malignant cells recognize and invade the sea urchin membrane, which not only indicates the usefulness of the invertebrate membrane in experimental studies but "suggests that molecules participating in basement membrane recognition and invasion have been functionally conserved during the time separating vertebrates from invertebrates" (Livant et al. 1995).

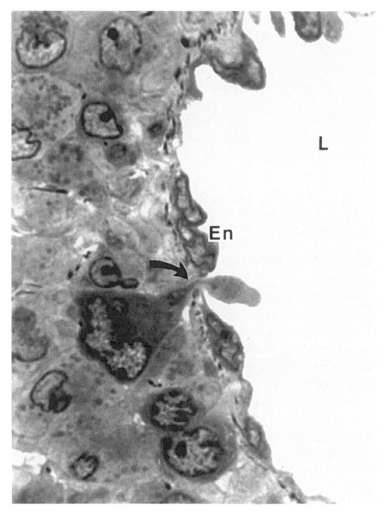

Figure 2–9. Cancer cells gain access to blood circulation by diapedesis. The cancer cell traverses a small opening (arrow) between endothelial cells (En) of the vessel to gain access to the lumen (L). (From Song, M.J., Reilly, A.A., Parsons, D.F., and Hussain, M. 1986. Patterns of blood-vessel invasion by mammary tumor cells. *Tissue Cell* 18:817–825. By permission of Churchill Livingstone.)

2.2e   Penetration of the vascular system

Cancer cells disseminate via capillary and lymphatic vessels. The thick wall of an artery is rarely penetrated. It is believed that cancer cells can enter newly formed capillaries more easily than preexisting capillaries because of defects in the new blood vessels, such as gaps between endothelial cells and a discontinuous or absent basement membrane (Ausprunk and Folkman 1977; Folkman 1993). Even if defective capillaries are not present, the entrance of tumor cells into vessels may be facilitated by the production of type IV collagenase (Liotta et al. 1979; Shields et al. 1984).[3] Neoplastic cells gain access to small vessels by digesting an opening in the outside wall of the vessel through which they migrate to the lumen (Figure 2–9). The process of slipping through the opening in the vessel is known as diapedesis (Engell 1955). Fortunately, most of these cells are damaged or destroyed upon entrance to the circulation, and only a minority of cells live to express their metastatic potential. Perhaps because of the short life expectancy of cancer cells in the circulation, cancer cells are not found in the blood of all cancer patients (Snellwood et al. 1969; Salsbury 1975).

2.2f   Cancer cells in the circulation

Cells thought to have originated in cancers were detected in the blood of a cancer patient well over a century ago (Ashworth 1869). With remarkable insight, the author wrote, "The fact of cells identical with those of the cancer itself being seen in the blood may tend to throw some light upon the mode of origin of multiple tumours existing in the same person." As many as three million cancer cells per gram of rat mammary carcinoma enter the circulation per day (Butler and Gullino 1975). What is the implication of cancer cells in the circulation? Quite simply, malignant cells in blood vessels are obviously a requisite for hematogenous dissemination. It is worth noting here that the density of capillaries in a tumor is positively correlated with metastatic behavior (Macchiarini et al. 1992; Bosari et al. 1992). Obviously, capillaries provide a avenue for entrance into the vascular system; the more avenues, the better is the chance for a successful entrance.

Although the significance of cancer cells in the circulation of a *particular* cancer patient is uncertain, surgeons nevertheless attempt to minimize the release of these cells during cancer operations. Although the circulatory system is a hostile environment for cancer cells, if the cells survive (and some obviously do), they may grow in the vasculature or they may be swept away to anatomically distant sites. The fate of the cancer patient cannot be foretold by the presence of circulating cancer cells (Griffiths et al. 1973; Salsbury 1975). Indeed, Tarin et al. (1984a, 1984b) reported that one-half of cancer patients with shunts to recirculate peritoneal fluid (which contained living cancer cells) to the general circulation had no evidence of metastatic cancer. Hence, vascular dissemination of cells in cancer patients does not necessarily result in metastasis (Tarin 1992). Although this is true, hematogenous dissemination and subsequent metastasis obviously cannot occur without blood-borne neoplastic cells.

---

[3] Different types of collagen are formed of different but related collagen molecules. Further, specific amino acids (proline, lysine, and cysteine) of the collagens are modified so that several types of collagen are recognized, namely types I through V. The collagen types are found in particular anatomic locations; for example, type I is found in skin, bone, tendon, and connective tissue; type IV is found in, among other places, the basement membranes of epithelial cells.

2.2g Arrest of circulating cancer cells (stasis)

Cancer cells in the circulation adhere to each other and to lymphocytes and platelets, forming emboli that may adhere to the inner surface of capillaries. A fibrin-containing thrombus (blood clot) forms, which stabilizes the embolus. Growth of the tumor cells and possible occlusion of the capillary follows. The stabilized cancer cells interact with the capillary endothelial cells, causing endothelial cell retraction with the exposure of the subjacent basement membrane. Metastasis is not an efficient process; some cancer cells may die within the thrombus, and the fibrin may dissolve, resulting in the subsequent loss of the stabilized tumor cell mass to the general circulation. It has already been emphasized how hazardous it is for cancer cells in the circulation.

Indirect evidence of the importance of thrombus formation ensues from the enhanced survivability of some patients on anticoagulant therapy (Michaels 1964; Ryan, Ketcham, and Wexler 1968; Rickles et al. 1988). Treatment with warfarin, an anticoagulant, enhanced survival time of terminally ill, small cell lung carcinoma patients (Zacharski et al. 1981; Zacharski 1984) but had little effect on survival of patients with other tumors (Zacharski 1984). Presumably the efficacy of an anticoagulant derives from the dissolution of the clot, resulting in expo-sure of the released cancer cells to the hazards of the blood circulation. Salsbury (1975) wrote that cancer cells "may do less harm to the host in the circulation than elsewhere."

The role of a fibrin enclosure of a tumor cell nidus (cluster) and lysis of that enclosure is complex. The fibrin enclosure may act as a barrier to undefined and unknown noxious factors and stabilize the trapped tumor cells, resulting in enhanced invasion. Accordingly, digestion of the enclosure (fibrinolysis) may decrease invasion. Alternatively, fibrinolytic activity may enhance invasion. Digestion of the fibrin barrier may increase cell detachment and active invasion. Thus, fibrinolysis may either inhibit or enhance metastatic behavior depending on the particular tumor. The role of fibrinolysis in invasion and metastasis is still being studied (Costantini et al. 1991).

2.2h Extravasation, growth of metastases, and metastasis of metastases

Extravasation of cancer cells requires cell motility and transgression of the basement membrane of the capillary. By active migration, the cancer cells move between the retracted endothelial cells, through the breached basement membrane, and into the intercellular matrix of the connective tissue (Figures 2–2 and 2–10). Resumption of neoplastic growth with the proliferation of tumor cells may either be immediate or after a variable period of dormancy (Chapter 1). The nidus of cancer cells is restricted in growth to a lump 0.5 to 1 mm in diameter due to the limited nutrient supply available by diffusion from blood. Without

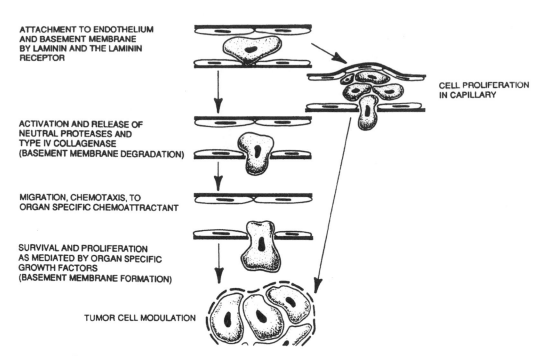

ATTACHMENT TO ENDOTHELIUM
AND BASEMENT MEMBRANE
BY LAMININ AND THE LAMININ
RECEPTOR

CELL PROLIFERATION
IN CAPILLARY

ACTIVATION AND RELEASE OF
NEUTRAL PROTEASES AND
TYPE IV COLLAGENASE
(BASEMENT MEMBRANE DEGRADATION)

MIGRATION, CHEMOTAXIS, TO
ORGAN SPECIFIC CHEMOATTRACTANT

SURVIVAL AND PROLIFERATION
AS MEDIATED BY ORGAN SPECIFIC
GROWTH FACTORS
(BASEMENT MEMBRANE FORMATION)

TUMOR CELL MODULATION

Figure 2–10. Diagram of extravasation. A lodged malignant cell, or a small population of cells, induce the retraction of vessel endothelial cells and digest the basement membrane. This permits exit from the vessel with subsequent formation of a metastatic colony. (Reprinted with persmission from Terranova, V.P., and Maslow, D.E. 1991. Interactions of tumor cells with basement membrane. In F.W. Orr, M.R. Buchanan, and L. Weiss, eds., *Microcirculation in Cancer Metastasis,* pp. 23–44. Copyright © 1991 CRC Press, Boca Raton, FL.)

angiogenesis, the center of the tumor cell colony becomes necrotic (D'Amore 1988). However, new capillaries are recruited into the metastases that will survive (Zetter 1990). Vascularization of the new tumor colony allows blood to deliver essential oxygen and nutrients and permits the removal of toxic metabolites from the metastatic mass. Growth follows. With growth, the whole process repeats, and the metastatic colony will give rise to additional metastatic colonies; thus, metastases metastasize.

2.3   Are there genes that control metastasis?

Some cancers are highly metastatic – others, like ovarian cancer, exhibit little tendency to disseminate through the bloodstream. Is it possible to discriminate between these extremes in neoplastic cell behavior by the expression of a particular gene? Identification of such a gene would be useful to prognosticate which cancer patient already has metastasis at diagnosis and which patient is vulnerable to early spread of the neoplasm with an unfortunate outcome. Knowing which neoplasms metastasize presents the possibility of identifying cellular processes that govern metastasis and offers the hope of exploiting that knowledge for the control or cure of cancer. It is not surprising, therefore, that genes which control metastasis have been sought. Although individual genes important for the process of metastasis have been identified, bear in mind that metastasis is a complex series of events, and it is likely that the phenomenon is under the control of a series of genes which when activated in a coordinated manner result in the inappropriate cell behavior of invasion and metastasis.

A gene with a long evolutionary history may provide partial understanding of the metastatic process. Reduced expression of this gene (in a murine cell line) is correlated with metastasis. Thus, the gene would be classified as a tumor-suppressor gene. The gene, *nm23*, is generously expressed in normal mice and in melanoma clones with a low metastatic potential but is poorly expressed in clones with a high metastatic potential (Steeg et al. 1988; Leone et al. 1991). The gene is not unique to mice. It has been detected in rat mammary carcinomas as well as other rodent cancers. It is also present in human breast cancer where relatively high *nm23* mRNA levels (compared to other metastatic cancers) are reported to be correlated with well-differentiated cancer, lack of lymph node metastasis, and longer disease-free survival (Hennessy et al. 1991). Because of these and other observations, there is increased interest in this putative metastasis-inhibiting gene.

Curiously, the *nm23* gene is very similar to bacterial (Mundoz-Dorado, Inouye, and Inouye 1990), slime mold (Wallet et al. 1990), *Drosophila* (Biggs et al. 1990) and rat (Kimura et al. 1990) nucleoside diphosphate (NDP) kinase genes. Certainly, the retention of molecular similarity across such phylogenetically diverse and ancient groups argues strongly for an important role for *nm23*. NDP kinase genes (and thus presumably *nm23*) are involved in the synthesis of nonadenine-containing nucleoside triphosphates by the transfer of phosphate from ATP to the appropriate diphosphate. Perhaps significantly, with respect to invasion and metastasis, NDP kinase genes play a role in microtubule assembly as well as intracellular signaling. As noted in the section on cell motility earlier in this chapter, microtubules are implicated in directional motility. Further, because the mitotic spindle is structured of microtubules, and because NDP kinases may not be functioning properly in malignant cells, abnormalities in mitosis resulting from spindle malformations may lead to heterogeneity, and with selection, further tumor progression. Perturbations in cellular signal transduction may also generate tumor progression. More recently, *nm23* has been identified as a factor that enhances transcription of the c-*myc* oncogene (Chapter 5) (Postel et al. 1993). How increased activity of c-*myc* may be related to inhibition of metastasis is unknown. In contrast, c-*myc* is related negatively to differentiation, which in its turn (i.e., loss of differentiation in tumor cells) is related positively to tumor aggressiveness (Chapter 1).

There are exceptions to what would be expected based on the breast cancer studies described here. For example, a study was made of *nm23* gene expression in normal gut mucosa and neoplastic colon tissue. Transcripts of *nm23* were detected in the normal tissue, but expression was even greater in the cancer tissue (Haut et al. 1991). Similarly, *nm*23 expression is positively correlated with metastasis in human pancreatic cancer (Nakamori et al. 1993). Is the expression of *nm*23 directly related to metastasis, or does expression of the gene have a function in cell biology that is only incidentally related to metastatic behavior? Much remains to be learned about the role of *nm*23 in cancer biology and the inhibition of metastasis.

In this context, putative metastasis-inhibiting genes (there are now two *nm*23 genes) most likely have important roles to play in normal biology. Many normal cells translocate during development (see section 2.5). It may be that *nm*23, or a related gene, is down-regulated during normal cell migratory behavior. The gene would be activated when the wandering cells arrive at their correct anatomic location. If this speculation were true, then the reduced transcription of *nm*23 in cancer cells manifesting metastasis is not behavior uniquely malignant but rather behavior that is normal, but inappropriate with respect to chronology or anatomic location.

Other genes have been identified as potentially important in metastasis. One family of genes that encodes for a glycoprotein believed to be involved in adhesion is designated CD44 ("CD" stands for "cluster designation," used to describe surface antigens of leukocytes). The CD44 exons are transcribed in different combinations (a phenomenon known as alternative splicing), depending on cell activity. Matsumura and Tarin (1992) reported that they could distinguish between metastatic tumors of the breast or colon versus those with no metastasis on the basis of the overproduction of specific variants of the CD44 genes. They suggested that overexpression of the differently spliced genes could be a useful tumor cell marker (Chapter 1) and assist in the identification of tumors with metastatic potential using surgical biopsy. Although CD44 (and other molecular markers) have been suggested as candidates for prognosis indicators (Tarin 1996), there is at least one exception to the notion that CD44 is a general requirement for metastasis (Driessens et al. 1995).

Greatly augmented metastatic potential was obtained by the insertion of high molecular weight human DNA, obtained from metastases, into mouse tumor cell lines. Only a relatively small fragment of DNA was taken up by transfected cells, narrowing the search for the genes responsible for metastasis. Clones of transfected cells were assayed for metastatic potential by injection into mice. Some mice cells transfected with DNA from metastatic tumors showed increased metastasis. Mouse cells transfected with DNA from nonmetastatic or nonmalignant cells did not display enhanced competence for metastasis. Study of transfected cells with high metastatic potential offers the opportunity to isolate and characterize DNA sequences that induce metastatic behavior (Hayle et al. 1993).

Recently, a gene located on the short arm of human chromosome 11 has been associated with the suppression of metastasis in prostate cancer. The gene is designated KAI1 (*kang ai* is Chinese for "anticancer") and its expression may interfere specifically with prostate cancer metastasis (Dong et al. 1995).

2.4 Soil and seed hypothesis of Paget

Cancer cells that gain access to the vascular system are disseminated throughout the body. If tumor cells were to arrest, escape from the vasculature, and grow as secondary colonies at *random* anatomic sites, then there would be no pattern of specific target organs associated with particular metastatic neoplasms. This is not what happens. It has long been recognized that prostatic carcinoma metastases are commonly found in bone, small cell carcinoma of the lung has a predilection for the brain, and neuroblastoma metastases frequently occur in the liver. About a hundred years ago, Paget studied the postmortem records of 735 patients who died of breast cancer. He noted that the spleen, which has a generous arterial supply, was infrequently the site of breast cancer metastasis (2%), in contrast to the liver, with less arterial blood, which had metastatic breast cancer in a greater proportion (33%) of the autopsies. Paget suggested that cancer cells are not unlike seeds, which, when cast upon the earth, grow only in hospitable soil. These observations became the basis of the "soil and seed" hypothesis (Paget 1889; see also Willis 1973; Tarin 1992).

An alternative view to this hypothesis, which does not necessarily exclude the concept that certain tissues or organs are particularly favorable for the growth of some cancers, has been suggested. It simply states that organ preference is a function of entrapment of cancer cells in the first capillary bed encountered. Tumor cells released into capillaries, venules, or lymphatics and transported to the heart may be caught in the sieve of the pulmonary capillaries. Tumors that release cells to the hepatic portal vein may colonize the liver. Subsequently, the hepatic or pulmonary colonies may release cells that express their affinity for a particular organ or tissue (Sugerbaker 1981; Tarin 1992). Therefore, the location of a metastasis may not be organ specific but may be related to blood-borne cells being trapped in the first capillary bed encountered. Tumor cells may also have homing mechanisms for tissue-specific endothelium (Fidler 1984), and may actively seek out certain tissues very similar to lymphocytes.

The pattern of organ involvement in metastasis probably depends on both the mechanical sieving effect of capillary beds *and* the presence of specific environmental conditions in certain organs that favor the adhesion of tumor cells in a capillary at a particular site, the escape from that capillary, and the subsequent growth of a metastatic colony.

2.5 Is metastasis limited to malignant cells?

If one assumes that the malignant tumor cell is inappropriately expressing a preexisting normal cell program of physiological invasion, then the fundamental difference between normal and malignant cells is regulation.

Kohn and Liotta 1995

Although Ewing (1916) considered metastasis the most convincing evidence of cancer (and few would disagree with him), many if not all aspects of metastatic behavior can be observed and studied in normal cells. Frost and Levin (1992) correctly stated, "There is nothing that a metastatic cell can do that is not a routine task for normal cells." The reason for listing attributes of normal cells that (seem to) mimic malignant behavior is twofold. First, the inappropriate behavior of cancer cells does not derive from cell properties that develop de novo but is probably related to the expression of otherwise normal cell attributes at an inappropriate time or place. Second, in at least some instances, the normal cell analogue may be more convenient to study than the same characteristic in malignant cells. Normal cell activities that are similar to metastatic phenomena of cancer cells include (but are not limited to) the following.

Motility is a characteristic of many normal cells as it is, of course, also of malignant cells. Embryonic and adult cell types that exhibit cell motility are neutrophilic, eosinophilic, and basophilic granulocytes, lymphocytes, monocytes and macrophages of the blood (Armstrong 1984), neural crest cells (Lofberg, Ahlbars, and Fallstrom 1980), myogenic stem cells of the somites involved in the formation of the wings and legs of chick embryos (Jacob, Christ, and Jacob 1978, 1979), and primordial germ cells (Kamimura, Kotani, and Yamagata 1980).

Primordial germ cells (PGCs; early progenitors of sex cells found in the embryo) in chordates do not have their embryologic origin in the gonad; rather, because of cell motility and other activities, they translocate from their site of origin to the gonad. In mammalian embryos, the PGCs develop first in the posterior yolk sac near the allantois. The PGCs subsequently translocate via the hindgut to hindgut mesentery, from there to the dorsal body wall, and then to the germinal epithelium adjacent to the developing mesonephros (Witschi 1948; Godin, Wylie, and Heasman. 1990). One could use Paget's venerable soil and seed aphorism to describe the PGCs as seed and the germinal epithelium as soil. PGCs that reach the hospitable germinal epithelium flourish and ultimately give rise to functional gametes after their metastasislike trek.

Another instance of the translocation of yolk sac cells involves hematopoietic stem cells. These embryonic cells are liberated into the circulation where they are subsequently "seeded" to embryonic blood-forming organs (e.g., bone marrow, spleen, thymus) (Moore and Owen 1967). This is an example of the blood-borne dissemination of cells that flourish and grow after arrival in hospitable soil. Related to natural developmental translocation of blood stem cells is bone marrow transplantation in adult humans. The bone marrow, containing hematopoietic stem cells, is transplanted directly to the recipient's circulation where the inserted blood-forming cells have access to all of the organs of the body. Despite this fact, the hematopoietic stem cells localize and repopulate only the host's bone marrow where sustained hematopoiesis occurs (Tavassoli and Yoffey 1983; Tavassoli and Hardy 1990). The "homing" of the hematopoietic progenitor cells to bone marrow precludes the need to transplant the cells surgically into the marrow. Bone marrow cell transplantation into the vasculature with subsequent population of the bone marrow is an elegant normal cell analogue of malignant metastasis.

Neural crest cells have their origin at the dorsal surface of the neural tube in the embryo. Several migratory pathways permit mitotic descendants of neural crest cells access to the skin where they become pigment cells, to the adrenal gland where they become medullary cells, laterally to the spinal cord where they give rise to spinal ganglia, and to other sites where they give rise to other normal tissues (Perris and Bronner-Fraser 1989). Similarly, the parathyroid gland cells translocate from the dorsal parts of the third and fourth pharyngeal pouches in the human embryo to the thyroid gland (Moore and Persaud 1993).

Although the examples given thus far are from normal biology, there are pathologic but nonmalignant examples of conditions that mimic metastasis. One such condition is endometriosis in which the lining of the uterus, the endometrium, grows in inappropriate locations in the body. The misplaced uterine tissue may be found in the ovaries, uterine tubes, uterine ligaments, and other places (Fox and Buckley 1992). Just as the structure and function of a malignant metastasis are true to the tissue of origin (e.g., metastatic malignant thyroid nodules accumulate iodine not unlike normal thyroid), endometrial tissue that has migrated to a distant site retains its original behavior, and cyclic bleeding occurs at menstruation. Of course, a significant difference of metastasis, when compared to endometriosis, is the capacity for continued growth and expression of other aspects of the malignant phenotype.

Perhaps it would be instructive to consider several specific steps in the metastatic cascade that occur in normal embryos (Figure 2–11). During early development, the chicken (also mouse and human) embryo consists of two discoidal layers. The upper layer, the epiblast, is epithelial and rests on a basement membrane. Mesoderm (the "middle layer") formation occurs when epiblast cells migrate toward the center of the blastodisc (Spratt 1946) and become less differentiated (would it be appropriate in the present context to designate these cells as "anaplastic"?). Loss of differentiation occurs with a down-regulation of the cell adhesion molecule E-cadherin. Earlier in this chapter, E-cadherin was described in the context of the loss of cell differentiation, cell detachment, and metastasis (Mareel et al. 1993). The embryonic premesoderm cells detach and invade through the primitive streak (Vakaet, VanRoelen, and Andries 1980; Vakaet 1984; Harrisson, Andries, and Vakaet 1988) after loss of the basement membrane barrier, a process analogous to carcinoma cell migration across the digested basement membrane. One might wish to study chick embryos because of the strict chronologic predictability of the appearance of migratory cells, cells which detach from each other and pass through the digested basement membrane. Primitive streak and mesoderm formation highlight what has been stated before, namely, that activities of malignant cells have their counterparts in normal cells. In this instance, those activities in the normal cells (i.e., in the chicken embryos) are easily studied and the embryos are inexpensive and widely available.

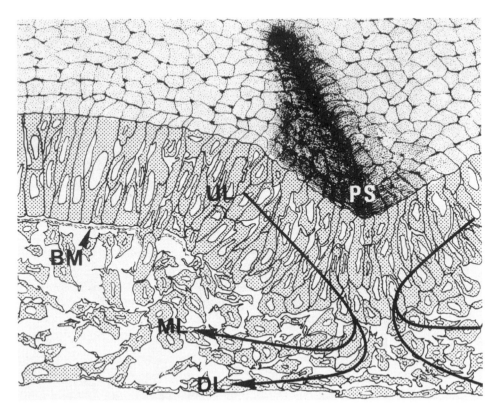

Figure 2–11. Diagram of morphogenetic movements that give rise to the primitive streak, mesoderm (middle layer), and some cells of the endoderm (deep layer) of the chick embryo. Note that cells lateral to the primitive streak (PS) in the upper layer (UL) are highly differentiated and rest on an intact basement membrane (BM). Cells become progresssively more anaplastic with loss of the basement membrane at and near the site of the primitive streak (PS). The cells that detach and migrate through the primitive streak and form the middle layer (ML), as well as those which contribute to the deep layer (DL), have a reduced expression of the cell adhesion molecule E-cadherin. This example from normal embryology parallels steps in the metastatic cascade in which cancer cells digest the basement membrane, become anaplastic, and detach and migrate from the primary tumor ultimately to form metastatic colonies. A down-regulation of expression of E-cadherin takes place during the steps in the malignant process. Primitive streak formation illustrates that normal cells share many activities of cancer cells. (Reprinted from Harrisson, F., Andries, L., and Vakaet, L. 1988. The chick blastoderm: Current views on cell biological events guiding intercellular communication. *Cell Diff* 22:83–106, with kind permission from Elsevier Science Ireland Ltd., Bay 15K, Shannon Industrial Estate, Co. Clare, Ireland.)

Cancer cell and normal cell transgression across the basement membrane in vivo have an in vitro analogue. Migration through a reconstituted basement membrane matrix was proposed as a method to discriminate between normal and malignant cell types – it was postulated that the former could not but the latter could migrate through the reconstituted basement membrane. Recently, however, it was shown that a number of malignant cell types failed this bioassay, whereas certain other normal cells readily invaded (Noel et al. 1991). The lesson to be learned here may be that many cell characteristics are shared by both normal and neoplastic cells.

Fertilization in humans occurs outside of the uterus. The developing embryo does not reach the uterus until it has cells competent for implantation into the endometrial lining. Implantation is a process in which the endometrium is invaded by aggressive embryonic trophoblast cells (Steer 1971; Yagel et al. 1990). These cells have the capacity to digest the endometrial lining and its connective tissue matrix as well as the walls of maternal arterioles. Were this aggressive invasive activity not to occur, implantation would fail and pregnancy would not ensue. The trophoblast releases proteolytic enzymes, which are implicated in this normal but highly invasive activity (Billington 1965; Kirkby 1965; Strickland, Reich, and Sherman 1976; Denker 1980). Again, compare this normal phenomenon of proteolytic enzyme activity with its malignant counterpart discussed earlier in this chapter.

Dissemination of malignant cells often occurs via a vascular route. There are examples of vascular dissemination of normal cells. Just described is the trophoblast of pregnancy, which has the competence to digest maternal arterioles, thereby providing access to the maternal circulation. Hence, trophoblast cells do intravasate (enter the vasculature) and are not uncommonly found in the blood of pregnant women where they are disseminated throughout the body presumably causing no damage. The circulating trophoblast cells may become entrapped in the lungs by the sieve effect of the pulmonary capillaries of the pregnant woman (Moore and Persaud 1993). Do any of these activities of trophoblast cells remind you of equivalent activities of cancer cells?

Primordial germ cells of the chick embryo (and other bird embryos) intravasate, disseminate, and extravasate in the process of moving from the germ crescent to the gonad primordia via the vascular system (Figure 2–12) (see Swift 1914). Although these cells do not have the malignant phenotype and do not form tumors, they are nevertheless behaving much as many metastatic tumor cells do.

Many more examples could be cited to support the contention that metastasis of malignant cells has similarities to phenomena observed in normal cell biology. Perhaps these examples are adequate to remind you that tumor cells do not have extraordinary capabilities – they are cells with ordinary capabilities expressed in an inappropriately extraordinary manner.

2.6   How do we know a metastasis to the liver is not a primary neoplasm of the liver?

A skeptic may remark that a malignancy in the brain is a brain tumor and a malignancy in the liver is a liver tumor. How does one distinguish between a primary brain tumor and a metastatic growth in the brain that had its origin elsewhere?

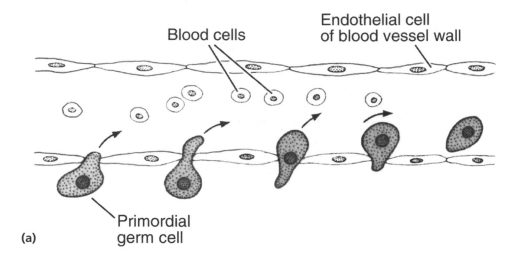

Blood cells

Endothelial cell
of blood vessel wall

Primordial
germ cell

(a)

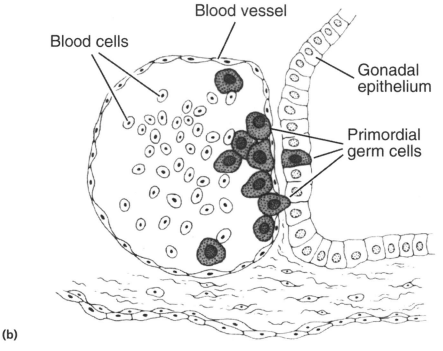

Blood vessel

Blood cells

Gonadal
epithelium

Primordial
germ cells

(b)

Figure 2–12. Primordial germ cells in the chick embryo gain entrance to a vessel by means of diapedesis (a). The primordial germ cells egress from the vessel at the gonadal epithelium (b). This translocation of normal cells involves vascular access; compare (a) here with the intravasation of tumor cells of Figures 2–1, 2–3, and 2–9. Also note the correspondence of primordial germ cell egress from a vessel (b here) with tumor cell extravasation in Figures 2–1 and 2–10. The sometimes organ-specific pattern of metastasis, known as the soil and seed hypothesis of Paget, has a normal analogue with vascular dissemination of primordial germ cells, that is, primordial germ cells egress vessels at the gonadal epithelium and not randomly at other sites. (From Gilbert, S.F. 1991. *Developmental Biology,* 3rd ed. Sunderland, MA: Sinauer Associates, Inc.)

Because stem cell determination is not lost during acquisition of the malignant phenotype (Chapter 1), the metastatic tumor faithfully preserves the cell morphology of the primary tumor regardless of its anatomic location (Chapter 1). Thus, we can distinguish a metastatic tumor cell colony in a particular organ from a primary tumor in that organ because the metastatic tumor often retains histologic characteristics of its tissue of origin – the primary tumor does this too and it thus may be a relatively simple matter to distinguish one from the other (on the basis of microscopic histology). This is not to say that the origin of all malignancies can be ascertained. Highly anaplastic cancers occur, making it difficult to determine in what tissue the tumor originated. However, in less anaplastic malignancies, metastatic tumors are recognized because of their morphologic similarity to the primary tumor.

2.7   Why study metastasis?

If about two-thirds of cancer patients already have cancer that has metastasized at the time of diagnosis, what good is there in knowing the biology of metastasis? In the words of Frost and Levin (1992), "knowledge of how the metastases developed offers little insight about control." Many believe the issue to be addressed is effective treatment, not more biology. The academic scholar who loves knowledge for knowledge sake would find this attitude puzzling. An analogy from another discipline concerns those who would seek to understand why an archeologist studies ancient native culture when contemporary populations of those people may be hungry. The archeologist studies previous cultures because they are interesting per se and solutions to today's problems may lie in an examination of the past. Similarly, a cell biologist may find metastasis interesting and this interest may have salubrious results. It is rarely possible to prognosticate the value of basic research prior to its being done. Understanding the subtleties of invasion and metastasis may lead to clinical applications not possible to anticipate with present knowledge.

Malignant neoplasms are comprised of a mosaic of cells with differing characteristics (called tumor cell heterogeneity). One of the differences is the competence to produce metastatic colonies. Perhaps only a very small minority of cancer cells have this competence. Treatment of that subpopulation of tumor cells clearly should be the target of therapy. It may become possible to treat the minority of cells with lethal metastatic potentialities without killing the mass of less destructive nonmetastatic cells, thereby leaving normal host cells less damaged than they would be with chemotherapy that does not target metastatic cells specifically.

What are some other practical applications of the study of metastases? Markers (discussed in Chapter 1) may be found to be associated with particularly aggressive tumors with a high potential for metastasis (Tarin 1996). However, all such tumors within that category may not be similarly aggressive. Would it not be useful to advise a patient who has a less aggressive cancer of the great likelihood that he or she probably will have many years of symptom-free good health and to go about living as usual? In contrast, sure knowledge obtained from a marker for metastasis that the patient has a poor prognosis allows him or her time to make rational decisions for family and self. That information, although not related to a cure, is useful and may well develop from the study of genes associated with metastasis.

Although two-thirds (some say up to 75%) of cancer patients have metastases at the time of presentation, one can take heart that about one-third *appear* not to. Metastasis can be abrogated, in theory, at any point in the cascade. With greater knowledge, it may become possible to repress *potential* metastasis in those patients who appear to be free of that dreaded complication. Temporal (or permanent) inhibition of metastasis in a patient free of secondary tumors would have benefits for both the patient and physician.

Millions of dollars have been spent looking for a cure for cancer. Most of the "cures" (Chapter 8) kill cancer cells, and, in the process, kill a significant number of normal host cells, thus proving to be highly toxic. It would be a boon to patients of cancer if a method could be devised to contain a primary tumor – to cage it – so it could not spread. Further, for those unfortunate two-thirds who already have metastatic cancer, containing their metastases while undergoing therapy is not a worthless goal. Perhaps restraint of metastasis of metastases will provide a somewhat larger window of time for effective therapy of existing tumor burden.

2.8 Summary

Metastasis accounts for much of the lethality of cancer and is feared for good reason. However, this most malignant aspect of cancer has been shown to depend on a number of interrelated steps known as the metastatic cascade. Interruption or significant perturbation of any of the steps has the potential for abrogating malignant spread. The steps of the cascade seem to be understandable and therefore controllable because the pathway to malignant dissemination follows the route of normal cell behavior, but with the trek sidetracked by untimely and disadvantageous cell function. The business of the cancer cell biologist is to set the pathway straight, and, by doing so, it is at least within the realm of possibility that more effective therapy will emerge.

3

Carcinogenesis

Alan O. Perantoni

3.1 Introduction

Our environment has been described as a "sea of carcinogens" awash with a variety of chemicals and, to a lesser extent, with oncogenic viruses and high-energy radiations, all of which may contribute significantly to cancer incidence in humans. Although much of our attention has been focused on the proliferation of synthetic chemicals and their waste by-products, an examination of these substances has revealed that, in fact, most chemicals are not carcinogenic. Testing by the U.S. National Toxicology Program (NTP) (Huff et al. 1988) and the International Agency for Research in Cancer (IARC) (Tomatis 1988) has shown that only one-fourth to one-third of those substances even suspected of being carcinogenic on the basis of their chemistries actually are cancer-causing agents. Thus, despite the initial bleak portrayal, these findings lead us optimistically to the possibility that cancer might be reduced or even eliminated through the identification and removal of carcinogens from the environment, assuming the numbers of carcinogenic substances are limited and that we are primarily responsible for their introduction into the environment.

Although industrial contributions have captured considerable attention and may, in fact, account for a small share of cancers (perhaps 1% to 4% of cancers in the United States) (Tomatis 1988), studies now underscore the potential of naturally occurring, or endogenous carcinogens (Gold et al. 1992): substances which, by their nature, cannot be eliminated. This chapter focuses on several aspects of carcinogens and the carcinogenic process, including the classes of carcinogens, the kinetics and dynamics of carcinogen-organism interaction, theories of carcinogenesis, government regulation of carcinogenic substances, and the continuing argument of risk versus benefit in regulation.

3.2   What is a carcinogen?

Simply stated, a carcinogen is any substance or agent that significantly increases tumor incidence. Operationally, that is, in a regulatory sense, this means any substance that increases tumor incidence at any dosage level by any route of administration in any species but usually in tester strains of laboratory rodents (Zwickey and Davis 1959). Rodents, however, tend to exhibit greater sensitivities to carcinogens than humans, which can lead to overestimates of carcinogenicity for humans, as will become clear from the discussion of carcinogenesis. Conversely, testing procedures may overlook other incomplete carcinogens, that is, agents, like promoters, which are incapable of inducing tumors by themselves but can accelerate the passage of altered cells through individual stages of the carcinogenic process. Because all substances that have been identified as human carcinogens by epidemiologists are also carcinogenic in bioassay animals (Wilbourn et al. 1986), animal studies are currently the most conservative and only definitive means for testing. Most human carcinogens have been identified either through epidemiologic studies of occupational risks or the inadvertent induction of tumors in humans receiving drug therapy for a medical condition. Due to space limitations, only the major classes of carcinogens are discussed here. For more information on individual or groups of chemicals, the IARC Monograph series provides continuous comprehensive updates on the various classes of carcinogens.

Although discussion here of the various types of carcinogens focuses on those described as "complete," to better understand what this means in terms of tumor induction, a description of the putative stages of the carcinogenic process is necessary.

3.3   Carcinogenesis as a multistage process

The consensus among cancer researchers is that carcinogenesis is a multistage process (Cohen and Ellwein 1991; Harris et al. 1992; Barrett 1993). This is based on observations from several perspectives: the latency of tumor development following carcinogen exposure, the apparent sequence of pathologic events described in the evolution of several types of tumors, age-specific tumor incidence, the existence of cancer-prone families, and now the molecular and genetic lesions associated with the various pathologic events of tumorigenesis. As mentioned in Chapter 1, carcinogenesis is often characterized by four sequential stages: initiation, promotion, progression, and malignant conversion (Figure 3–1). In this process, initiation occurs when a carcinogen interacts with DNA, producing a strand break

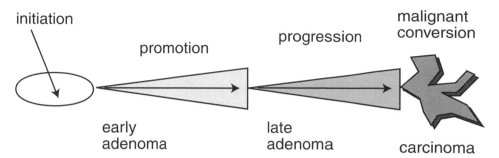

Figure 3–1. Carcinogenesis, a multistage process, begins with the genetic event of initiation followed by selective expansion of altered cells during promotion to form benign early adenomas. In the absence of continued promotion, the adenomas regress and disappear. With a second genetic event, a small number of promoted adenomas progress to form late adenomas, some of which may then undergo malignant conversion.

or, more often, an altered nucleotide called an adduct. Then, if the genome is replicated before repair enzymes can correct the damage, a DNA polymerase may misrepair the damaged sequence and permanently fix a heritable error in the genome. For adducts, this entails the insertion of an incorrect nucleotide in the DNA strand opposite the damage. The vast majority of such misincorporations are probably neutral to the cell. However, if the alteration occurs in a sequence that encodes a growth regulatory protein, for example, such an error may, under certain circumstances, provide that cell with a selective growth advantage. Those circumstances include exposure to a class of compounds known as promoters. Promoters are believed to preferentially select for or stimulate proliferation of initiated cells to form multiple benign tumors or hyperplastic lesions, and they represent the second stage of carcinogenesis. The effect of promotion is completely reversible, and withdrawal of the promoter results in the disappearance of the expanding clones of cells or hyperplastic nodules. Promoters, therefore, are not considered genotoxic, that is, able to induce heritable alterations or mutations. They are also poor inducers of malignant conversion (discussed later).

Finally, a second genetic event or series of events is proposed that allows some permanent selective growth advantage to initiated cells or somehow increases the probability a cell will become neoplastic. This stage, called progression, provides the impetus for conversion from benign adenomas to infiltrative and finally metastasizing neoplasms. These alterations could arise from additional exposures to a carcinogen, spontaneous mutations due to the natural infidelity of enzymes involved in replication, or genomic instabilities induced by initiating mutations. Regardless of the mechanism, the outcome is an irreversible change in the cell that allows expression of the neoplastic phenotype. In light of this theory, a "complete" carcinogen is an agent capable, through single or multiple exposures, of providing an initiating insult, promoting a selective growth advantage for altered cells, and generating lesions during progression that result in malignant conversion.

3.4 Chemical carcinogenesis

The English surgeon Percivall Pott is generally regarded as the father of carcino-genesis studies for his astute recognition that an environmental agent was responsible for tumor induction in chimney sweeps in London (Pott 1775). In a series of papers, he characterized a high incidence of scrotal cancer associated with this occupation and suggested that deposits of soot in the scrotal area were causative. As a consequence, other European nations legislated bathing and clothing requirements for chimney sweeps and, thus, virtually eliminated scrotal cancer as an occupational hazard (Butlin 1892). This clearly demonstrated not only that environmental substances could induce tumor formation but also that by identifying and eliminating the etiologic agent from the immediate environment, tumor incidence could be significantly reduced.

Others later established associations between exposures to soot and cancer incidence for a variety of professions, for example, gardening (Earle 1808) and tar manufacturing (Bell 1794), but more than a century passed before an animal model was developed to directly test the idea that chemicals cause tumors and to permit purification of the etiologic agents. Subsequently, Yamigawa and Ichikawa (1918) demonstrated that chronic application of coal tar to the ears of rabbits could induce skin carcinomas. Thus, chemists could examine the carcinogenicity of purified fractions from tar products. Their detailed study established a clear multistage process consisting of early irritation and proliferation of cells, papilloma formation at the site of exposure in treated animals, neoplasia, and even metastases in several animals. Finally, Cook et al. (1933) described the successful extraction of a crystalline substance responsible for the carcinogenicity of coal tar on rabbit skin. The compound benzo[a]pyrene (BP), a member of a class of carcinogens called polycyclic aromatic hydrocarbons (PAH), has since been studied extensively as a metabolism-dependent chemical carcinogen and has proven invaluable in elucidating the mechanisms for carcinogen activation. Since the discovery of BP, hundreds of other occupational or environmental chemical carcinogens have been identified (Table 3–1), including a variety of organic substances (e.g., PAH and aromatic amines), as well as a number of environmentally important inorganic metals and minerals, such as nickel and asbestos. Despite the diversity of chemistries, more than 95% of the various carcinogenic chemicals fall into one of three major categories: alkylating agents, aralkylating agents, or arylhydroxylamines (Figure 3–2). Although these agents differ in the type of residue transferred to create a DNA adduct, they share certain characteristics. Most notably, they are either intrinsically reactive with DNA or can be metabolically activated to a stable DNA-reactive form. These so-called electrophiles bond with the electron-sharing atoms of the DNA nucleotides, such as ring nitrogen or exocyclic oxygen atoms (Figure 3–3), to form stable altered nucleotides or adducts (Hemminki 1994).

Because of their ability to react with DNA and cause misincorporation of

Table 3–1. *Known and suspected carcinogens identified by the National Toxicology Program*

Known chemical carcinogens

aflatoxins
analgesic mixtures containing
  phenacetin
arsenic
asbestos
azathioprine
benzene
benzidine
bis(chloromethyl)ether

1,4-butanediol dimethyl-sulfonate
chlorambucil
(2-chloroethyl)-3-(4-methyl-cyclo-
  hexyl)-1-nitrosourea
chromium
conjugated estrogens
cyclophosphamide
diethylstilbestrol
erionite

melphalan
methoxsalen with ultraviolet light
mustard gas
2-naphthylamine
radon
thorium dioxide
vinyl chloride

Chemicals reasonably anticipated to be carcinogens

acetaldehyde
2-acetylaminofluorene
acrylamide
acrylonitrile
2-aminoanthraquinone
o-aminoazotoluene
1-amino-2-methyl-anthraquinone
amitrole
o-anisidine hydrochloride
benzotrichloride
beryllium
bischloroethyl nitrosourea
bromodichloromethane
1,3-butadiene
butylated hydroxyanisole
cadmium
carbon tetrachloride
ceramic fibers
chlorendic acid
chlorinated paraffins
1-(2-chloroethyl)-3-cyclohexyl-1-
  nitrosourea
chloroform
3-chloro-2-methylpropene
4-chloro-o-phenylenediamine
C.I. basic red 9 monohydro-
  chloride
cisplatin
p-cresidine
cupferron
DDT
2,4-diaminoanisole sulfate
2,4-diaminotoluene
1,2-dibromo-3-chloropropane
1,2-dibromoethane (ethylene
  dibromide)
1,4-dichlorobenzene
3,3'-dichlorobenzidine
1,2-dichloroethane
dichloromethane
1,3-dichloropropene
diepoxybutane
di(2-ethylhexyl)phthalate
diethyl sulfate

diglycidyl resorcinol ether
3,3'-dimethoxybenzidine
4-dimethylamino-azobenzene
dimethylcarbamoyl chloride
1,1-dimethylhydrazine
dimethyl sulfate
dimethylvinyl chloride
1,4-dioxane
direct black 38
direct blue 6
epichlorohydrin
estrogens (not conjugated):
  estradiol-17ß
  estrone
  ethinylestradiol
  mestranol
ethyl acrylate
ethylene oxide
ethylene thiourea
ethyl methanesulfonate
formaldehyde (gas)
glasswool (respirable size)
glycidol
hexachlorobenzene
hexachloroethane
hexamethyl phosphoramide
hydrazine
hydrazobenzene
iron dextran complex
kepone¨ (chlordecone)
lead acetate and lead phosphate
lindane
2-methylaziridine
4,4'-methylenebis(2-chloroaniline)
4,4'-methylenebis(n,n-dimethyl-
  benzenamine)
4,4'-methylenedianiline
methyl methanesulfonate
n-methyl-n'-nitro-n-nitrosoguanidine
metronidazole
michler's ketone
mirex
nickel
nitrilotriacetic acid

nitrofen
nitrogen mustard hydrochloride
2-nitropropane
N-nitrosodi-N-butylamine
N-nitrosodiethanolamine
N-nitrosodiethylamine
N-nitrosodimethylamine
N-nitrosodi-N-propylamine
N-nitroso-N-ethylurea
4-(N-nitrosomethyl-amino)-1-
  (3-pyridyl)-1-butanone
N-nitroso-N-methylurea
N-nitrosomethyl-vinylamine
N-nitrosomorpholine
N-nitrosonornicotine
N-nitrosopiperidine
N-nitrosopyrrolidine
N-nitrososarcosine
norethisterone
ochratoxin a
4,4'-oxydianiline
oxymetholone
phenacetin
phenazopyridine hydrochloride
phenoxybenzamine hydrochloride
phenytoin
polybrominated biphenyls
polychlorinated biphenyls
polycyclic aromatic hydrocarbons:
  benz[a]anthracene
  benzo[b]fluoranthene
  benzo[j]fluoranthene
  benzo[k]fluoranthene
  benzo[a]pyrene
  dibenz[a,h]acridine
  dibenz[a,j]acridine
  dibenz[a,h]anthracene
  7H-dibenzo[c,g]carbazole
  dibenzo[a,e]pyrene
  dibenzo[a,h]pyrene
  dibenzo[a,i]pyrene
  dibenzo[a,l]pyrene
  indeno[1,2,3-cd]pyrene
  5-methylchrysene

(continued)

(Table 3–1 continued)

procarbazine hydrochloride	selenium sulfide	thiourea
progesterone	silica, crystalline (respirable size)	toluene diisocyanate
1,3-propane sultone	streptozotocin	o-toluidine
beta-propiolactone	sulfallate	toxaphene
propylene oxide	2,3,7,8-tetrachlorodibenzo-p dioxin	2,4,6-trichlorophenol
propylthiouracil	tetrachloroethylene (perchloro-	tris (1-aziridinyl)phosphine sulfide
reserpine	ethylene)	tris (2,3-dibromopropyl) phosphate
saccharin	tetranitromethane	urethane
safrole	thioacetamide	4-vinyl-1-cyclohexene diepoxide

Source: This list was derived from the 7th Annual Report of the NTP, and evaluations of individual chemicals can be obtained on the Net at ntp-server.niehs.nih.gov

nucleotides in the genome, many, but not all, carcinogens are also mutagenic (Shelby and Zeiger 1990). One of the most sensitive and rapid assays for mutagenicity employs strains of *Salmonella typhimurium,* which are readily mutable to growth independence from histidine supplementation of growth media. In the presence of a mutagen (such as a mutagenic carcinogen), an inactive enzyme involved in bacterial histidine synthesis can be mutated to an active form, allowing the bacterium to grow in a medium that lacks histidine. The problem with this assay is that bacteria do not contain the drug or carcinogen-metabolizing enzymes found in mammalian cells, so carcinogens requiring metabolic activation depend on the addition of metabolizing enzymes for mutagenicity.

Many of the metabolizing enzymes exist in the endoplasmic reticulum, so inclusion of microsomal preparations from rodent livers in bacterial assays is often sufficient to permit activation. Other carcinogens, however, appear to be nongenotoxic by current means of testing, yet increase tumor incidence. This may be due in part to the shortcomings of testing procedures. Asbestos, for example, induces large chromosomal alterations and not point mutations, the end points for the bacterial mutation assays (Jaurand 1989). Pure promoters like phenobarbital are nontransforming and nonmutagenic by themselves but can enhance tumorigenicity several fold in combination with weakly transforming doses of a carcinogen. For others, like amitrole, current assay systems may lack the enzymes for metabolizing carcinogens to reactive electrophilic forms (Kraus et al. 1986).

Although many carcinogens exhibit mutagenic activity, as just described, not all carcinogens have been shown to be mutagenic, nor are all mutagens carcinogenic. Perhaps the best example involves certain nucleotide analogues that serve as potent mutagens in bacterial assays but are generally regarded as noncarcinogenic in animal carcinogenicity tests.

The most frequently studied carcinogens include the N-nitroso compounds and aflatoxins (alkylating agents), the polycyclic aromatic hydrocarbons (aralkylating agents), and the aromatic amines and aminoazo dyes (arylhydroxylamines). With the exception of certain N-nitroso carcinogens, these substances are all

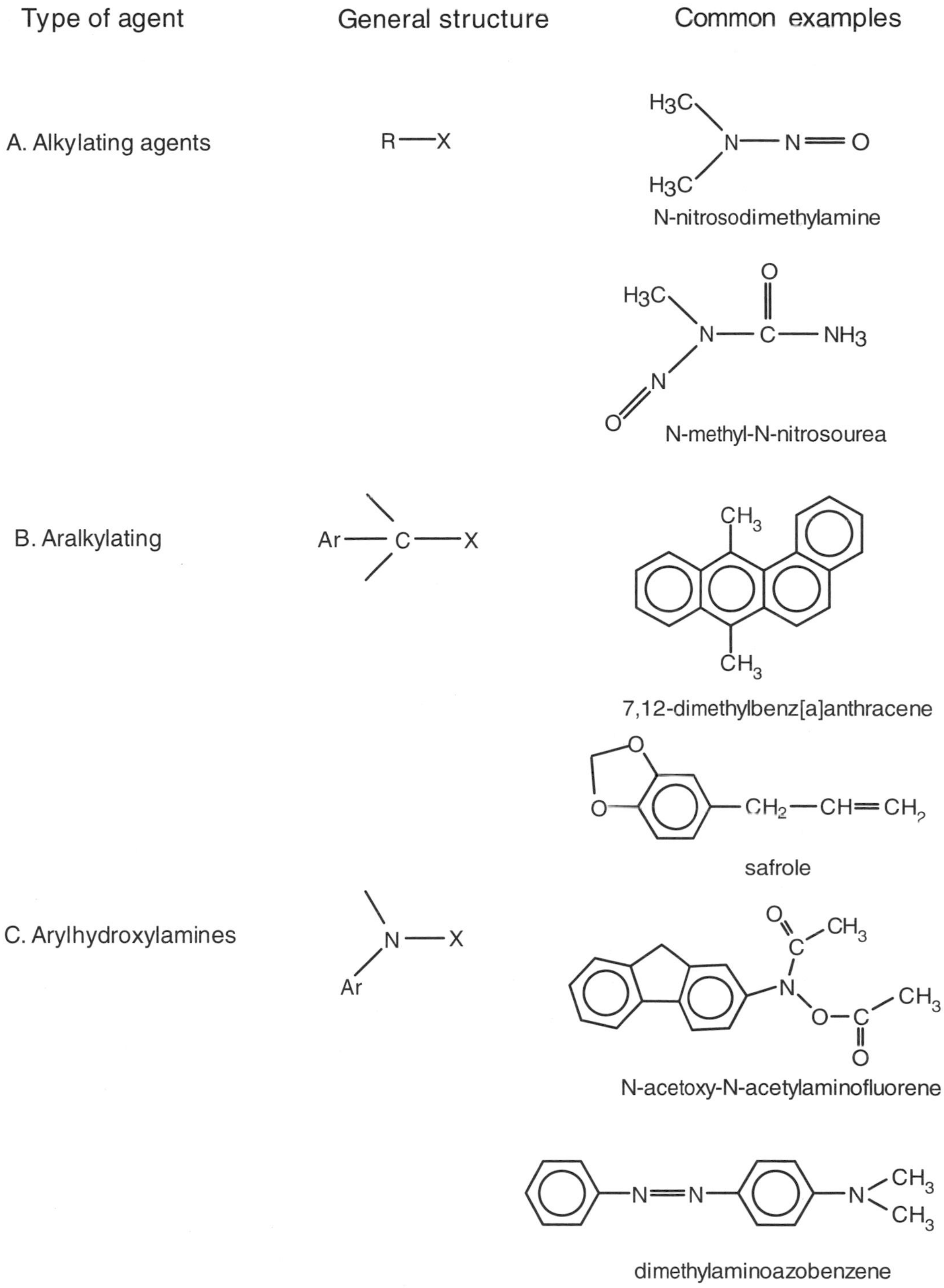

Type of agent	General structure	Common examples

A. Alkylating agents R—X

H3C\
N—N=O
H3C/

N-nitrosodimethylamine

H3C\
N—C—NH3 (with O double bond on C)
N
O

N-methyl-N-nitrosourea

B. Aralkylating Ar—C—X

CH₃
(ring structure)
CH₃

7,12-dimethylbenz[a]anthracene

CH₂—CH=CH₂

safrole

C. Arylhydroxylamines N—X with Ar

C=O CH₃
N
O—C—CH₃ (with O double bond)

N-acetoxy-N-acetylaminofluorene

N=N
N<CH₃ CH₃

dimethylaminoazobenzene

Figure 3–2. Classification of the major types of chemical carcinogens. The majority fall into three categories based on their donor groups in adduct formation: alkylating, aralkylating, and arylhydroxylamines.

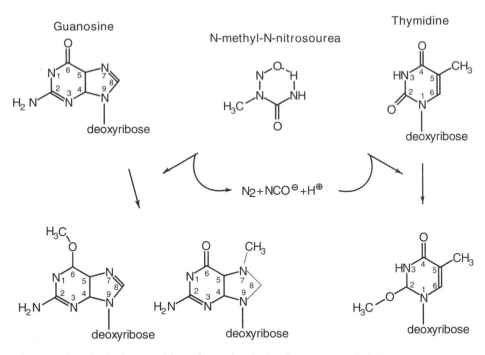

Figure 3–3. Principal DNA adducts formed with the direct-acting alkylating agent N-methyl-N-nitrosourea (MNU). The carcinogen MNU donates a methyl group primarily to the ring N-7 and the exocyclic O-6 in guanosine and the exocyclic O-2 of thymidine to cause mutations in the genome.

metabolized by the P-450 mixed-function oxidase enzymes (or P-450 cytochromes), a superfamily of drug-metabolizing microsomal enzymes, to electrophilic intermediates that react with DNA bases. Although structure/activity studies of these compounds have resolved details of cellular mechanisms for metabolic activation, they have been less useful in defining events of carcinogenesis. For this, studies of direct-acting agents, such as the N-nitrosoureas, have clarified the chemistry of carcinogen-target interactions.

3.4a   Organic compounds

Alkylating agents – chemicals that transfer alkyl groups, often methyl or ethyl groups, to nucleotides to form DNA adducts (Figures 3–2 and 3–3).

The N-nitroso compounds, especially the nitrosamines, are perhaps the most insidious and therefore potentially most hazardous of the various carcinogens. Epidemiologic studies have provided no definitive link of members of this class of carcinogens to human cancer, with the exception of the use of the chloroethylnitrosureas, chemotherapeutic agents for gastrointestinal cancers (Boice et al. 1983).

However, the potent carcinogenicity of several of these compounds in every species tested (including nonhuman primates; Magee and Barnes 1956; Kelly et al. 1966), the hepatotoxicity of certain nitrosamines in workers exposed occupationally (Kimbrough 1983), and the ubiquitous nature and volatility of these chemicals strongly suggest that they pose a significant health hazard to humans (Bartsch et al. 1992).

Activation of these compounds often requires biotransformation either enzymatically by oxidation (as for dimethylnitrosamine) or directly by alkali-mediated hydrolysis (as for the direct-acting carcinogen methylnitrosourea). In either case, a methyl group (CH_3^-) or an ethyl group ($CH_3CH_2^-$), depending on the chemical, is available for the modification of a DNA base. Methylation occurs predominantly at exocyclic oxygen moieties or ring nitrogens (Figure 3–3).

The problem with these compounds is not only their ubiquitous distribution in the environment but also the ease with which they are generated endogenously. Formation of N-nitroso compounds by the nitrosation of secondary amines occurs fairly readily under acidic conditions, for example, in the stomach, and bacteria can catalyze similar reactions under nonacidic conditions, for example, in the gut. Thus, compounds like dimethylnitrosamine can be readily produced in the body, but whether or not this represents a significant health risk has not been established.

Numerous naturally occurring toxins have also been characterized, many of which may function as natural pesticides or antibiotics for plants and fungi and also may exhibit carcinogenic activity (Ames, Magaw, and Gold 1987). Over three hundred mycotoxins have been identified, some of which are strong mutagens and carcinogens. The potent rodent liver carcinogen aflatoxin B_1, produced by the common mold *Aspergillus flavus,* can heavily contaminate grains, vegetables, and nuts on which the mold thrives. Aflatoxin is especially a problem in areas of the world where methods of food preservation are deficient. Its general distribution and potency has been considered serious enough in the United States to warrant limitations on levels in such food products as peanut butter.

Aflatoxin was first identified as the hepatotoxic substance in contaminated feed in England that devastated the domesticated turkey population in 1960 (Blount 1961) and was subsequently shown to induce liver tumors in several species (IARC 1992). Its carcinogenicity for humans is suggested in epidemiologic evaluations of African populations with a high incidence of liver tumors (Alpert et al. 1968) and in the mutational spectrum of genetic lesions found in the p53 suppressor gene from hepatocellular cancers (Hsu et al. 1991). However, the multiplicity of factors identified in these populations (notably, chronic infections with hepatitis B) obscures the role of aflatoxin B_1 in human liver tumorigenesis. Metabolic activation of aflatoxin B_1 involves oxidation of the aflatoxin molecule by the P-450 mixed-function oxidases to generate an intermediate which reacts with a cyclic nitrogen in guanine, forming the adduct 8,9-dihydro-8-(N7-guanyl)-9-hydroxyaflatoxin B_1 (Essigmann et al. 1982).

Aralkylating agents – chemicals that transfer aromatic or multiringed compounds to a nucleotide to form an adduct (Figure 3–2).

Polycyclic aromatic hydrocarbons (PAH), the principal group of aralkylating agents, remain an occupational problem in several industries. As previously mentioned, exposure to the hydrocarbons in soot increased the incidence of scrotal cancer in British chimney sweeps. More recently, the crude oils used in cotton spinning (Heller 1930) or tool setting (Waldron, Waterhouse, and Tessema 1984) were reported to cause scrotal cancer in workers whose pants became saturated with the oils. Structurally similar compounds have been found in combustion products (Grimmer and Misfeld 1983) and as such are added continuously to the environment. This includes products such as benzo[a]pyrene and the potent 7,12-dimethylbenz[a]anthracene, which are generated in cigarette smoke and on charcoal-grilled meats (Lijinsky and Shubik 1964). These compounds readily induce tumors in laboratory animals, causing rapid tumorigenesis in rat mammary tissue following ingestion (Huggins, Grand, and Brillantes 1961), the major route of exposure for humans.

Metabolism of PAH has been investigated extensively, and certain generalizations can be made from these studies. The reactive metabolite appears to be a diol-epoxide, generated by two consecutive cycles through the P-450 mixed-function oxidase system (Figure 3–4). The intermediates are sufficiently stable to allow their passage to the cell nucleus but also highly reactive with DNA to cause mutations (Jerina et al. 1991).

Arylhydroxylamines – chemicals that transfer aromatic amines to nucleotides to form adducts (Figure 3–2).

The carcinogenic activity of the aromatic amines, a major group of arylhydroxylamines, was established from epidemiologic studies of workers in the dyestuff industry. Occupational exposures to aniline dyes caused a high incidence of bladder cancer (Rehn 1895), and the etiologic agents most responsible were eventually identified as 2-naphthylamine and benzidine (Case et al. 1954). In a study of workers involved in the distillation of 2-naphthylamine, nearly all heavily exposed individuals subsequently developed bladder cancer (Case 1969). Similarly, 2-naphthylamine was implicated in the high incidence of bladder tumors in the manufacturing of rubber (Case and Hosker 1954). Bioassays for carcinogenic activity in rats and dogs showed these compounds to be active primarily in the bladder following dietary exposures (Radonski 1979).

The arylhydroxylamines, like PAH, require metabolism by the P-450 mixed-function oxidase enzymes, and commonly undergo N-oxidation to generate reactive intermediates. In the case of 2-naphthylamine, the chemical is first enzymatically oxidized to generate an N-hydroxyl intermediate, which may be further metabolized in liver or kidney to form a stable glucuronide conjugate that is

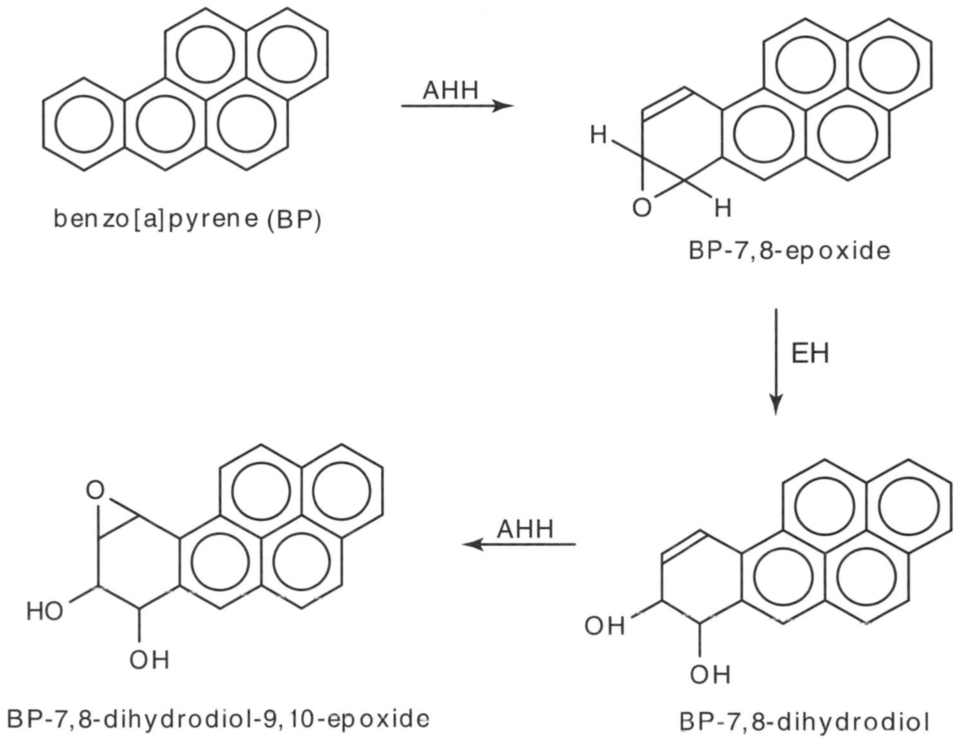

Figure 3–4. Metabolic activation of the aralkylating agent benzo[a]pyrene (BP). Activation is mediated by the P-450 microsomal enzymes and specifically arylhydrocarbon hydroxylase (AHH), which oxidizes the carbon-carbon double bond to form an epoxide. Another enzyme, epoxide hydratase (EH), destroys the epoxide, and AHH further metabolizes the BP to form the presumed reactive intermediate.

passed to the bladder (Figure 3–5) (Kadlubar, Miller, and Miller 1977). In the bladder, the final activated electrophile, presumed to be a nitrenium cation, is formed, which can react with DNA in bladder epithelia (Orzechowski, Schrenk, and Bock 1992).

3.4b   Inorganic compounds and asbestos

Certain inorganic metals and minerals exhibit carcinogenic activities or are associated with elevated risks for cancer in humans. These include arsenic, nickel, chromium, and asbestos. The unequivocal effect of asbestos on tumorigenesis and its extensive presence in the environment such as in cement construction materials, insulation, and fireproofing, have established it as a significant health hazard.

The term "asbestos" encompasses a variety of silica fiber types. Although one form of asbestos, the serpentine magnesium-containing chrysotile ($Mg_6Si_4O_{10}(OH)_8$), represents more than 90% of the mined asbestos in the

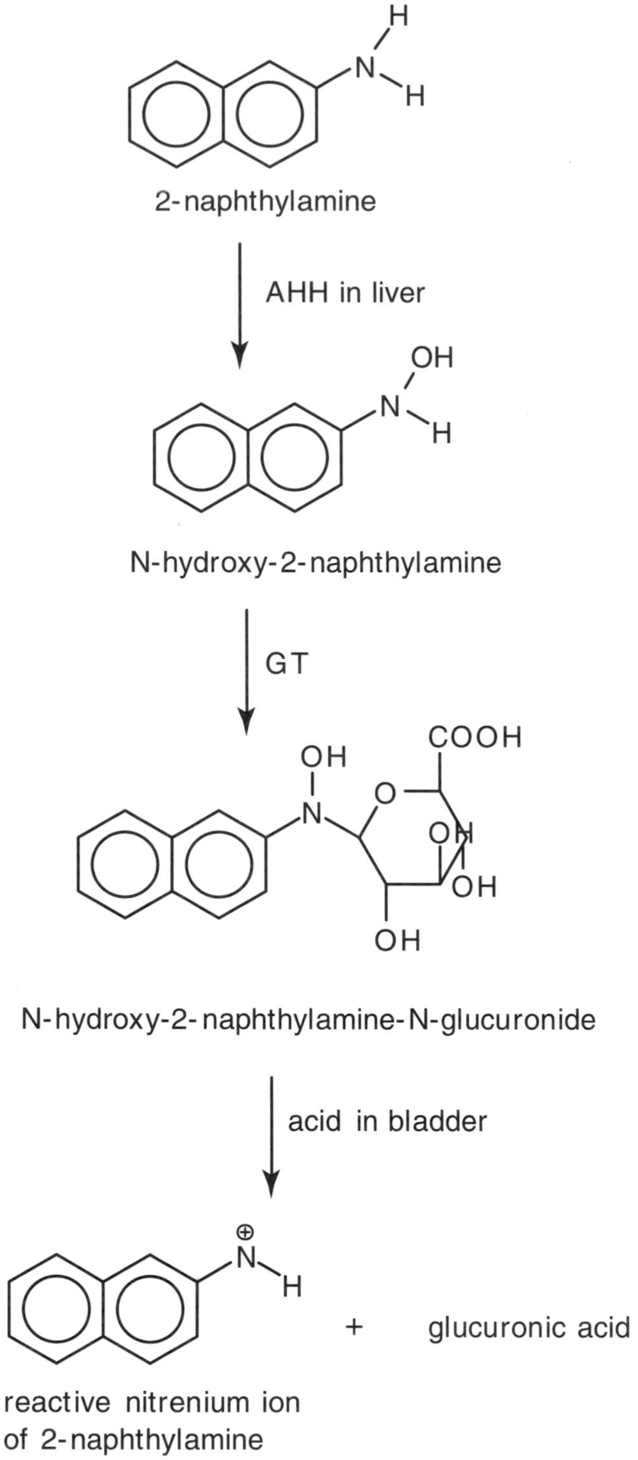

Figure 3–5. Metabolism of the arylhydroxylamine 2-naphthylamine. AHH activity in the liver oxidizes the nitrogen to generate the N-hydroxy derivative. The conjugating enzyme glucuronide transferase (GT) adds the sugar residue to the carcinogen. This stabilizes the compound until it reaches the bladder, where acid conditions cause the formation of the carcinogenic nitrenium ion.

United States, carcinogenic activity is generally associated with the amphibolic iron-bound crocidolite $(Na_2(Fe^{3+})_2(Fe^{2+})_3Si_8O_{22}(OH)_2)$ or calcium-containing tremolite $(Ca_2Mg_5Si_8O_{22}(OH)_2)$. The various mineral forms of asbestos generally reflect differences in fiber structure, differences that affect the ability of the fibers to be retained in the lungs upon inhalation. For example, chrysotile fibers often occur as clusters of curly fibers that penetrate the lungs inefficiently and are cleared far more rapidly than the amphiboles. Conversely, the amphiboles (a large group of structurally similar silicate minerals) such as crocidolite exist as individual rods that readily penetrate deep into the lungs and may remain there for several months to years (Mossman et al. 1990).

From epidemiologic studies, it is apparent that all individuals carry a significant number of asbestos fibers in their lungs and that includes infants as well (Haque and Kanz 1988). However, workers exposed to asbestos in the mining, milling, or manufacture of insulation carry a much greater lung fiber burden than the general population and have nearly a tenfold greater risk for lung cancer. In addition, cigarette smoking acts synergistically with asbestos to enhance lung cancer risk ten times above incidences for asbestos exposure alone and five times for smoking alone (Saracci 1987).

Asbestos workers present with either lung carcinomas or mesotheliomas of the pleura or peritoneum. As many as 20–25% of heavily exposed workers develop lung cancer (Lemen, Dement, and Wagoner 1980). Because the occurrence of mesotheliomas is extremely rare and almost totally associated with asbestos exposure (Wagner et al. 1960), epidemiologists were readily able to establish asbestos as the etiologic agent. Although all types of asbestos fibers can cause chromosomal aberrations (Sincock, Delhanty, and Casey 1982; Hei et al. 1992) as well as lung tumors in rats (Davis et al. 1978), studies of fiber biodistribution in asbestos workers with mesothelioma suggest that the lung burden of amphiboles is significantly greater than chrysotile fibers (MacDonald and MacDonald 1987; Wagner et al. 1988) as would be predicted on the basis of the fiber structures described here. These studies therefore are consistent with the idea that the amphiboles present a significantly greater health risk than the chrysotiles. Although epidemiologic evidence seems to indicate that the crocidolite fibers are more potent inducers of mesothelioma than chrysotile fibers, evidence for a similar circumstance in lung cancer induction is far less compelling. The fact that fiber size, shape, and composition varies in different asbestos deposits causes problems interpreting the epidemiologic studies. Sufficient evidence, however, suggests that both crocidolites and chrysotiles may be relevant to lung cancer induction in asbestos workers (Stayner, Dankovic, and Lemen 1996).

The absolute stability of the asbestos fibers and their ability to travel long distances in the air as well as epidemiologic and toxicologic evidence support the tight controls currently placed on the manufacture and use or removal of amphibole-containing products. Furthermore, the growing literature implicating the chrysotiles in tumor induction reinforces the concept that both classes of asbestos should be treated with equal caution, and in fact the Occupational Safety and Health Administration has revised standards of acceptable occupational exposure levels to include all forms of asbestos.

What of nonoccupational exposures to asbestos? Low levels of fibers persist in outdoor air, which probably accounts for the fact that all individuals possess a small but significant lung burden. Interestingly, levels of airborne fibers are no greater in buildings with damaged asbestos materials than are found outdoors. Consequently, medical experts have suggested that the elevated levels of fibers, which can occur with removal, may actually increase the health risks to the asbestos (Mossman, Kamp, and Weitzman 1996). Because the cancer risk from those materials already in place could conceivably be negligible, the costly issue of asbestos removal from buildings warrants further investigation before significant resources are committed to its elimination.

3.4c   Naturally occurring chemicals

Finally, and perhaps of greatest importance, is the category of generally unexplored naturally occurring chemical carcinogens, which have only recently received attention. Although the U.S. government regulates synthetic compounds and food additives, there is currently no control over natural substances such as pesticides or antibiotics produced by plants. It has been estimated that humans are exposed to 10,000 times more of these natural pesticides than synthetic pesticides, and, in fact, these natural pesticides may constitute as much as 5% to 10% of a plant's dry weight. Although most of these compounds remain untested, preliminary studies indicate that they are not without risk. Of the 57 substances evaluated thus far, 29 are carcinogenic in rodents (Gold et al. 1992). In addition to the multiplicity of chemicals, some of these compounds may occur naturally in volumes as high as parts per thousand, whereas exposure to synthetic pesticides is generally in parts per million or less. Thus, as Ames (1987) has asserted, "nature is not benign." Granted, this area requires further investigation to assess the risks from these substances, but they deserve greater attention in light of public pressure to eliminate beneficial synthetic compounds, which may in fact contribute little to our total carcinogenic burden.

3.5   Radiation

3.5a   Ultraviolet radiation

Although most agree that, of the exogenous agents, chemicals are responsible for the greatest share of cancer, chronic exposure to ultraviolet light as a consequence of increased leisure time or to radon because of improved standards of household insulation have refocused attention on the various forms of radiation. Ultraviolet light is continuously bathing our environment. Although predominantly from sun, artificial lighting ensures, to a lesser but significant extent, that this exposure will be continuous because both fluorescent and the increasingly popular tungsten-halogen fixtures emit significant amounts of UV radiation. As the ozone is depleted from the stratosphere, prospects are that exposure levels to ultraviolet irradiation will also increase.

Ultraviolet radiation includes wavelengths between 200 and 400 nm (visible light ranges from 400 to 700 nm) and is often subdivided into three regions: UV-A, 320 to 400 nm; UV-B, 280 to 320 nm; and UV-C, 200 to 280 nm. Biologic effects are elicited primarily with UV-B radiation, which induces the acute symptoms of sunburn and the adaptive responses to exposure of hyperpigmentation and skin thickening. Because stratospheric ozone and layers of dead skin effectively absorb UV-C radiation, exposure in this region appears to be rather limited. Little is known regarding the biologic activity of UV-A radiation, although levels do not vary with time of day or season, suggesting a lack of association with factors that play a role in skin cancer induction. There are, however, indications that it can effect tissue injury in conjunction with UV-B radiation or certain photosensitizing chemicals (Urbach 1993).

Circumstantial experimental and epidemiologic evidences have implicated UV radiation in the induction of both basal and squamous cell carcinomas, the most common but generally curable forms of cancer (Urbach 1993; Marks 1996b). As discussed in Chapter 7, although nearly 100% curable with early detection and treatment, these skin cancers have been observed with increasing frequency as leisure time has expanded. Epidemiologic studies have shown that tumor incidence correlates positively with circumstances that elevate cumulative skin exposures to UV radiation. Thus, tumor incidence increases with decreasing latitude (Elwood et al. 1974). Furthermore, tumors arise predominantly in weakly pigmented individuals or ethnic groups (Chuang et al. 1990) and generally appear on those body surfaces that receive the greatest exposure, such as the head and neck (Haenszel 1963). Also, individuals in occupations that require greater outdoor exposure clearly have a higher tumor incidence (Vitaliano 1978).

Ultraviolet radiation catalyzes the formation of pyrimidine cyclobutane dimers (Beukers and Berends 1960) and 6–4' photoproducts, both of which are formed between adjacent thymine bases and can cause GC-to-AT transition mutations in DNA if not repaired (Figure 3–6). The involvement of dimer formation in carcinogenesis is strongly supported by studies of the genetic defect xeroderma pigmentosum, a complex of disorders characterized by deficient excision repair of ultraviolet-induced pyrimidine dimers and a high skin cancer incidence (Kraemer, Lee, and Scotto 1987). In addition, in vitro studies have shown that the UV action spectrum for transformation of hamster embryo cells (Doniger et al. 1981) or human embryonic fibroblasts (Sutherland et al. 1981) is consistent with that for UV induction of dimer formation.

Malignant melanoma, the highly aggressive cutaneous cancer of melanocytes, is

cyclobutane dimer   6-4' photoproduct

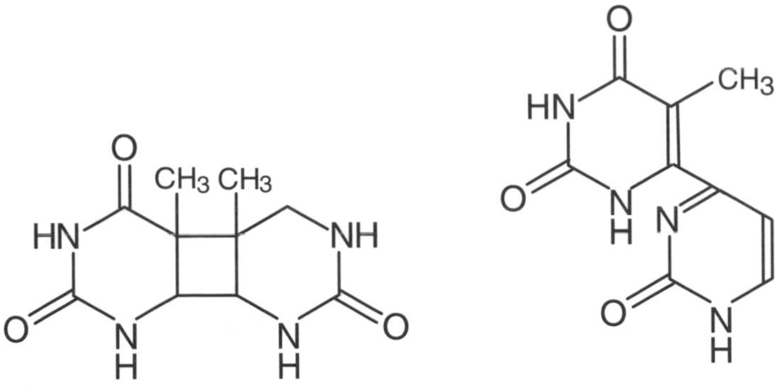

Figure 3–6. Ultraviolet light causes the formation of two principal adducts in DNA: the cyclobutane dimers and 6–4 photoproducts of thymidine. Both products require excision repair processes to correct.

of growing concern because it affects primarily young adults, is increasing in incidence faster than any noncutaneous cancer, and is often lethal. Although melanoma incidence is similarly associated with exposure to sun (IARC 1992), the relationship is more complex than for other skin tumors and seems to involve intense, intermittent exposures. For example, melanoma incidence in individuals with outdoor occupations is actually lower than for those receiving intermittent exposures (Lee and Strickland 1980; Garland et al. 1990). However, the incidence does increase with decreasing latitude (Armstrong 1984) and predominates in white-skinned populations (Muir et al. 1987) as do other skin cancers. Other risk factors include the extent of intermittent exposure to sun, the numbers of dysplastic nevi (possible precursors of melanoma), and susceptibility to sunburn; frequency of sunburn shows no correlation (Armstrong 1988). For risk factors associated with skin cancer, see Chapter 7.

3.5b   Ionizing radiation

Ionizing radiation is a well-established human carcinogen first noted by experimentalists during the development of Roentgen's cathode tube, the basis for the x-ray machine. Not recognizing its health consequences, early radiologists often used their own hands to focus the electron beam in primitive x-ray machines, resulting in the frequent induction of skin cancer (Frieben 1902). Later, luminescent dial painters in watch factories suffered a high incidence of osteosarcomas, a tragedy attributed to radium ingestion when painters orally formed pointed tips on paintbrushes (Martland 1931). The radium was localized to and retained by cells in the bone due to its ability to mimic calcium, thus making bone the immediate target of radium's high-energy alpha emissions.

Ionizing radiation has been clearly linked to the excess cancer cases in populations exposed to nuclear detonations. An intensive epidemiologic investigation of the 76,000 atomic bomb survivors who were exposed to neutron and gamma radiation in Hiroshima and Nagasaki has revealed an increased risk of cancer mortality from leukemias especially, and to a lesser but significant extent from tumors of the digestive tract (especially esophagus, stomach, and colon), lungs, liver, bladder, female breast, and ovaries (Shimizu et al. 1989; Shimizu, Kato, and Schull 1990). Interestingly, the increases occurred in direct proportion to normal increases in cancer with aging; that is, tumor latency was apparently not abbreviated, but the number of observed versus expected cases increased. It is also noteworthy that all measures of germline genetic damage in this population have thus far demonstrated no increases attributable to parental exposures, and, therefore, the offspring of survivors of ionizing radiation are apparently not at additional risk to any adverse genetic effects (Neel et al. 1990). Study group estimates now project that a total of about a thousand deaths in the 76,000 survivors will be attributable to atomic bomb irradiation (Lenihan 1993). These observations demonstrate the relative carcinogenic impotence of ionizing radiation when not tissue localized and suggest that current low-level environmental or health-related exposures may be of little consequence.

On the other hand, tissue localization of an ionizing radiation source can have adverse health consequences. In an inadvertent human exposure during an aboveground U.S. nuclear test, an unexpected wind shift carried high levels of the radioactive isotope ^{131}I over the Marshall Islands. Subsequent epidemiologic analysis of the indigenous population on one of the heavily exposed islands revealed thyroid tumors in more than 50% of children under the age of 10 at exposure (Conrad, Dobyns, and Sutow 1970). Because thyroid tumors were extremely rare in this population, the high tumor incidence could be attributed to localization of this moderate-energy gamma emitter in the thyroids of exposed children.

The mechanism of carcinogenesis from ionizing radiation is believed to involve indirect formation of mutagenic oxygen free radicals. Due to the tissue penetrance of certain types of ionizing radiation, oxygen free radicals can be generated at the DNA level by ionizing the shell of hydration surrounding the DNA, thus making it a readily available target for these highly reactive and extremely short-lived molecules. Once formed, the reactive oxygen species (.OH [hydroxyl radicals], H_2O_2 [hydrogen peroxide], $^1O_2$ [singlet oxygen], or O_2^- [superoxide radicals]) can induce more than thirty different DNA adducts as well as DNA-protein crosslinks (Feig, Reid, and Loeb 1994). The free radical–generated mutations may result directly from the specific adducts formed, or they may be caused indirectly by free radical–induced alterations in the DNA polymerases. In either case, the spectrum of mutations formed is determined in part by the activity of the specific polymerase that reads through the damage, that is, what base it preferentially incorporates opposite the adduct.

The various forms of ionizing radiation described are carcinogenic when presented at unusually high doses but what about the chronic low levels to which most are exposed routinely, for example, chest x-rays, dental exam x-rays, endogenous isotopes, and cosmic or terrestrial irradiation? Although many efforts have been made to quantify tumor incidence from low-dose exposures, the weakness of ionizing radiation as a carcinogen and the qualitative similarity between spontaneous and radiation-induced classes of tumors have hampered such attempts. The sizes of experimental groups needed to reach significance are enormous and beyond the capabilities of most institutions. To circumvent these problems, epidemiologists have extrapolated risks based on data from occupational or public exposures and made estimates by inference. The problem with this approach is knowing the shape of the curve to apply when extrapolating from higher dosage portions of exposure/ incidence curves to low-dose exposure. Nonthreshold linear dose-response curves may lead to an overestimate of cancers but are a conservative approach to risk estimates because one assumes that any amount of exposure has an effect. This model appears to be appropriate for nonleukemic cancers in atomic bomb survivors (Shimizu et al. 1990). A nonthreshold linear-quadratic representation, however, is accepted as the most accurate predictor of risk for leukemias due to ionizing radiation, again based on results from studies of atomic bomb survivors (Shimizu et al. 1990). Under either scenario, low-dose exposures are predicted to increase tumor incidence, and, as a result, federal regulations mandate that exposures to radiation must be kept "as low as reasonably achievable."

3.5c Endogenous ionizing radiation

Finally, although impossible to evaluate the hazard, we are continuously exposed to endogenous levels of ionizing radiation estimated to constitute 11% of the annual average total radiation dose to a person in the United States (NCRP 1987). Such isotopes as ^{40}K, ^{14}C, ^{3}H, and ^{226}Ra are assimilated internally and may even exist in DNA. One might argue that due to their proximity and continuous impact on cellular targets, these endogenous isotopes are of greater consequence to cellular transformation than is exposure to exogenous sources.

3.6 Radon

More recently the volatile isotope radon, a decomposition product of uranium, has been recognized as a significant environmental carcinogen. Its carcinogenic potential was first suggested from studies of uranium miners, who showed a nearly 50% incidence of lung cancer in a single region of Czechoslovakia (Holaday 1969). This was attributed to the inhalation of high levels of the radioactive decomposition products of radon, although other factors such as cigarette smoking or exposure to silicates were not considered. In a population of predominantly nonsmoking Navajo uranium miners, lung cancer incidence was also significantly increased (Samet et al. 1984), but again silicate contamination or respirator use was not monitored. In unusual circumstances, radon exposures to miners have approached the relatively low levels commonly reported for homeowners (Tirmarche et al. 1993),

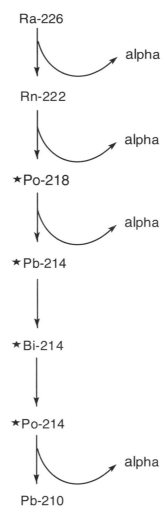

Figure 3–7. Radioactive decay of radon. The inert gas radon decomposes to form several particulate daughters, including the two polonium isotopes.

or conversely, high levels in homes have even exceeded those of uranium miners (Ennemoser et al. 1993, 1994). In each case, lung cancer incidence was increased significantly, suggesting that exposure is not without risk in populations receiving significant levels.

The problem as such is not with radon itself, which is an inert gas, but instead with its particulate daughters (e.g., polonium-218 and polonium-214), which can be deposited and accumulated in the lungs (Figure 3–7). These isotopes are high-energy alpha emitters with considerable potential for ionizing damage, but, due to their short effective range and inability to penetrate the outer layer of skin, they are inconsequential outside the body. Thus, they are harmful only when inhaled, and their effects are localized to the lungs or bronchi.

Because radon exposure represents more than 50% of the total ionizing radiation to which individuals in the United States are subjected, its risks cannot be ignored (Hart, Mettler, and Harley 1989). However, considerable uncertainty in risk assessment does exist due to numerous biologic variables and the questionable reliability of the mining data on which risk estimates have been based. Even most attempts to evaluate residential radon risk have ignored important variables such as occupation, smoking habits, family cancer history, diet, and residential history. Several ongoing studies have included these parameters, but based on the National Research Council's predicted relative lung cancer risk for women of 1.2 to 1.5 from household radon exposure, these current studies may not be of sufficient size to test the hypothesis that radon is a residential risk factor (Neuberger 1992).

Animal studies seem to support the epidemiologic conclusion that radon/radon daughters are carcinogenic, but the results are by no means clear cut. Hamsters exposed to radon over their lifetimes develop lung tumors infrequently (Cross 1988). In rats, low exposures result in increased lung tumor incidence; high doses induce hyperplasia but are not carcinogenic (Chameaud, Masse, and Lafuma 1984). This so-called inverse exposure-rate effect has also been observed in studies of neutron-induced transformation, but its causation remains a matter of speculation. Furthermore, only doses much higher than those to which even uranium miners are exposed are carcinogenic to dogs (Cross et al. 1982), suggesting that the inverse effect is not universal. Thus, although epidemiologic studies of occupational exposures indicate that radon/radon daughters or conditions associated with the presence of radon are risk factors for human lung cancer, the complex results from animal studies and the numerous uncontrolled parameters in human studies prevent extrapolations to the low levels generally experienced from indoor exposures. In fact, at the low radon levels typically found in houses, only a small number of lung cells would experience genetic damage, and the likelihood of even a second event would be minuscule (Jostes 1996). Therefore, a threshold dose rate may be required for radon induction of lung tumors. Although the EPA asserts that radon causes a significant share of lung cancers in the United States, and therefore levels of exposure should be controlled in the home, the jury is still out on the health consequences of such exposures.

3.7   Viral carcinogenesis

The belief that viruses participate in carcinogenesis has now gone full circle from the days in the early 1970s, when the NCI director declared that cancer would be cured in five years, presumably by a vaccine to tumor viruses, to the mid-1970s when retrovirologist and Nobel laureate Howard Temin proposed that the only tumor induced by a virus was the nonmalignant wart, to the discovery of cellular oncogenes incorporated into the genomes of retroviruses. The literature is replete with associations of viral sequences in certain tumors (e.g., Epstein-Barr virus and

Burkitt's lymphoma or hepatitis B virus and liver cancer), but direct proof of their involvement in carcinogenesis has remained somewhat elusive. Perhaps the best case has been demonstrated for the human papilloma viruses (HPV) and anogenital cancers, especially cervical cancer. Epidemiologic studies (discussed in Chapter 7) have long implicated a venereal component in cervical carcinoma (Kessler 1976), but initial attempts to correlate the now reputed etiologic agent HPV with these neoplasms were unsuccessful due primarily to the multiplicity of HPV viral types. Some 65 different HPVs have been reported (DeVilliers 1989), of which 10 are categorized as "high-risk" types due to their involvement in cases of severe dysplasia and neoplasia (Werness, Munger, and Howley 1991). Recent examination of cervical carcinomas now has shown that nearly 85% of these cancers contain "high-risk" HPV sequences (Riou et al. 1990). Molecular fragmentation of the "high-risk" forms has revealed two viral oncoproteins that interact with one of the cellular growth suppressor gene proteins pRb or p53 (Munger et al. 1989; Scheffner et al. 1990), important regulatory factors for cell cycle progression that are discussed in Chapter 5. The HPV oncoproteins are capable, either independently or in cooperation with other transforming genes, of immortalizing (Hawley-Nelson et al. 1989) or transforming (Phelps et al. 1988) a variety of cell types in vitro. Although these observations in no way prove that HPVs play a role in human cervical cancers, the epidemiologic reports combined with the molecular findings provide compelling evidence that they function in the neoplastic process and, on the basis of their late detectability, probably do so during the tumor progression stage.

Clearly, viral participation in carcinogenesis is a rare occurrence and therefore probably impacts few individuals; however, viral studies have provided among the most important observations in carcinogenesis of the 1980s and 1990s through the discovery of oncogene sequences in tumor retroviruses. These are regulatory genes that normally function within vertebrate cells but have been modified and incorporated into retroviral genomes, converting the viruses to oncogenic forms. They are discussed in considerable detail in Chapter 5.

3.8   Endogenous carcinogenesis

Investigations of etiologic agents in carcinogenesis have focused primarily on exogenous effectors, but what about endogenous processes? Might they account for events in carcinogenesis? Despite the evolution of highly accurate proofreading functions in the nucleus, the cellular processes involved in DNA synthesis and repair are inherently mutagenic and therefore may present a risk for carcinogenesis. Additionally, the multistep nature of carcinogenesis and the aneuploidy common in tumor cells during the latter stages of tumorigenesis have led to speculation of a mutator phenotype. This is a genetic sequence, which when mutated could increase the rate of spontaneous mutations throughout the genome. Such a mutation would thus destabilize the genome and theoretically would allow the fixation of the multiple genetic alterations frequently observed in neoplasms, especially the more aggressive forms.

Loeb (1989; Cheng and Loeb 1993) describes several mechanisms whereby genomic instability associated with tumor progression or carcinogenesis itself could occur without participation of an exogenous carcinogen. First, DNA is inherently unstable in aqueous environments, as witnessed in the relatively high frequency of depurination, that is, purine release from its deoxyribose moiety (3×10^{-11} events/base/s) in the genome (Lindahl and Nyberg 1972). These apurinic sites may increase the likelihood that misincorporation of a base during repair or replication will produce a spontaneous mutation. Apurinic sites are created by a repair enzyme that recognizes an altered base or adduct and eliminates it from the DNA strand. This produces a gap in the sequence during replication, leaving the DNA polymerase some latitude in determining what nucleotide should be incorporated opposite the apurinic site. Such sites can therefore lead to the misincorporation of nucleotides and the fixation of mutations in the sequence.

Another potential source of endogenous transforming mutations may be from certain biochemical processes, for example, respiration and phagocytosis, which generate significant quantities of oxygen free radicals and are estimated to yield 10,000 alterations in DNA per human cell per day (Richter, Park, and Ames 1988). This frequency far exceeds rates of spontaneous mutations, but it is based on measurements of free radical–modified nucleotides in urine. These measurements probably exaggerate the frequency of spontaneous mutation because the modified nucleotides are presumably released during phagocytosis in a nonnuclear environment laden with free radicals.

Transforming mutations may also result from the infidelity of the enzymes responsible for DNA repair and replication. Although base complementarity, base selection by DNA polymerases, 3′ to 5′ proofreading activity, and postreplicative mismatch correction together ensure the high level of accuracy of replication, the latter three processes offer potential targets for the mutator phenotype. For example, a single base substitution in the proofreading function of *E. coli* DNA polymerase III increases mutation rates by 10,000-fold (Fowler, Schaaper, and Glickman 1986). Similarly, mutations in mismatch repair genes enhance mutation rates as much as 100-fold (Cox 1976). Somatic or germline mutations in these genes then might predispose individuals or entire families to neoplasia. It is noteworthy that mutations derived from the lack of DNA polymerase fidelity in bacteria are predominantly GC-to-AT transition mutations (Schaaper and Dunn 1987). Similarly, spontaneous mutations observed in cultured mammalian cells are predominantly GC-to-AT transitions (De Jong, Gorsovsky, and Glickman 1988) as are the mutations identified most frequently in human tumor studies of activated *ras* oncogenes (Burmer and Loeb 1989).

Conceivably, therefore, carcinogenesis is caused by processes inherent in cells and not necessarily from exposures to exogenous agents. That is not to say that occupational exposures to carcinogens do not accelerate this tumorigenesis. There is no question that exogenous chemicals can significantly increase the risk of cancer; however, the major mechanism of carcinogenesis might be completely internal, especially for certain embryonal tumors that arise in the pediatric population and, in all likelihood, develop during gestation when exposures to exogenous chemicals should be limited.

3.9 Metabolism of xenobiotics

The kinetics of organismal processing of carcinogens and the dynamics of carcinogen-target interaction, that is, the mutagenic events involved in carcinogenesis, have until recently, with the discovery of oncogenes, received the greatest attention in cancer research. This is due, at least in part, to the number of toxicologists-pharmacologists involved in the cancer effort. In general, the principles of toxicology and xenobiotic processing are similarly applicable to studies of carcinogens. The dose of a carcinogen that reaches a transforming target depends on the kinetics of its processing by the organism: absorption, distribution, metabolic activation, detoxification, and excretion of the carcinogen. All of these processes in combination modulate the carcinogen's effectiveness. The dynamics of interaction of an "ultimate" carcinogen, that is, the active electrophile, with DNA targets determine the formation and permanent fixation of transforming lesions.

3.9a Host defenses

Despite the formidable task of devising a protective scheme against the variety and quantity of xenobiotics to which we are exposed, evolution has provided humans with a fairly effective shield. Regardless of form or method of presentation of foreign substances, cellular mechanisms at several levels modify and often eliminate the risks of chemical injury, including carcinogenesis.

To begin, the skin effectively blocks the entry of most carcinogens. The layer of keratinized dead skin cells can trap and eliminate, upon sloughing, many external electrophiles. It is also thick enough to prevent certain types of ionizing radiation such as alpha particles from reaching living tissue, except through inhalation. Ingested carcinogens experience the intense acidity of the stomach, which hydrolyzes and inactivates certain chemicals, such as those with diazo groups. Stomach acids, however, may actually aid in the production of other carcinogens like the nitrosamines, as described previously in this chapter.

Those substances that survive the stomach may find targets in the gut, but cell turnover is so rapid as to prevent any initiated cells in the gut, other than stem cells, from having a lengthy latency period. Additionally, selective absorption further restricts carcinogen internalization. Most carcinogens that are absorbed into cells enter by passive diffusion, and as such their lipid solubility determines their effectiveness at entry. It also influences the ability of a cell to store these carcinogens because the lipophilic substances can be sequestered for long periods in fatty tissues. Thus, the lipid solubility of the polycyclic hydrocarbon 7,12-dimethyl-benz[a]anthracene (DMBA) may in part explain its potency as a mammary carcinogen because it may be stored and concentrated in the fatty tissue of the mammary gland. There are also carcinogens that structurally mimic important biologic molecules and are therefore similarly transported and concentrated. Azaserine, a diazo amino acid that structurally resembles the amino acid glutamine, is actively concentrated by amino acid transport pathways in the pancreas and kidney where it exerts its carcinogenicity (Longnecker and Curphey 1975).

Following absorption, the next protective tissue barrier is the liver where a superfamily of hemoproteins has evolved for the monoxygenation of lipophilic substances to form more hydrophilic compounds. In mammals, the P-450 mixed-function oxidases, enzymes that convert chemicals to more water-soluble forms for distribution and excretion, may be placed into two major functional classes: (1) steroid- and cholesterol-synthesizing (metabolism of endogenous compounds), and (2) foreign substance or xenobiotic-metabolizing (metabolism of exogenous chemicals). Both classes are generally located in the cellular microsomal membranes, that is, the endoplasmic reticulum, and are expressed at the highest levels with the greatest complexity of function in the liver. Their expression is not restricted to the liver, however, and various isoforms can be detected in several tissues. Certain forms are expressed only in specific nonhepatic tissues. Because of their broad but overlapping substrate specificities, the xenobiotic-metabolizing P-450s represent a formidable barrier to a remarkable range of inhaled or ingested lipophilic substances.

Approximately 36 P-450 gene families have been identified, of which at least 12 occur in all mammals (Nelson et al. 1993). This multiplicity of P-450 enzymes is thought to have resulted directly from selective pressures imposed on evolving eucaryotes by diversification in the plant community, which elaborated a variety of novel toxic natural pesticides (Nelson and Strobel 1987). Indeed, the majority of P-450 isoforms function specifically in the metabolism and elimination of xenobiotics present in the air or in our diet. Unfortunately, as a result of cellular efforts to eliminate some of these toxins, the cell may in some cases metabolically activate them to highly mutagenic and/or carcinogenic forms. Thus, while protecting an organism from large numbers of toxic environmental chemicals, a small number may be inadvertently metabolized by the P-450s to harmful forms.

3.9b   Inducibility of xenobiotic metabolism

The discovery of the P-450 proteins resulted from studies of an hepatic microsomal pigment, which produced a characteristic absorbance maximum at 450 nm when reduced and bound by carbon monoxide (Garfinkel 1958; Klingenberg 1958). Then, in seemingly unrelated studies, chronic administration of the barbiturate phenobarbital to rats was shown to reduce the sedative effect of phenobarbital by stimulating its own metabolism (Remmer 1962). These separate observations were subsequently unified by the demonstration that P-450 proteins play a role in drug and carcinogen metabolism (Cooper et al. 1965). Critical to these findings was the fact that the P-450s are inducible. In this case, induction of the P-450s is similar to the classic process of induction observed in bacteria for the *lac* operon. Essentially, enzyme activity is increased by elevating the levels of transcription and translation from the gene that encodes the enzyme. For the P-450 drug-metabolizing enzymes, the chemicals generally induce levels of the same P-450s that metabolize them. Thus, phenobarbital stimulates its own metabolism in liver by increasing the expression of a specific P-450 isoform. By examining P-450 induction with a spectrum of chemicals, induction categories have been identified: the polycyclic aromatic hydrocarbons, ethanol, peroxisome proliferators, glucocorticoids, and phenobarbital. Each group represents a P-450 isoform that its member chemicals induce preferentially and by which they may be metabolized (Okey 1990). Therefore, induction by a specific chemical will generally enhance the metabolism of several compounds within its grouping.

Ah receptor and induction. The chemical specificity of inducer compounds probably reflects the ability of the chemical to interact with a specific cytoplasmic receptor. This has been effectively demonstrated for the Ah (aromatic hydrocarbon) receptor, a cytosolic protein with high affinity for PAH. The Ah receptor was discovered using the potent P-450 inducer 2,3,7,8-tetrachlorodibenzo-p-dioxin (TCDD), a chemical that enhances the metabolism of PAH. With a radio-labeled form of TCDD, Poland et al. (1976) found a receptor protein in mouse liver cytosol with an extremely high affinity for TCDD. Further studies have suggested that this receptor belongs to the steroid receptor superfamily and, as such, upon complexing with its ligand, is translocated to the nucleus where the complex enhances transcription of its specific P-450. In the cytoplasm, the receptor forms a complex with another protein called hsp90 (heat shock protein 90), which permits the protein to interact with the chemical ligand (Wilhelmsson et al. 1990). Once the chemical is bound to the receptor, hsp90 is released, allowing the receptor to be translocated to the nucleus by a transporter protein (Hoffman et al. 1991). The hsp90 also masks the DNA-binding site in the receptor, so once removed, the receptor-transporter complex can interact with target sequences in the genome.

The Ah receptor is not the only receptor for activating the P-450s. Inducers of peroxisome proliferation (for oxidation of long-chain fatty acids) and its associated P-450 activate a member of the steroid receptor superfamily as well, one of which has now been cloned (Issemann and Green 1990). As for phenobarbital, although it acts through transcriptional stimulation, its receptor remains uncharacterized.

P-450 induction and carcinogenesis. How does P-450 induction relate to carcinogenesis? The effects of induction can be profoundly beneficial or detrimental.

Amino azo dye–induced carcinogenesis is markedly reduced in rodents first treated with the weak PAH carcinogen methylcholanthrene, which apparently stimulates P-450-mediated azo dye metabolism (Conney 1982). Furthermore, differences in levels of carcinogen metabolism occur in strains of mice with dissimilar abilities to induce P-450 activity. Studies of Ah responsive (inducible) and nonresponsive (noninducible) mice have shown that nonresponsive mice have lower levels of the Ah receptor and a weaker affinity of the inducing chemical for the Ah receptor. Tumor incidence increases in responsive mice when the carcinogen is administered locally and in nonresponsive mice when administered systemically (Okey et al. 1986). The explanation for this may be that responders with inducible liver P-450 can clear carcinogens from their system during the first pass of the chemical through the liver, but nonresponders cannot (Okey 1990). Conversely, high levels of inducible tissue P-450s in the responders may only result in greater levels of carcinogen activation in locally exposed tissues where inactivating pathways may be deficient.

For humans, inducibility has also been linked with carcinogenesis, at least in the smoking population. In an examination of arylhydrocarbon hydroxlyase (AHH), a catalytic activity for PAH associated with P-450 isoform IA1, lymphocytes from patients with lung cancers were found to have the highest AHH activities (Kouri et al. 1982). Although interindividual or familial variations for P-450 activities do exist and, in theory, could significantly impact tumor incidence, clear involvement in carcinogenesis has not been established. For example, expression of a certain variant of the P-450IA1 gene correlated with lung cancer susceptibility in a Japanese population (Nakachi et al. 1991; Kiyohara, Hirohata, and Inutsuka 1996), but the same association was not observed in Norwegian lung cancer patients (Tefre et al. 1991). Further efforts are necessary to clarify this issue in human populations.

3.9c   Metabolic activation of chemical carcinogens

Most chemical carcinogens require metabolic activation to elicit their carcinogenicity. This may be due to the fact that metabolism-independent chemicals can react with a multitude of nucleophilic molecules outside of the cell or cell nucleus and therefore cannot reach the DNA unless present in very high concentrations. Conversely, metabolism-dependent chemicals as procarcinogens can be distributed systemically as nonreactive substances, enter cells in a variety of tissues, and become activated to their "ultimate" reactive carcinogenic forms by the cellular P-450 proteins. The classic studies of the Millers (1970) demonstrated for the first time that unreactive chemical carcinogens or procarcinogens can be metabolized to more reactive intermediates or ultimate carcinogens. Studies of the carcinogen 2-acetylaminofluorene (AAF) had established a species-specific pattern of carcinogenesis in which rats, mice, hamsters, and rabbits were susceptible to its carcinogenic effects but guinea pigs were not (Figure 3–8). However, an N-hydroxy deriv-

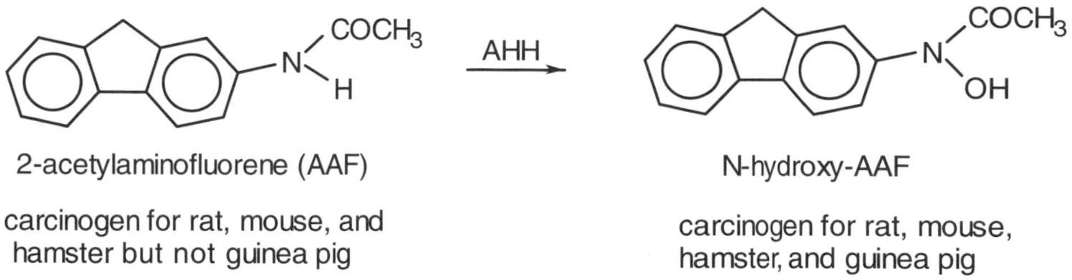

Figure 3–8. Species-specific metabolism of 2-acetylaminofluorene (AAF). AAF is a carcinogen in rat, mouse, and hamster but not guinea pig because it lacks the ability to oxidize AAF. Studies of these differences provided the first evidence that some carcinogens require metabolic activation.

ative of AAF induced tumors even in the guinea pig, showing that the guinea pig lacked the ability to metabolize AAF to its N-hydroxy form. This intermediate did not react directly with DNA, however, suggesting further metabolism was necessary for activation and that guinea pig cells retained the metabolic capability to process the N-hydroxy intermediate to its ultimate carcinogenic form.

Although most environmentally important carcinogens require metabolic activation, some occupational substances are direct-acting or metabolism-independent compounds that can induce tumors systemically. Although highly reactive substances often decompose too rapidly to be distributed beyond the site of administration, other compounds like N-ethyl-N-nitrosourea are sufficiently stable to penetrate transplacental barriers and induce, at high incidence, tumors of the central and peripheral nervous systems in fetal rats (Druckrey, Ivankovic, and Preussman 1966). Thus, the direct-acting carcinogens cannot be ignored.

3.9d Inactivation of chemical carcinogens

In addition to the role of activation in susceptibility to carcinogenesis, inactivation plays a similarly important function in determining susceptibility. During P-450 induction, several conjugating or detoxifying enzymes are simultaneously induced: liver cytosolic glutathione S-transferase, liver microsomal epoxide hydrase, acetyltransferase, and glucuronide transferase. These enzymes collectively add polar groups to P-450 reaction products, increasing their water solubility or hydrophilicity and allowing their elimination by excretion. In liver, these enzymes are present at high levels, like the P-450s; however, tissues other than liver generally do not maintain the same capabilities, at least quantitatively. Thus, as described earlier, Ah-responsive mice may be susceptible to local carcinogenesis because high levels of carcinogenic intermediates are generated by the P-450s in the absence of appropriate conjugating or detoxifying enzymes.

Although the desired effect of xenobiotic conjugation is the detoxification and removal of hazardous compounds, this mechanism is also not foolproof. For example, liver-mediated glucuronidation of the bladder carcinogen 2-naphthy-lamine actually stabilizes the intermediate until it reaches the bladder where it is further metabolized to its ultimate carcinogenic form (Figure 3–4).

3.9e   Systemic distribution of chemical carcinogens

Once a carcinogen has passed through the liver, it is available for distribution throughout the organism and metabolic activation by P-450 enzymes in individual tissues. For transport, a variety of serum proteins are available. Lipophilic compounds can be sequestered by proteins like albumin and shunted to adipose tissue where they can be temporarily stored unmodified. Certain metals such as cadmium may be bound by either transferrin or metallothionein and concentrated in tissues with a capacity for iron storage. As already mentioned, it seems remarkable that even highly reactive substances like N-ethyl-N-nitrosourea are disseminated uniformly throughout the tissues not only in the host but also in fetuses exposed transplacentally. It appears then that distribution may not be a limiting factor in carcinogen exposure, especially in circumstances where the liver is unable to clear the first pass of carcinogen from circulation. This would allow unreacted chemicals to reach tissues without the liver's detoxification mechanisms.

3.9f   Mechanisms for carcinogen suppression

The fact that most environmental agents require metabolic activation provides opportunities for manipulating this process to suppress carcinogenic activity. To this end, a group of chemopreventative blocking agents has been identified that function at one of three levels in inhibiting carcinogen-target interactions of metabolism-dependent compounds. These include inhibitors of metabolic activation, enhancers of detoxification pathways (conjugating enzymes, discussed previously), and scavengers of electrophilic intermediates (the ultimate carcinogens) (Wattenberg 1992).

As described in the chapter on epidemiology (Chapter 7) and in Appendix A, dietary components can provide significant protection against certain forms of cancer. Dietary factors seem to modulate the risks of colorectal, lung, cervical, and stomach cancers, for rates are lowest in those groups that consume the highest levels of fruits and vegetables (Hirayama 1979; Boyle, Zaridze, and Smans 1985). Examining these foods has led to the identification of several natural substances that block carcinogen-target interactions by interfering with processes described earlier. Diallyl sulfide, a member of the organosulfur compounds found in garlic and onions, the citrus-derived monoterpenes D-limonene, and the product of caraway seeds, D-carvone, all reduced forestomach tumor and pulmonary adenoma formation in diethylnitrosamine-exposed mice apparently by inhibiting carcinogen activation (Wattenberg, Sparnins, and Barany 1989). Other organosulfur compounds inhibit benzo[a]pyrene-induced tumorigenesis in mice by enhancing the activity of the conjugating enzyme glutathione S-transferase (Sparnins, Barany, and Wattenberg 1988). Thus, diallyl trisulfide and allyl methyl disulfide inhibit carcinogenesis by stimulating detoxification pathways.

The scavengers of reactive carcinogens, notably the antioxidants, have drawn considerable attention since the tocopherols in wheat germ oil were found to inhibit carcinogenesis from coal tar (Davidson 1934). Several naturally occurring antioxidants, such as vitamins C and E and beta-carotene, have been shown in animal experiments to inhibit the formation of chemically induced tumors (Steinmetz and Potter 1991). Vitamins C and E also restrict the endogenous formation of nitrosamines. The plant flavonoid ellagic acid reduced benzo[a]pyrene-diol-epoxide-initiated lung tumor formation (Chang et al. 1985) and 3-methylcholanthrene- and DMBA-induced skin tumor incidence when applied orally to rodents at the time of carcinogen exposure. Similarly, beta-carotene inhibited colon tumor formation induced with dimethylhydrazine. On the basis of experimentation with the ultimate carcinogenic metabolite of benzo[a]pyrene, it appears that these antioxidants elicit their effects at the level of the ultimate carcinogen. Despite the epidemiologic associations of antioxidants with reduced colorectal cancer and the supportive experimental evidences described, the application of antioxidant therapies to populations at high risk for cancers has thus far have been somewhat disappointing. Clinical trials with vitamins C and E and beta-carotene to prevent tumors in patients previously treated for colorectal adenomas have failed to inhibit polyp formation (Greenberg et al. 1994).

3.10 Modulation of carcinogenesis

After a carcinogen has been taken up, processed, and transported to its target (presumably DNA), the carcinogenic process may be further modified by a number of dynamic processes inherent to the host or its tissues. Such properties as DNA repair processes, cellular proliferative or transcriptional status, immunologic state, stage of tissue development, extent of growth inhibition, and hormonal status enhance or suppress carcinogenicity by affecting the fixation and maintenance of a carcinogenic lesion.

The interaction of a carcinogen with DNA results in an hydroxyl, alkyl-, aralkyl-, or arylamine-modified nucleotide, a frameshift mutation due to the intercalation of a ringed compound into the DNA strand, or strand breaks that may cause sequence additions or deletions. For example, 8-hydroxypurines commonly result from base interactions with an oxygen free radical (Kasai et al. 1986). These oxidative adducts cause the misincorporation of dATP (Shibutani et al. 1991) in the complementary DNA strand opposite the adduct. The adduct also destabilizes neighboring bases so multiple mutations in adjacent bases can occur (Kuchino et al. 1987). For the metabolism-independent alkylating agent N-ethyl-N-nitrosourea (ENU), two minor adducts, O^2- and O^4-ethyldeoxythymidine, persist in DNA following carcinogen exposure, causing misincorporation of dTTP opposite the modified base (Bhanot et al. 1992) and generating the T to A mutations commonly observed in ENU-exposed tissues (Perantoni et al. 1987, 1994). Therefore, the ability of a chemical to behave as a carcinogen may reflect the persistence of the adducts it generates in DNA.

Examination of adduct formation by the other classes of carcinogens has established preferred sites for carcinogen-base interactions (Figure 3–3). The alkylating agents, in general, attack exocyclic oxygens and ring nitrogens, especially N-7 in the purines. The aralkylating agents react preferentially with exocyclic nitrogens, and the arylamines form adducts primarily with either ring or exocyclic nitrogens in the purines. The qualitative distribution of adducts formed therefore is inherent in and restricted by the chemistry of the class of carcinogen.

The fixation of mutations in the genome depends on a dynamic equilibrium between cell replication and DNA repair processes. Prior to repair, normal DNA replication can produce mismatches where DNA adducts occur, generating point mutations. Thus, the most sensitive cell cycle phase to mutagenesis is the S phase or DNA replicative phase. The clearest example of replicative sensitivity to carcinogenesis involves the transplacental induction of neurogenic tumors in rats. ENU is 50 times more effective at inducing these tumors in the fetus than in the adult animal, presumably because there is substantially more cell replication in fetal neuroectodermal tissue than in the adult (Druckrey et al. 1966).

Transcriptional activity also profoundly influences the extent of mutagenesis by a carcinogen. Excision repair accounts for virtually all adduct repair in DNA and is accomplished by two different mechanisms: a slow inefficient general genome repair mechanism primarily for untranscribed DNA and a rapid transcription-coupled repair process involving an ATP-dependent excinuclease (excision nuclease) activity (Huang et al. 1992). The general repair process depends on the recognition of DNA damage by a DNA surveillance complex. For transcription-coupled repair, the damage is identified by a stalled RNA polymerase. In studies of UV repair in cultured mammalian cells, excision of UV-induced dimers in a transcriptionally active gene was found to occur preferentially in the transcribed or coding sequence (Mellon, Spivak, and Hanawalt 1987). Similarly, repair of methylated adducts induced by N-methyl-N-nitrosourea (MNU) occurred significantly faster in transcriptionally active sequences (LeDoux et al. 1991). Through the study of genetic complementation groups in patients with deficiencies in DNA repair processes, for example, xeroderma pigmentosum (XP), a series of proteins has been cloned that function both in excision repair and as transcription factors. The study of XP has been especially useful in this regard because the genetic disorder involves a cutaneous hypersensitivity to ultraviolet radiation and a predisposition to all forms of skin cancer. Furthermore, the genetic lesions are predominantly associated with excision repair mechanisms. Based on these studies, the following model for repair has been proposed. When an RNA polymerase complex encounters an adduct in the DNA that it is transcribing, elongation ceases. The stalled enzyme then recruits "excinuclease" proteins to the damaged site to remove the altered region. Several of the participating proteins involved in this complex process have been cloned and partially characterized: one of which is a helicase to unwind the

DNA for repair and at least two are endonucleases to nick the DNA on both sides of the adduct (Drapkin, Sancar, and Reinberg 1994; Sancar, Siede, and van Zeeland 1996).

Several other factors have been implicated in pathways of carcinogenesis and are mentioned briefly here. Considerable effort has been focused on immune surveillance, a postulated process mediated by the immune system for the selective elimination of aberrant cells. It was hoped that an understanding of this process would lead to anticancer therapies that could enhance surveillance to neoplastic cells and perhaps even lead to the development of antitumor vaccines. More than twenty years of immunologic research have failed to clearly demonstrate that immune surveillance modulates human tumorigenesis or even that immune suppression can increase the incidence of human tumors. The exception might be cases in which a virus has been implicated as an etiologic factor because viral antigens might provide a basis for an immune response in infected tumor cells.

The stage of tissue development modifies carcinogenesis in a variety of ways, but perhaps most significantly by production of a high proportion of cycling stem cell targets for carcinogens in undifferentiated tissues, and the presence or absence of carcinogen-metabolizing enzymes in differentiating tissues. As already discussed, rat neurogenic tissues are considerably more sensitive to carcinogenesis in fetal development than in adult tissues, probably because of the high proportion of replicating stem cells. Similarly, the less differentiated proliferative cells in the lower layers of skin are more susceptible to carcinogenesis than the highly keratinized differentiated nonproliferative upper layers. In the case of metabolism-dependent carcinogens, the undifferentiated proliferating tissues generally express lower levels of P-450 proteins. This reduces tissue sensitivity to carcinogenesis unless the carcinogen is metabolized elsewhere and transported to the target tissue as occurs in transplacental exposures.

Factors involved in growth suppression are discussed in Chapter 5 but probably markedly influence carcinogenesis at multiple levels. The suppressive factors are thought to affect primarily initiated cells by inhibiting the formation and progression of autonomous colonies. It is clear that initiated cells can remain quiescent in the suppressive environment of normal cells. The classic example of this is the suppression of the malignant phenotype of mouse embryonal carcinoma cells upon implantation into a normal mouse blastocyst. These tumor cells, when placed in an embryo, can be distributed throughout the animal, yet their tumorigenic behavior is suppressed and they function normally in their respective tissues (Mintz and Illmensee 1975).

In summary, several factors modulate carcinogenesis once the carcinogen has reached its target and initiated the process of tumor formation. These factors include properties inherent in the chemicals themselves, the tissues, the repair processes, and the cellular environment surrounding the altered cells. The combined influences of these various factors determine whether or not a cell will begin its journey on the road to independence.

3.11 Tumor promotion

The carcinogens described thus far all have genotoxic effects on cells; that is, they can directly cause mutations. As already mentioned, at least one class of compounds involved in carcinogenesis shows little if any genotoxicity. These are the so-called promoters, which, mostly by yet undefined mechanism(s), are believed to stimulate growth or block differentiation preferentially of initiated cells. The existence of this group of substances was originally suggested in early skin painting experiments with mice when it was observed that chemical irritation enhanced the rate of skin tumor formation (Berenblum and Shubik 1947). In these experiments, mouse skin was first exposed to a PAH carcinogen that induced few skin tumors and then to repetitive skin paintings of croton oil, a noncarcinogenic plant extract that markedly increased tumor incidence. The active component of the oil was identified as 12-O-tetradecanoylphorbol 13-acetate (TPA) (Hecker 1967), a substance now used routinely in tumor promotion studies because of its extreme potency. Following a single "nontumor-inducing" dose of an initiating carcinogen, repeated exposures to TPA induced the appearance of multiple proliferative lesions or adenomas of the skin. Maintenance of the early lesions required the continuous presence of the TPA, and withdrawal of TPA resulted in their complete regression and disappearance.

The phenomenon of promotion has also been observed in other organ systems. The classic rat liver carcinogenesis studies of Peraino, Fry, and Staffeldt (1971) identified phenobarbital as an effective promoter of hyperplastic nodules and eventually of hepatocellular carcinoma. More recently, rat renal cortical and transitional cell tumors were also found to be promotable with barbital (Diwan, Ohshima, and Rice 1989).

The chemicals used in promotion protocols are generally of little environmental consequence to humans. However, certain chemicals that function as promoters in bioassay systems have been used therapeutically on humans. For example, the skin promoter benzoyl peroxide is frequently employed in the treatment of certain skin ailments. Perhaps of greater importance are possible endogenous promoters. No intensive study of these substances has been undertaken, but certainly growth factors and hormones are likely candidates. TPA mimics the mitogenic activity of epidermal growth factor (EGF), which induces proliferation in a wide variety of cell types. It is possible therefore that EGF or its family members (e.g., transforming growth factor-α) function as promoters for certain initiated cells. Hormonal status or responsiveness also influences tumorigenesis by providing selective growth-promoting stimuli. For example, prostate tumor cells are often initially androgen dependent, and tumor growth can be slowed by orchidectomy, which reduces testosterone levels by 90%. Because of their role as inducers of proliferation, these substances may be important endogenous promoters of tumorigenesis. Elucidation of endogenous promoters therefore may lead to novel therapies for tumor management.

Although the biologic mechanisms responsible for promotion have not been fully characterized, examination of various promoting agents has established that hyperplasia is an absolute prerequisite for promotion. Based on studies of specific target tissues for promoters or genetic susceptibility for promotion (Sisskin, Gray, and Barrett 1982; Loury et al. 1987), all promoters induce cell proliferation. Induction of proliferation, however, is not sufficient for promotion; for example, acetic acid, a potent inducer of hyperplasia in skin, is an inefficient promoter (Slaga, Bowden, and Boutwell 1975). Exactly what the additional requirement might be is unclear. For promotion in mouse skin, two events have been described: one for conversion of an initiated cell to a cell that can be selectively propagated and the second for expansion or propagation of the converted population (Drinkwater 1990).

If an initiating mutation in a gene enhances responsiveness to growth stimulation by a growth factor or inhibits the ability of a stem cell to terminally differentiate or become apoptotic, then a promoter such as TPA might induce proliferation selectively in those initiated cells. Alternatively, TPA, like other promoters, is cytotoxic. If the normal cells surrounding the initiated cells are killed during exposure, the initiated cells might be released from regulation and additionally stimulated by the various growth factors secreted during tissue regeneration and healing. Finally, promoters at noncytotoxic levels also inhibit junctional intercellular communication (Trosko et al. 1982), providing another mechanism whereby initiated cells can escape regulation by surrounding normal tissues.

Although any of the mechanisms just described could account for the selective growth or loss of normal growth regulation of initiated cells, there is another side to the promotion story. First, promoters like TPA can have genotoxic effects and are weakly carcinogenic, perhaps as a result of oxygen free radical formation during exposure (Cerutti 1985). Secondly, promoters can enhance carcinogenesis, albeit weakly, even when applied prior to the initiating carcinogen. This would seem to negate the concept of clonal expansion of initiated cells during the promotion stage. Thus, unanswered questions remain on the mechanisms of promotion and its role in carcinogenesis despite its widespread acceptance. Because discussions of molecular interactions of tumor promoters include signal transduction pathways and possible oncogene involvement, this aspect of promotion is discussed in Chapter 5.

3.12 Tumor progression

Once a cell has been initiated and that population of cells specifically expanded by promoters, the altered cells can either regress following promoter removal or experience another genetic event to facilitate tumor cell progression. The specific events involved in the progression phase are poorly understood primarily because of the multiplicity of changes that occur, many of which do not participate in malignant conversion. With the exception of certain types of leukemia, tumor cells at this stage generally show considerable heterogeneity and become aneuploid. The heterogeneity results from pressures on cells to differentiate, which even malignant cells can do, and from the selective growth pressures just described. One might predict even greater heterogeneity during progression, but instead actually observe a decrease in heterogeneity as populations of tumor cells undergo selection. Additionally, the genome becomes unstable, causing chromosomal alterations with increasing frequency during tumor progression. These lesions do not accrue randomly, and tumors often show remarkable uniformity in their chromosomal aberrations. In nearly every category of tumor, specific primary and secondary chromosomal alterations are consistently observed and may contribute to the progression process.

As the progression phase ends, tumor cells have converted to the neoplastic phenotype. They are invasive, sometimes metastatic, and highly autonomous. They can now erode their surrounding normal tissue barriers, penetrating cellular layers as well as the basal laminar matrices. They can escape both physical and growth regulatory restraints.

3.13   Alternative pathways for carcinogenesis?

Most cancer researchers support the initiation/promotion/progression/conversion concept of carcinogenesis, and, indeed, certain molecular events in tumor formation associate with specific phases of the process at least in promotable tissues such as skin (Yuspa 1994). Vogelstein proposed that tumors result from the accumulation of genetic events and has documented a series of genetic alterations, both dominant and recessive, in tumorigenesis of the colon (Fearon and Vogelstein 1990; Kinzler and Vogelstein 1996). Although the sequence of events may not be absolute, clearly certain events must precede others in tumor formation, a finding consistent with this multistage model.

3.14   Federal regulations

Considering the variety and quantity of synthetic chemicals added to the environment annually, what type of protection are we provided? Does the government do anything to regulate exposures to potentially hazardous substances? For foods, drugs, and cosmetics, the Food and Drug Administration had until recently been assigned the task of enforcing the Delaney Clause of the 1958 Food Additive Amendment of 1958, U.S. Congress. The clause mandated that any color/additive substance intended for human consumption should be prohibited from use if the substance is found by what is now the Department of Health and Human Services to induce tumors in humans or experimental animals, that is, a zero-tolerance policy. The National Toxicology Program (NTP) within the National Institute of Environmental Health Sciences has been charged with the responsibility of verifying the safety of this group of chemicals. NTP, through animal testing, establishes a basis for the FDA's determination of carcinogenic risk in human terms. The problems with assigning risks on this basis are twofold: the estimation of carcinogenic potency in humans from studies of rodents known to be considerably more sensitive to many carcinogens than humans, and extrapolation from high doses employed in animal experimental protocols to considerably lower exposures experienced by humans in their environment. The risk assignments have long been misleading because they apply only to additives; naturally occurring carcinogens are regulated differently or not at all. As a result, an additive projected to produce 1 cancer per 1 million lifetimes would have been eliminated from use despite its benefits, and a carcinogen like aflatoxin that is produced by a mold in a food product would be allowed at levels that could increase tumor incidence by 700 cancers per 1 million lifetimes (Ames et al. 1987). The FDA attempted to redefine the Delaney Clause to allow additives with risks of less than 1 cancer per 1 million lifetimes to be legally *de minimis,* that is, too little to trifle with; however, the courts rejected this concept on the basis that Congress must act to modify its own legislation (Curran 1988). Because of this, orange dye no. 17 with its estimated risk, at worst, of 1 cancer in 19 billion was banned from use (FDA 1986).

No specific legislation similar to the Food Additive Amendment had been enacted to protect the public from environmental carcinogens, although the Environmental Protection Agency was assigned the responsibility of protecting the public's health. In 1988, recognizing that most pesticides currently in use would be banned under a strict interpretation of the Delaney Clause, the EPA established its own minimum risk factor of 1 cancer per 1 million lifetimes and applied this standard to problems of contamination involving the inadvertent addition of carcinogens to the environment, such as through agricultural use. The EPA also makes risk assessments on the basis of bioassays performed by the NTP. When the pesticide ethylene dibromide was found to be carcinogenic in animal studies, it was the EPA's problem to assess the public risk and eventually remove it from the food chain in a timely manner after weighing any health risks that might result from its removal. The EPA was not initially bound by the Delaney Clause. However, the National Resources Defense Council challenged the EPA's methods in the courts, forcing the zero-tolerance policy on the EPA as well as a ban on the use of several commonly applied pesticides (Corliss 1993).

The chief criticism of the Delaney Clause arose from its inability to deal with advances in technology (Ashby 1994). When the clause was originally passed, levels of chemical detection did not exceed one part in ten thousand, but currently carcinogens are detectable to one part in a billion, a 100,000-fold increase in sensitivity for zero tolerance. Finally, in 1996, the U.S. Congress passed the Food Quality Protection Act, resolving the disputes between environmentalists and the food industry over the use of pesticides. In summary, it sets a standard of "reasonable certainty that no harm will result from aggregate exposure" to pesticides on both processed and raw foods, provides standards for all risks (not just cancer), establishes tougher standards for safe levels specifically in infant and children's foods, considers the benefits of pesticide use in determining tolerances (e.g., when a pesticide prevents greater health consequences), calls for a complete reevaluation of existing tolerance levels to occur within ten years, provides for periodic review of all pesticides, mandates a process for the accelerated consideration of "safer" pesticides, and prevents states from establishing tolerances that differ from those of the EPA. All provisions considered, this act provides consumers with what appears to be a much more balanced and rational standard of protection than previously legislated.

Guidelines for protection in the workplace were established in the Occupational Safety and Health Act of 1970. In this document, the Occupational Safety and Health Administration (OSHA) was charged with defining health and safety standards for the workplace and enforcing those standards. Its mission to reduce exposures to hazardous substances has in some cases virtually eliminated certain occupational cancers. Vinyl chloride, a polymerizing agent in plastics manufacturing that induced a high incidence of the rare cancer angiosarcoma, is no longer a health problem in industry despite its continued use because of OSHA standards. Establishing appropriate industry standards has been a difficult problem for this organization, but efforts are being made by the current administration to devise risk-assessment guidelines for all federal agencies.

3.15 Summary

Carcinogenesis is a multistage process initiated and/or promoted by any of a variety of external chemicals, radiations, or endogenous processes. The ability of a carcinogen to induce tumor formation is intrinsic to the nature of the carcinogen and its ability to interact with a host and modify target DNA. Until the 1980s, cancer investigators focused on a characterization of the various carcinogens and their gross interactions with the host, for example, the metabolism of chemicals and DNA-adduct formation. This period also resulted in the development of model systems for studying tumor formation in most tissues. The rapid development of molecular technologies in the 1980s ushered in a new era in carcinogenesis studies, allowing researchers to focus on the specific targets of carcinogens, the transforming lesions. The discovery of cancer-related genes has many scientists convinced that we are rapidly approaching an understanding of carcinogenesis. Furthermore, this knowledge has provided significant insight into the fundamental processes of cell growth regulation and differentiation as well. Chapter 5 addresses this aspect of carcinogenesis: the putative targets and how they function in carcinogenesis.

4

Cancer genetics

Robert G. McKinnell

The vast majority of human carcinomas and sarcomas have not been investigated fully because the cytological pictures are so complex that they almost defy analyses, even with the banding techniques.

T. C. Hsu 1979

[A]ll cancers exist in both hereditary and nonhereditary form . . . the heritable cancers are all uncommon.

A. G. Knudson 1985

For decades there has been no doubt that cancer is genetic, in the sense that transformation of a normal cell to invasive and malignant growth is due to changes in the DNA. But most cancer is genetic only at the level of the transformed cell, not in the germline of the patients.

M-C. King, S. Rowell, and S. M. Love 1993

4.1 Introduction

The field of genetics in the late twentieth century encompasses a wide area of research and includes within its ranks scientists of a diversity of specializations. Included in "cancer genetics" are studies that range from examining human pedigrees to investigating genes and their products. Although molecular biology is clearly a discipline of the late twentieth century, genealogy (the study of human pedigrees) is an ancient endeavor. For example, the familial clustering of breast cancer was known to the Romans of the first century (Lynch 1985a) (a more recent account of familial breast cancer is considered later). This chapter begins with a review of chromosomes as they relate to cancer and proceeds to consider the role of heredity in human cancer.

Chapter 1 discussed tumors that fail to differentiate properly and which, when properly treated, differentiate, and in some instances, give rise to benign and post-mitotic cells. One might argue that cancers with the potential to give rise to normal cell progeny must have an intact genome (i.e., the cancer cell DNA does not differ from normal cell DNA), or if genetic defects exist, they can be bypassed by appropriate treatment (Sachs 1987). However, it is proposed here that even these tumors with a normal or near-normal genome have an aberrant genetic component which consists of abnormal gene regulation. Obviously, certain genes are inappropriately expressed in these and all cancer cells, and whether the inappropriate gene expression is due to mutation or some epigenetic phenomenon, the fact remains that the genome is not properly programmed. Thus, if this premise is accepted, we can say that all cancer has a genetic basis.

4.2   Chromosomes and cancer

4.2a   Aneuploidy

"Aneuploidy" designates an abnormal number of chromosomes. Aneuploid cells have either too many or too few chromosomes, whereas euploid cells have a normal chromosome number. Missing chromosomes may result in incomplete genomic DNA and the likely loss of critically important genes. Extra chromosomes may lead to unbalanced gene expression. Interest in aneuploidy as it relates to normal development and cancer has a long history. Theodor Boveri (1862–1915) studied the effect of aneuploidy in sea urchins. He observed that when sea urchin eggs were fertilized with two sperm instead of one, an additional cleavage center was formed. As a result, the sea urchin chromosomes were distributed to the daughter cells unequally, thus creating aneuploidy. The abnormalities of development were related to the abnormal distribution and combination of chromosomes. Boveri concluded that the improper combination of chromosomes, resulting from the fertilization with an extra sperm, was the cause of the "ruinous" larval development of the sea urchins (Boveri 1907). If the fate of a population of dividing embryonic cells is profoundly affected by aneuploidy, could a population of dividing cancer cells be similarly afflicted because of chromosomal imbalance?

Boveri (1914) speculated that "malignant tumors might be a result of a certain abnormal condition of the chromosomes." However, he abandoned this notion, at least temporarily, because of the skepticism and hostility of his colleagues and other experts. Certainly, Boveri's early thoughts on the role of aneuploidy in embryonic development have been amply sustained in plants (for example, the genus *Datura*, Blakeslee 1934) and in animals (Fankhauser 1945). More recently, the abnormal development of certain nuclear transplant embryos (DiBerardino 1979, 1987, 1997) and the progeny of animals exposed to mutagens (McKinnell, Picciano, and Schaad IV 1979; McKinnell et al. 1980) has been shown to have a chromosomal basis. Boveri was clearly correct, and abnormal chromosomes are now known to be associated with abnormal development and cancer. A two-volume compendium, which catalogs cytogenetic data on cancer, was published with almost 2,000 pages devoted to more than 14,000 cases of chromosomal aberrations associated with cancer (Mitelman 1991). Boveri was vindicated – few would argue with his hypothesis in the late twentieth century.

The development of cancer cytogenetics has been slow despite Boveri's work. One reason for the sluggishness is the historic difficulty of determining the correct number of chromosomes in humans and other animals. Boveri and many other investigators believed the diploid number of chromosomes in humans was 48. Other less precise estimates put the diploid number variously at 8 to more than 50 (Kottler 1974). It was not until 1956 that the *correct* human diploid number of 46 was published. Chromosomes were counted in 265 dividing human embryo cells, and all but four cases showed the number 46 (Tjio and Levan 1956). Since that time, no deviation from the number 46 in normal diploid cells of humans has been reported by any competent observer.

4.2b   Euploidy does not preclude genetic change

Aneuploidy is unequivocal evidence of an altered genome. As already discussed, there was early interest in establishing the relationship between chromosomes and cancer. The real issue, however, is not an aberrant chromosome number as a cause of cancer but rather whether or not a change in the genome can be causally associated with cancer. Obviously, aneuploidy is one kind of genomic change. Genomic changes can occur that do not result in a changed chromosome number. It is not apparent when examining an euploid set of chromosomes, obtained from a malignancy, whether or not mutations have occurred which may have caused that malignancy. An example of this is a neoplasm that occurs in frogs, the Lucké renal adenocarcinoma. The chromosome complement of the frog neoplasm is virtually euploid – it is not known whether or not specific mutations have occurred that induce the malignant state.

The Lucké renal adenocarcinoma of the northern leopard frog, *Rana pipiens,* is caused by a herpesvirus (Chapter 6). *Rana pipiens* has a euploid diploid chromosome number of 26 (Figure 4–1) (DiBerardino 1962; Ellinger, King, and McKinnell 1975), and the chromosome number of the renal neoplasm is also 26 (Figure 4–2) (DiBerardino, King, and McKinnell 1963). Further, consistent differences in chromosome morphology (shape) have not been detected in the tumor chromosomes.

Triploid *R. pipiens* (3n = 39) are viable, and the renal carcinoma has been induced in these animals (McKinnell and Tweedell 1970; Lust et al. 1991). Tumor cells of triploid frogs also have the triploid number (39) of chromosomes (Figure 4–3) (McKinnell et al. 1991; Williams et al. 1993). The information about diploid and triploid tumors may lead one to suggest the frog neoplasm is an exception to the notion that cancers arise as a change in the complement (number) or morphology of

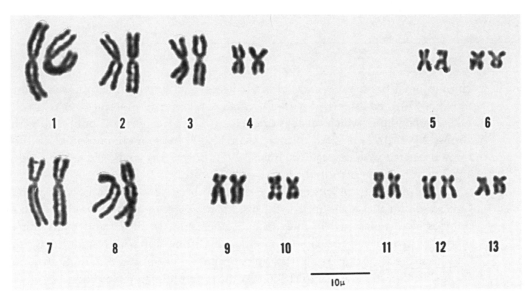

Figure 4–1. Chromosomes derived from normal adult tissue of *Rana pipiens.* The diploid number is 26. (From DiBerardino, M.A., King, T.J., and McKinnell, R.G. 1963. Chromosome studies of a frog renal adenocarcinoma line carried by intraocular transplantation. *J Natl Cancer Inst* 31:769–789.)

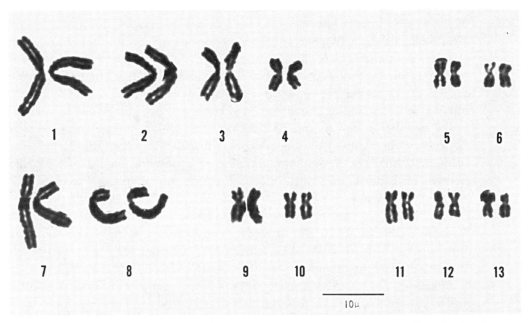

Figure 4–2. Chromosomes of a spontaneous renal adenocarcinoma allografted to the anterior eye chamber of a normal adult *Rana pipiens.* The chromosome number of this neoplasm is 26. (From DiBerardino, M.A., King, T.J., and McKinnell, R.G. 1963. Chromosome studies of a frog renal adenocarcinoma line carried by intraocular transplantation. *J Natl Cancer Inst* 31:769–789.)

Figure 4–3. Chromosomes of a virus-induced triploid renal adenocarcinoma with 39 chromosomes. These chromosomes are of a metaphase plate and have not been arranged in the form of a karyotype as illustrated in Figures 4–1 and 4–2. (From Williams III, J.W.W., Carlson, D.L., Gadson, R.G., Rollins-Smith, L.A., Williams, C.S., and McKinnell, R.G. 1993. Cytogenetic analysis of triploid renal carcinoma in *Rana pipiens*. *Cytogenet. Cell Genet* 64:18–22. By permission of Karger, Basel.)

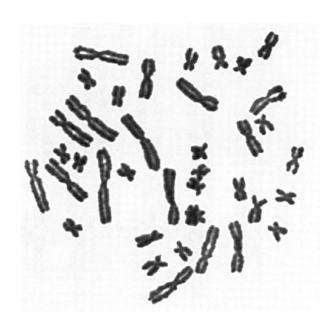

chromosomes. Other exceptions exist, of course. For example, there has been a failure, thus far, to detect karyotypic aberrations in the chromosomes of patients with Li-Fraumeni syndrome (Malkin et al. 1990) (see discussion in section 4.5b). Other exceptions have also been documented (Bayreuther 1960; Sasaki 1982).

Modern cytogenetics is concerned with structural rearrangements of chromosomes as well as numerical changes. There may be translocations, deletions, and inversions (see Le Beau 1997, for a glossary of cytogenetic terms related to cancer), but none of these structural anomalies necessarily involves a change in the chromosome number. How are the structural rearrangements detected? The primary mode of detection is by banding of stained chromosomes. Banding is the longitudinal cross staining that varies between chromosomes but is constant for each chromosome pair (Figure 4–4) (Drets and Shaw 1971; Patil, Merrick, and Lubs 1971). Banding permits the detection of subtle chromosomal structural change. A novel method for microscopic characterization of chromosomes involves labeling with a combination of fluorescent dyes such that *each* chromosome has its own distinct color (Schröck et al. 1996). This procedure will lead ideally to rapid detection of chromosomal aberrations associated with cancer and other maladies.

Let us now return to the frog neoplasm with its euploid set of chromosomes. Have these chromosomes been banded, or treated with fluorescent dyes, to reveal structural rearrangements? The answer is no, at least with respect to Giemsa banding and the fluorescent technique (King 1979). Without these procedures structural rearrangements are difficult to ascertain.

Even if banding were possible, it may not provide the final answer to the condition of frog chromosomes. As pointed out by Ruddon (1995) and Cannistra

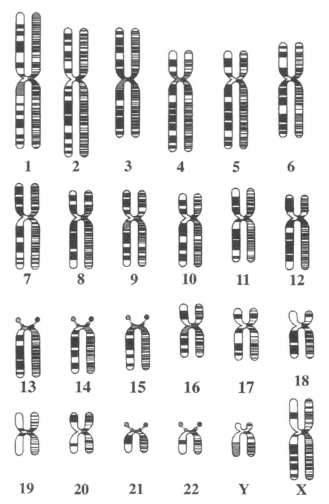

Figure 4–4. Banding pattern of human chromosomes. G-banding patterns of mid-metaphase are represented in the left chromatid; G-banding patterns of late prophase chromosomes are represented in the right chromatid. (Adapted from Yunis, J.J. 1976. High resolution of human chromosomes. *Science* 191:1268–1270. Copyright © 1976 American Association for the Advancement of Science.)

(1990), the average chromosome band is composed of about 5×10^6 nucleotide pairs. Structural aberrations (i.e., deletions or duplications) of 2×10^6 nucleotide pairs would probably go unnoticed. Over a thousand genes (each adequate to code for a protein of 50,000 daltons) could be lost or duplicated with no cytogenetic stigmata. *For this reason, failure to detect chromosomal changes does not prove the lack of genetic anomaly.* Even if Giemsa banding became possible with leopard frog chromosomes and no structural changes were detected, we would still have no final answer to the question of whether the chromosomes of the Lucké renal adenocarcinoma are normal. More sensitive procedures, such as the detection of changes in specific gene sequences, would be required. Hence, although the Lucké renal adenocarcinoma is clearly euploid, it remains unknown if genetic change related to neoplasia has occurred in these cells.

This discussion concerning the detection of cytogenetic aberrations in cancer cells should alert you to the fact that *failure to detect* altered chromosomes in a particular neoplasm does not rule out genetic change. Detection of consistent chromosomal aberrations associated with a particular malignancy, however, is irrefutable evidence of genetic change.

4.2c   Cancers with chromosomal aberrations

Chronic myeloid leukemia: The first malignancy associated with a specific chromosomal aberration. Improved cytogenetic techniques led to the identification of the correct chromosome number for normal humans (Tjio and Levan 1956) as noted, and the search was on for consistent (i.e., nonrandom) chromosomal anomalies associated with specific neoplasms.

In a remarkably short time, a consistent chromosomal aberration was found in the cells of patients with chronic myeloid leukemia (also known as chronic myelogenous leukemia, CML). An overabundance of neutrophils and monocytes with normal form and function characterizes the early chronic phase of CML. After several years, the bone marrow precursors of the neutrophils and monocytes become progressively more immature. As a consequence the bone marrow fills with blast (i.e., immature) cells. The blast phase of CML results in bone marrow failure and the eventual death of the afflicted individual (Cannistra 1990).

A minute (a tiny chromosome fragment) was found to "replace" one of the four small chromosomes (19, 20, 21, and 22) in some cells of seven patients with CML (Nowell and Hungerford 1960). Because both Nowell and Hungerford were affiliated with institutions in Philadelphia, the tiny chromosome fragment was named the Philadelphia chromosome and designated Ph[1]. More than 95% of patients with CML have the Philadelphia chromosome in their leukemic cells (Morrison 1994).

The seminal discovery of Ph[1] was made prior to banding procedures, and it seemed reasonable to ascertain whether the missing material of the small chromosome (equivalent to about 2×10^7 nucleotide pairs, or about 0.5% of the entire diploid genome, Rudkin, Hungerford, and Nowell 1964) was lost or translocated to another chromosome. Translocation was the case as shown in banded chromosomes a decade later (Rowley 1973). The missing material from Ph[1] (now shown to be an abnormal chromosome 22) was discovered to be *translocated* to the long arm of chromosome 9. Note that other chromosomal changes and variant translocations occur in CML in addition to the common reciprocal translocation between chromosomes 9 and 22. Furthermore, the 9 to 22 translocation is not uniquely associated with CML. About 20% of adult acute lymphoblastic leukemia (ALL), 10% of childhood ALL, and somewhat less than 5% of adult acute myelogenous leukemia (AML) also have the translocation (Cannistra 1990).

A cellular proto-oncogene (Chapter 5) known as Abelson (abbreviated c-*abl*) is known to be located on the long arm of chromosome 9 in normal human cells. However, in CML patients, it is found on the long arm of Ph[1] (the abnormal

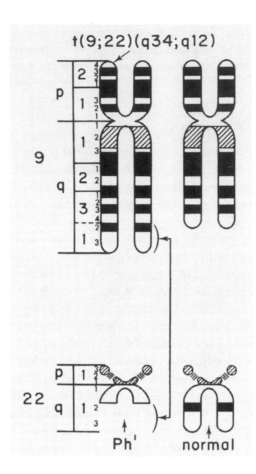

Figure 4–5. Diagrammatic portrayal of the reciprocal translocation in cases of chronic myelogenous leukemia. A portion of chromosome number 22 moves to chromosome number 9. The diminished chromosome number 22, known as a minute, is designated the Philadelphia (Ph[1]) chromosome. (From Whang-Peng, J., Lee, E.C., and Knutsen, T.A. 1974. Genesis of the Ph chromosome. *J Nat Cancer Inst* 52:1035–1036.)

chromosome 22) as detected with c-*abl* and v-*abl* hybridization probes (de Klein et al. 1982). This means that the CML patients with the Ph[1] chromosome have a reciprocal translocation with no obvious loss of genomic material. Part of chromosome 22 breaks and moves to a break site on the long arm of chromosome 9, and there is a reciprocal exchange with a very small amount of chromosome 9 moving to the break site of chromosome 22 (Figure 4–5) (Rowley 1973). It would seem that the evolution of understanding the significance of the Ph[1] chromosome follows closely the evolution of advances in chromosome technology. In this case, it has permitted the identification of a specific gene (c-*abl*) that may be involved in the malignant process.

Most CML patients have the Ph[1] chromosome, but not all do. The breakage site on chromosome 22 has a gene known as *breakage cluster region* (*bcr*) where the small amount of chromosome 9 attaches. A hybrid gene forms that ultimately produces a hybrid protein, part *abl* protein specified by the c-*abl* proto-oncogene from chromosome 9 and part *bcr* protein specified by the *breakage cluster region* gene from chromosome 22. The hybrid protein is translated in blood cells, and some cell lines, and thought to have activity as a tyrosine kinase (tyrosine kinases are enzymes that phosphorolate tyrosine residues and are important for growth factor receptors and the expression of oncogenes leading to malignancy). Some CML patients, although lacking the Philadelphia chromosome, nevertheless express the c-*abl* gene on chromosome 22. The gene has moved without the usual Philadelphia reciprocal translocation (Morris et al. 1986; Verma 1990). There are two important lessons to be learned here. First, the really important aspect of genetic change in CML is probably the presence and expression of c-*abl* on chromosome 22. The second lesson relates to a seemingly normal karyotype. As indicated in the discussion on exceptions to Boveri's hypothesis, chromosomes may appear normal in a stained metaphase spread examined with the clinical light microscope, but are in fact genetically aberrant. In the present situation of CML with a supposedly normal karyotype, chromosome 22 was shown to contain the *abl* gene, and, although appearing normal, it is a "masked Ph¹ chromosome."

Chromosome changes occur in other leukemias including acute nonlymphocytic leukemia, chronic lymphocytic leukemia, and acute lymphoblastic leukemia. Gains, losses, and rearrangements of chromosomes in these hematologic malignancies are reviewed by Verma (1990), Mitelman (1991), and Le Beau (1997).

Burkitt's lymphoma. Burkitt's lymphoma, the most common malignancy of African children, is named for the English surgeon D. P. Burkitt, who first described the neoplasm. This B-lymphocyte malignancy commonly afflicts young individuals, who frequently present with massive jaw neoplasms that grossly distort the face (Figure 4–6). The initial report of the cancer originated from Kampala, Uganda, but other patients were reported from a diversity of African locations. Further, the adrenals, kidneys, liver, and other organs may also be involved (Burkitt 1958). The common (endemic) form of this tumor is African and associated with the Epstein-Barr herpesvirus (EBV) (Chapter 3), but a less common (sporadic) form of the lymphoma occurs in North America. Both forms are associated with a chromosome translocation (Gaidano and Dalla-Favera 1997).

Cytogenetic analysis of cells obtained from several Burkitt's lymphoma patients revealed a loss of genetic material in the distal region of the long arm of chromosome 8 and an extra band on the long arm of chromosome 14. This observation was interpreted as a translocation (Zech et al. 1976). The 8:14 translocation is the most common, but 8:22 and 2:8 variant translocations occur also. The genetic material involved in the translocations that occur from chromosome 8 include the c-*myc* proto-oncogene. The precise mode of action of c-*myc* is unknown, but it is believed to be involved in cell proliferation (Chapter 5). In the 8:14 translocation, the c-*myc* gene moves to the locus on chromosome 14 that codes for the immunoglobulin heavy chain. Variant translocations 8:22 and 2:8 result in c-*myc* juxtapositioned to immunoglobulin light chain gene loci. The translocations result in deregulation and abnormal expression of c-*myc,* an event which would seem to be a critical step in the pathogenesis of this most common neoplasm of African children (Magrath 1990).

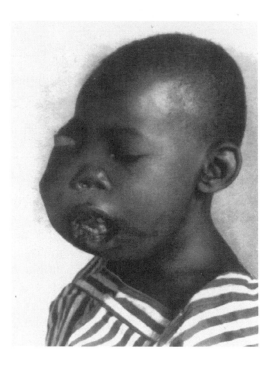

Figure 4–6. Child manifesting the severe facial distortion characteristic of some Burkitt's lymphoma patients. (From Burkitt, D. 1958. A sarcoma involving the jaws of African children. *Br J Surg*, 46:218–225. By permission of Blackwell Science Ltd.)

4.3 Chromosome damage, mutation, and vulnerability to cancer

Genoclastic agents (agents that cause chromosome damage) have been identified as substances which can cause cancer, and many, and perhaps most, carcinogens are mutagens (Ames et al. 1973). Further, various kinds of radiation also increase the risk for cancer. It is not surprising, therefore, to learn that chromosome damage occurs with ultraviolet radiation (Chapter 7 and 8), x-radiation, and other forms of radiation (Upton 1982). The survivors of the atomic bombs dropped at Hiroshima and Nagasaki have been exposed to high levels of ionizing radiation and have an increased risk for cancer, and they also manifest chromosomal anomalies (Miller 1966). Uranium miners, a group that suffers from increased risk for pulmonary cancer, are exposed to radiation, and the chromosomal aberrations they carry are a sensitive biologic indicator of their exposure to that radiation (Brandom et al. 1978). Of course, observations of chromosome damage do not establish that the cancers under consideration are genetic; that is, the observed chromosome damage may be independent from the cancer, which may arise for entirely other reasons, but the observations of genomic damage associated with increased cancer risk is an area that continues to elicit the interest of cell biologists.

Research into cancer prevention has been thought to be potentially the most beneficial area for cancer investigation (80% to 90% of cancer has been thought due to environmental insult and thus potentially preventable). Why discuss this

issue in a chapter on cancer genetics? The reason relates to "spontaneous" mutation. A natural rate of mutation (due to natural causes) seems to be unrelated to
factors in the environment that can be controlled by reasonable human effort.
Although 80% to 90% of cancer may be produced by human-generated or
human-controllable environmental insult, the remaining basal rate of 10% to
20% may *not* be preventable (Knudson 1985). Some humans, whose remains persist as fossils, clearly had cancer. These ancient individuals were not exposed to
smokestack emissions, automobile exhausts, preserved foods, and ubiquitous pesticides. Nevertheless, they became afflicted with cancer. Although knowledge concerning the genesis of cancer gleaned from the study of mutagenic agents may be
useful in understanding the cell alterations leading to malignancy, and perhaps
ultimately to the prevention of much environmentally induced cancer, it is
unlikely this knowledge will ever eliminate a basal low rate of cancer due to "spontaneous" mutation.

4.4   Hereditary cancers

There probably is a genetic component to the origin of all cancer (see earlier discussion). More than fifty forms of hereditary cancer are known, so many, in fact,
that it has been suggested every cancer exists in both a hereditary (genetic) form
and a noninherited (sporadic) form (Knudson 1985). Obviously, all of these cannot be considered in an introductory textbook. Only a selected few have been cho-
sen here because of their prevalence or because they are well known as examples of
genetically transmitted risk factors for cancer.

4.4a   Retinoblastoma

Although retinoblastoma (Rb) is a relatively rare cancer, it is nevertheless the most
common eye neoplasm of children. The cell type that evidences malignancy is
embryonic, indicated by the suffix of the term. Because immature retinal cells
become progressively less common as an individual ages and terminally differentiated retinal cells cannot become malignant, the cancer is rarely encountered in
adults.

The malignancy, if untreated, moves from the eye along the optic nerve to the
brain, metastasizes to other organs, and eventually causes death (Cowell 1989).
Slightly more than one-third of Rb cases are inherited. The remaining cases are
sporadic (not hereditary). Significant differences are noted between inherited and
sporadic Rb. The age of onset of the inherited type is approximately 10 months,
on average 8 months earlier than the sporadic type. Another difference is that
tumors of both eyes occur only with inherited Rb. Further, individuals with the
inherited type are also vulnerable to cancers of other tissue types.

A mutation or deletion in the long arm of chromosome 13 is associated with the pathogenesis of Rb (Sparkes et al. 1983). Every cell in the body of an individual with inherited Rb has the mutation or deletion in one homologue of chromosome 13. Children who are heterozygous for the chromosome 13 abnormality are at risk for developing Rb with the occurrence of another (second) mutation or deletion. The tumor arises when cells of the retina carry mutations or deletions in the long arm of both homologues of chromosome 13 (Lemieux et al. 1989). The conditions that lead to the second mutation or deletion (that is, the conditions that yield homozygosity for the genetic change in the cells of one eye), are such that there is significant risk of developing the malignancy in the other eye. This is why both eyes frequently become afflicted in hereditary Rb. Similarly, both alleles in the same retinal cell must be affected in nonhereditary retinoblastoma, but in contrast with hereditary Rb, the likelihood of this occurring in both eyes is so low that for all practical purposes it does not occur, and Rb in both eyes is thus diagnostic for the hereditary type.

Thus, hereditary Rb does not occur until a second mutation or deletion occurs. This second event led to the two-hit theory of Knudson. A statistical study by Knudson suggested that in the inherited form, one mutation is present prezygotically and found in all cells of the body. A second random mutation occurs at the same locus in the homologue postzygotically. Tumor initiation occurs with the second mutation, hence the "two-hit" theory (Knudson 1971). Retinoblastoma is also discussed in Chapter 5.

4.4b   Wilms' tumor

Another childhood cancer that shares some characteristics with retinoblastoma is Wilms' tumor (WT). It is a malignancy of the metanephrogenic blastema and as such contains several kinds of cells including blastema (embryonic) cells, kidney epithelial cells, and connective tissue cells. This most common kidney malignancy of childhood is generally diagnosed in childhood between the ages of 2 and 3 (Petruzzi and Green 1997). Several congenital defects are associated with Wilms' tumor (Li 1988), which constitutes about 5% of pediatric cancer (350 new cases) each year in the United States.

The similarities of WT with retinoblastoma include the following: WT occurs in both nonhereditary and hereditary forms. Nonhereditary (sporadic) WT accounts for about two-thirds of the cases; the remainder are hereditary. As in retinoblastoma, the two-hit theory of carcinogenesis may be applied to the hereditary form of this childhood cancer (Knudson and Strong 1972; Maurer et al. 1979). It is believed that one homologue of chromosome 11 has lost a functioning gene at one or more sites in its short arm in the hereditary form of the disease (Francke et al. 1979; Mannens et al. 1987). Malignancy occurs with the mutation or loss of the gene at the same locus on the other chromosome 11. The more common noninherited form of WT generally involves only one kidney. The less frequent hereditary form of WT, as in retinoblastoma, may be bilateral with both kidneys involved.

In both retinoblastoma and Wilms' tumor, mutation or gene loss at the appropriate locus in both chromosomes is necessary for the neoplasm to be expressed. In the inherited form of the malignancies, one mutated allele is found in all of the cells of a patient. A second mutation at the same locus in cells of the target tissue is required for the disease to appear. Sporadic (noninherited) disease requires two separate mutations at the appropriate locus. In both cases, the inherited form is less common than sporadic cancer.

4.4c Hereditary conditions that increase cancer risk

You may ask whether there are diseases related to chromosome damage that manifest an increased risk for cancer. The answer to that question is yes. For example, individuals with xeroderma pigmentosum do not repair radiation-induced damage to DNA (Grossman 1997). This genetic disorder, an autosomal recessive disease that occurs infrequently in the United States (about 1 in 250,000), does not result in the inheritance of cancer, but rather the individuals inherit a deficiency in DNA repair that means they are extraordinarily vulnerable to skin cancer (Cleaver 1968, 1994; Feinberg and Coffey 1982). Similarly, Bloom's syndrome (a condition with growth retardation, sensitivity of skin to light, elf-like features, and a vulnerability to repeated infections), Fanconi's anemia (congenital anemia with dwarfism, mental retardation, and other defects), and ataxia telangiectasia (abnormal dilation of small blood vessels on exposed skin with cerebellar atrophy and confinement to a wheelchair before the end of the second decade of life) are autosomal recessive genetic disorders characterized by chromosomal fragility with resultant chromosomal breaking and rearrangements. Not surprisingly, patients with these disorders are also vulnerable to selected cancers (Meyn 1995).

4.5 Familial cancer syndromes

Families with an unusual clustering of cancer incidence, for example, colon cancer in males and ovarian and breast cancer in females, are known from contemporary reports (Lynch et al. 1981) and from antiquity. Familial colon and breast cancer are discussed here. Because of the great abundance of prostate cancer and the increased longevity of males in the United States and much of western Europe (which is related to the abundance of the cancer), the recent report of a prostate cancer gene has aroused significant interest (Pennisi 1996). It is discussed briefly in section 4.5c.

4.5a Colon cancer

New cases of colorectal cancer will affect an estimated 131,200 individuals with a death toll of about 54,900 in the United States in 1997 (Parker et al. 1997). About 10% of cancer deaths are attributed to this neoplasm (Parker et al. 1997). Most (about 85%) cancers of the colon and rectum are thought to be sporadic, random, and not due to heredity. The remaining cases (15%) occur in families in which a first-degree relative (a parent, full sibling, or child) also has colorectal cancer. The family history of colorectal cancer may ensue from environmental similarities or coincidence or it may result from a hereditary condition. Two principal hereditary conditions lead to familial colon cancers: familial adenomatous polyposis (FAP) and hereditary nonpolyposis colon cancer (HNPCC). These hereditary forms of cancer have led to the contemporary intense study of the genetics of cancer of the colon and rectum.

Patients with FAP, an autosomal dominant trait, comprise less than 1% of the total of individuals with colorectal cancer. They are afflicted with many (hundreds to thousands) benign adenomatous polyps of the mucosa of the large intestine that, although not present at birth, appear as early as the first decade of life. The adenomatous polyps of FAP (also called familial polyposis coli, FPC) are precursors of colorectal cancer just as other (non-FAP) colorectal cancers are thought to arise from benign polyps (Fenoglio-Preiser and Hutter 1985; Fearon and Vogelstein 1990; Nishisho et al. 1991; Powell et al. 1992; Smith et al. 1993). Patients with FAP are likely to develop cancer by age 40 and will probably die of colorectal cancer before the age of 60 if untreated.

FAP is related to a mutation of a cancer suppressor gene, known as the adenomatous polyposis coli (APC) gene, located on the long arm of chromosome 5 (Bodmer et al. 1987). The APC gene specifies for a cytoplasmic protein, the expression of which is related to proliferation and maturation of colonic epithelial cells. Failure to produce proper APC protein thus results in enhanced proliferation of cells and failure of the maturation process. Mutation of the APC gene is associated with essentially all cases of FAP. Curiously, many individuals (40% to 50%) who are not afflicted with FAP but who have benign colonic polyps or the "sporadic" form of colon cancer also have a mutation or deletion of the APC gene. Thus, the APC gene obviously plays an important role in most cases of colorectal cancer. Although mutation of the APC gene is necessary for colorectal cancer, it is insufficient for the expression of the malignant phenotype. Carcinogenesis ordinarily results from a multistep process. Hence, other genetic changes are required for colorectal cancer. These changes involve the early loss of methyl groups in the DNA of adenomas perhaps leading to aneuploidy, mutation of the k-*ras* oncogene, deletions of the *p53* antioncogene, which is a tumor suppressor gene, and DCC (*d*eleted in *c*olon *c*ancer) antioncogene located on the long arm of chromosome 18 (Fearon and Vogelstein 1990; Kern and Vogelstein 1991; Knudson 1993). The DCC gene product is a protein with significant similarity to a cell adhesion molecule and may also affect cell differentiation. The multiplicity of genetic changes that can occur in FAP raises the question of in what order the changes occur. It is believed that malignancy occurs in FAP due to the accumulation of genetic changes rather than any particular order in changes.

Individuals with Gardner's syndrome, another autosomal dominant condition leading to intestinal polyposis, are afflicted also with abnormal growths and cancers at other sites, as well as bone abnormalities. Gardner's syndrome occurs at half the frequency of the relatively rare FAP. An autosomal recessive condition leading to adenomatous polyposis and tumors of the central nervous system is known as Turcot's syndrome. Gardner's syndrome, Turcot's syndrome, and FAP sometimes all occur in the same family, suggesting their respective etiologies may be in some way related (Fenoglio-Preiser and Hutter 1985).

Hereditary nonpolyposis colon cancer (HNPCC), hereditary colon cancer *without* adenomatous polyposis of the large intestine, occurs much more frequently than FAP, but both, as stated earlier, are far less common than sporadic colon cancer of unknown etiology (Lynch 1985b). HNPCC is characterized by multiple colon cancers at an early age.

Two genes have been isolated that account for the vast majority of HNPCC. Both are associated with DNA repair. Various kinds of errors occur in DNA replication in dividing cells. One kind of error is the mismatching of nucleotides in the complementary DNA strands. Failure of repair permits the retention and accumulation of thousands of DNA errors (mutations) located throughout the entire genome, and some of these mutations lead to cancer. A mutated gene designated hMLH1, located on chromosome 3 (Papadopoulos et al. 1994), is involved in about 30%, and another gene, hMSH2, located on chromosome 2, is related to approximately 60% of HNPCC cases (Peltomäki et al. 1993). The identification of two genes, both involved in DNA repair, which are damaged in hereditary (HNPCC) cancer, provides striking evidence of the relationship of the loss of normal gene repair and resultant mutation with the genesis of cancer (Cleaver 1994).

Another category of hereditary colon cancer is included in the term "cancer family syndrome" (CFS). CFS, which results from an autosomal dominant allele, is characterized by multiple colon cancers associated with an excess of other neoplasms, primarily endometrial, breast, gastric, and ovarian cancer. Patients with CFS develop their malignancies at an earlier age than individuals afflicted with comparable cancers that are not of genetic origin.

4.5b Breast cancer

An estimated 181,600 new cases of breast cancer in U.S. women are expected in 1997, and 44,190 women are expected to die of that malignancy in the same year (Parker et al. 1997). Prior to 1985, breast cancer claimed more lives of women than any cancer. Since that time, lung cancer has surpassed breast cancer as a cause of death, but breast cancer prevalence and mortality among women remain formidable. In the United States, women now have a lifetime risk of developing breast cancer of 1 in 8 (Miki et al. 1994). New breast cancer cases are expected to afflict about 1,000 men with about 300 deaths due to that neoplasm in 1997.

As would be expected from so common a malignancy, many risk factors have been identified, environmental as well as genetic. "Environment" here, interpreted in the broadest sense, includes any factor(s) that will lead to genetic alterations of somatic cells resulting in breast cancer. This includes most breast cancer. What is known of genetic factors (heredity) is discussed next.

Much of inherited susceptibility of breast cancer is due to an allele known as *BRCA1* (*BR* for breast, *CA* for cancer). Probably about half of all familial breast cancer cases (families in which there is a high breast cancer frequency) and 80% of families with early-onset breast cancer and ovarian cancer have mutations affecting the autosomal dominant, tumor-suppressor gene *BRCA1*. The gene is physically mapped to the long arm of chromosome 17 (Hall et al. 1990), and it and its protein product of 1,863 amino acids have been characterized (Miki et al. 1994). The wild-type *BRCA1* gene inhibits growth in vitro of many breast and cancer cell lines. Further, breast cancer–bearing mice, treated with a retroviral construct containing wild-type *BRCA1*, survived significantly longer than similar mice treated with mutant *BRCA1* (Holt et al. 1996). Another gene, *BRCA2*, located on the long arm of chromosome 13, accounts for almost as much inherited vulnerability to early-onset breast cancer as does *BRCA1* but appears not to be involved in ovarian cancer risk (Wooster et al. 1994; Marx 1996).

Identifying a breast cancer gene provides the potential for distinguishing individuals at high risk for that neoplasm. The identification of *BRCA1* allows for detection and characterization of mutations. Many mutations have been discovered scattered throughout the gene (31 separate mutations were known in late 1994), which may make screening of populations to detect individuals at high risk very difficult (Castilla et al. 1994; Simard et al. 1994; Friedman et al. 1994). Nevertheless, attention is on young women with breast cancer who have no family history of the disease. A significant proportion of these women have germline mutations in *BRCA1* (Langston et al. 1996; FitzGerald et al. 1996).

Another mutant gene related to breast cancer occurs at an even lower prevalence than those already discussed. It accounts for no more than 1% of breast cancers that arise in women prior to age 40. The rare cancers referred to here are the most common malignancy of a group of familial cancers, known as Li-Fraumeni syndrome (LFS), that includes, in addition to breast cancer, soft tissue sarcomas, osteosarcomas, brain tumors, leukemia, and adrenocortical carcinoma. The familial predisposition to cancer in LFS patients led to the notion that the condition was inherited (Li and Fraumeni 1969; Strong, Stine, and Norsted 1987; Li 1988). What then is known about the gene that results in familial vulnerability to breast cancer and other tumors in LFS? It has been postulated that mutations (or deletions) of the tumor suppressor gene p53 located on the short arm of chromosome 17 (Isobe et al. 1986; Malkin et al. 1990) leads to the LFS. The tumor suppressor gene p53 is discussed extensively in Chapter 5.

As the quotations at the beginning of this chapter indicate, breast cancer due to heredity is only a small proportion of breast cancer cases. Only 5% to 10% of all breast cancer and ovarian cancer cases will be women who inherit a gene-conferring cancer susceptibility. Women with inherited susceptibility generally become afflicted with breast cancer at an earlier age than those without inherited susceptibility. Breast cancer associated with *BRCA1* mutations has an incidence that peaks in women between 41 and 50 years of age (Simard et al. 1994). Breast cancer incidence peaks one or two decades later among women who do not have detectable *BRCA1* mutations (i.e., women with sporadic breast cancer). Early-onset cancer is particularly notable among young women. About 25% to 35% of women who have breast cancer between the ages of 20 and 29 have inherited susceptibility. There is no age group at which *most* breast cancer is of individuals with inherited susceptibility. However, given the prevalence of this malignancy (with, as indicated, 181,600 breast cancers annually), the minority of women who have inherited susceptibility become a relatively large group when compared with other genetic diseases.

Why do women with inherited susceptibility generally have breast cancer at a younger age? It is believed that breast cancer, as so many other cancers, ensues from not one mutation but from several; that is, cancer evolves from a multistep process (King, Rowell, and Love 1993). Inasmuch as every cell in the body of a person carrying an allele for breast cancer susceptibility already has one of these mutations, it follows that fewer added genetic changes are necessary to result in malignancy. This can be understood by comparison with retinoblastoma (see earlier discussion). Recall that the inherited form of retinoblastoma occurs at an earlier age than the sporadic form. Only one additional mutation is thought to be required for the inherited form of the malignancy, whereas two mutations are required for the non-inherited form of retinoblastoma. Thus, the inherited form likely will occur earlier (Knudson 1971).

4.5c   Prostate cancer

The estimated 334,500 new cases of prostate cancer with about 41,800 deaths among males in the United States in 1997 (Parker et al. 1997) has been reduced by about 24% (Haas and Sakr, 1997; Wingo, Landis, and Ries, 1997). It is the most common internal cancer of males despite the recent decline. For that reason, as noted earlier, there has been great interest in the report of a prostate cancer gene located on the long arm of chromosome 1. The gene is designated *HPC1* (hereditary prostate cancer 1). As always, the cancer that occurs in people with a genetic predisposition seems to be less common than nonhereditary ("sporadic") cancer, and prostate cancer is not an exception. The recently identified gene accounts for about a third of hereditary prostate cancer, or about 3% of all prostate cancer (Smith et al. 1996).

4.6   Summary

This chapter began with the premise that essentially all cancer has a genetic component. However, when specific cancers known to be hereditary are considered, it is estimated they comprise no more than 2% of all malignancies. What is the difference? Cancer is a disorder of cells, and cells cannot function in the absence of the cell genome and its controlling factors. Therefore, all cancer involves the genome and the regulation of that genome, and thus all cancer has a genetic component. But when we consider specific cancers with known hereditary transmission, we find they are all uncommon. At least with present knowledge, it seems most malignancies are of the sporadic, nonhereditary type. Even so, an enhanced understanding of cancer will surely ensue with a greater understanding of specific genomic changes that occur in cancer and the resultant inappropriate gene regulation, which is the hallmark of cancer cells.

4.7   A postscript concerning genetic services for familial cancer patients

Certainly, there is much in the popular press concerning genetic and familial cancer. Where is a cancer patient, or a family member of that patient, to learn specific information about inheritance of the cancer afflicting that individual or family? Unfortunately, reliable information is not all that easy to obtain. A survey was made of 41 cancer centers that received National Cancer Institute cancer center support grants. Only about half of the responding National Cancer Institute cancer centers provide "some" genetic services. The other half currently provide no genetic services for familial cancers – a "surprising finding" (Thompson et al. 1995). For further information, you may wish to call the American Cancer Society, Inc. (1800 ACS 2345) or the National Cancer Institute (1 800 4 CANCER).

5

Cancer-associated genes

Alan O. Perantoni

5.1 Introduction

Over the past 15 years, more than a hundred genes have been identified that can convert nontumorigenic tissue-cultured test cells from rodent cell lines to a transformed phenotype, that is, foci of piled-up and crisscrossed cells that grow in soft agar and form tumors when explanted into immunocompromised rodents. These dominant transforming genes, or oncogenes, encode proteins involved in signal transduction or cell cycle regulation. When overexpressed or structurally altered, these oncogenic proteins selectively induce the proliferation of cells that express them. More recently, studies of families predisposed to specific types of cancers have yielded what is at present a small but important group of recessive tumor suppressor genes. Loss of suppressor function by deletion or point mutation of both gene copies allows cells to proliferate unregulated or with reduced restraints. The discovery of both oncogenes and suppressor genes will almost certainly prove pivotal in our understanding of the mechanisms of carcinogenesis, for at last scientists may focus on the genetic targets of carcinogens and thus finally link the details of carcinogen activation, adduct formation and repair, and neoplastic conversion and metastasis with definable molecular events.

5.2 What is an oncogene?

An oncogene is an altered form of a normal cellular gene called a *proto-oncogene*. It encodes a regulatory protein with dominant transforming properties; that is, a single copy of the altered sequence can transform a cell and the normal sequence cannot block its transforming ability. The term "oncogene" is really a misnomer because no

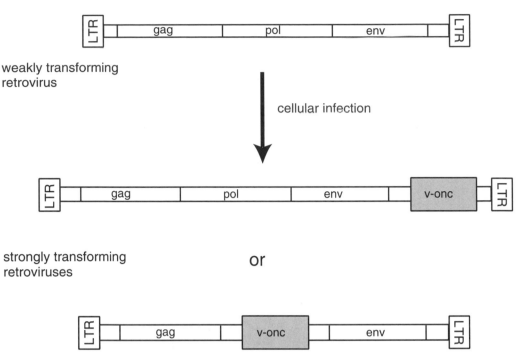

Figure 5–1. Structure of retroviruses before and after host integration. Prior to infection, a retrovirus contains three essential genes: *gag, pol,* and *env.* With these genes the viruses are only weakly transforming. By integrating into a host genome, the virus occasionally removes copy of a host sequence, which if oncogenic, can convert the retrovirus to a highly transforming virus.

one believes these genes evolved specifically for the induction of cancer but rather that the normal proto-oncogenes have been altered by one of several mechanisms to cause transformation. It is not surprising, then, that proto-oncogenes function in the regulation of normal cell growth and/or differentiation.

The principal method for identification of oncogenes has been through the isolation of the transforming genes in oncogenic *retroviruses,* although certain oncogenes have been discovered only in preparations of mammalian tumor DNAs. Regardless of source, the end point of detection is the ability of DNA preparations to transform certain rodent cell lines when those sequences are incorporated by transfection, a process involving cellular uptake of purified precipitated DNA.

The first oncogenic virus, an RNA tumor virus or retrovirus, was described in 1911 by Peyton Rous as an unfilterable agent (one not removed by filtration) from an homogenate of a chicken tumor. The filtrate could induce sarcomas in chickens upon reintroduction (Rous 1911). The etiologic agent was subsequently shown to be a tumor virus (Rous sarcoma virus, or RSV) in a family of avian sarcoma viruses, the study of which eventually led to the discovery of retroviruses.

The retroviral genome exists as single-stranded RNA that must be reverse transcribed to generate the double-stranded sequences required for incorporation into a host's DNA. Characteristically, it contains three genes necessary for the production of viral particles: *gag, pol,* and *env* (Figure 5–1). The *gag* gene encodes the synthesis of the core protein that packs the viral RNA in the viral particle; the *pol* gene contains the sequence for production of the reverse transcriptase or RNA-dependent DNA polymerase; and *env* provides the glycoproteins of the viral envelope. The retroviral genome terminates at both ends in long terminal repeats (LTRs), which provide the sites for integration into host DNA and promote the transcription or expression of viral sequences in the host genome. Viruses in this configuration are weakly or chronically transforming in host cells; however, when isolated from transformed cells and reintroduced into normal cells, they become highly transforming (acute transforming) retroviruses. For RSV, a comparison of viral recombinants with acute transforming ability and viral mutants that had lost their transforming ability revealed a specific region of the genome associated with tumor induction. This region was distinct from the three defined viral sequences (Kawai and Hanafusa 1971; Bader 1972). The novel sequence was designated v-*src,* for viral sarcoma gene. It was subsequently isolated and surprisingly was shown to be a host-derived sequence c-*src* (cellular *src* gene versus viral *src* gene) (Stehelin et al. 1976). The virus had actually incorporated a gene from the host chicken cell that it had infected. Presumably, the acute transforming virus was formed by accidental recombination between the chronic transforming virus and the c-*src* sequence in the host genome. The v-*src* gene encoded the synthesis of a single phosphoprotein, p60 (Purchio et al. 1978), which Hunter and Sefton (1980) subsequently demonstrated to be a tyrosine kinase, an enzyme that specifically phosphorylates tyrosine groups in proteins. It was most fortunate that researchers began with RSV because this virus, unlike most others, incorporated the oncogene sequence downstream of the *env* sequence, so the additional gene did not interfere with viral replication. Most other retroviruses lose their ability to replicate during host integration because the oncogene sequence is often inserted within the *pol* gene. Helper viruses with normal *pol* function are then required to rescue the transforming virus with its associated oncogene.

Using the same approach as for RSV, several other v-*oncs* or viral oncogenes that have a normal counterpart in the eucaryotic cell (c-*onc* or proto-oncogene) have been isolated from chicken or mouse tumors. Studies of these virally transmitted cellular sequences, however, indicate they are not accurately reproduced by the virus during viral incorporation and that the arising sequence differences between viral (v-*onc*) and cellular (c-*onc*) forms actually account for the oncogenicity of the v-*onc* gene (Bister and Jansen 1986). Mutations, including base pair substitutions, deletions, and insertions, arise at high frequency during viral transcription and splicing. Because transforming mutations in the v-*oncs* are necessary for tumor formation, the mutations observed in viral oncogene sequences have provided some indication of what mutations in proto-oncogenes might be transforming when retroviral infection is not responsible for tumor formation (e.g., in chemically induced rodent tumors or in human tumors).

The significance of retroviral oncogenic sequences as dominant transforming genes in normal cells remained largely unappreciated until Weinberg's (Shih et al. 1981) and Cooper's (Krontiris and Cooper 1981) laboratories independently found that DNA taken from a human tumor could transform mouse cells in culture. Subsequently, the labs of Barbacid (Pulciani et al. 1982), Weinberg (Shih and Weinberg 1982), and Wigler (Goldfarb et al. 1982) separately cloned the transforming sequence and established its homology with members of the *ras* oncogene family, which had previously been isolated from an avian sarcoma virus. Since then, several other cellular oncogenes isolated from retroviruses have been shown to cause cultured mouse cells to assume a transformed phenotype in transfection assays.

5.3   Proto-oncogenes function in signal transduction, cell cycle regulation, or differentiation

Proto-oncogenes encode proteins that participate in the regulation of growth and/or differentiation of normal cells, and all are involved at various levels in signaling from the extracellular compartment to the nucleus. In the normal process of signaling, or signal transduction, the interactions of the cell with its extracellular environment are translated to the nucleus to elicit a nuclear response. Through a cascade of events, which will be described in detail, an external signal is transmitted to the nucleus, provoking a response often through specific modifications of transcription. Signaling or signal transduction may be initiated by any one of several proto-oncogene-encoded growth factors that interact with membrane-bound receptors (e.g., *int*-2 or *hst,* members of the fibroblast growth factor family, or *sis,* platelet-derived growth factor). Other proto-oncogenes encode the receptors themselves, such as *erb*B-1, the epidermal growth factor receptor. The *ras* family members are localized to the inner side of the cell membrane and participate in signal transduction from the membrane. The products of others, like *jun,* which encodes a transcriptional-activating factor, or *cyclin D,* which regulates cell progression through the cell cycle, are localized to the nucleus.

In addition to their positioning and functions, the regulatory importance of the proto-oncogenes is also inferred from their conservation in evolution. Homologous *ras* sequences have been identified from yeast and fruit flies to humans (Santos and Nebreda 1989) and apparently function similarly throughout. The redundancy of function found in families of certain oncogenes also provides assurance that their regulatory capabilities will be maintained within a cell. For example, the *ras* gene family is represented by three separate forms: H-*ras,* K-*ras,* and N-*ras,* which differ significantly in their introns, or noncoding regions, but differ little in the encoding portions of the genes.

Proto-oncogene involvement in differentiation has been well documented in studies of expression in developing tissues. During embryonic and fetal development of the mouse, for example, specific proto-oncogenes are expressed at high levels, whereas others appear and disappear in a temporal manner (Slamon and Cline 1984). For example, expression of c-*myc* peaks at day 17 of gestation in the whole mouse, whereas c-*src* increases maximally on day 14 and then diminishes significantly. Thus, even gross determinations indicate that proto-oncogenes are expressed differentially. When individual mouse tissues are evaluated in a stage-specific manner, differential proto-oncogene expression is even more apparent (Zimmerman et al. 1986). Expression of nuclear N-*myc* (Mugrauer, Alt, and Ekblom 1988) and the receptor tyrosine kinases *met* (Sonnenberg et al. 1993) and *ros* (Sonnenberg et al. 1991) are detected in the embryonic kidney specifically in cells undergoing tubulogenesis or branching morphogenesis, respectively, suggesting their functional involvement in those developmental processes. Expression of the proto-oncogenic *ret,* which encodes a membrane receptor for glial-derived neurotrophic factor, is limited to the branching epithelium in the developing kidney, and genetic targeting of this locus results in animals deficient in renal development (Schuchardt et al. 1994). Thus, appropriate expression of a proto-oncogene during development can be critical to normal tissue differentiation.

5.4   Genetic approaches to delineate proto-oncogene function

Studies that correlate proto-oncogene expression with specific developmental processes, although helping to identify significant macromolecules, provide little insight into the actual cellular functions of these genes. For this purpose, several technical innovations have created a window into the cell (Figure 5–2). One approach involves the use of genetic targeting by so-called knockout mutations for specific genes. These are mutations that result in the cellular loss of normal function of a particular gene product. In the case of proto-oncogenes with receptor function as shown, the genetic sequences can be engineered so they maintain their ability to bind ligands but fail to convey a signal upon binding, a so-called dominant negative. Therefore, when incorporated into embryonic stem cells or a cultured cell line, they sequester ligands from the normal receptors and block specific signaling pathways. Although genetic redundancy often prevents loss of proto-oncogene function despite the loss of a normal proto-oncogene product, occasionally germline targeting does allow a glimpse at the cellular function of these genes in the developing embryo. Such is the circumstance in studies of the neurotrophin-binding protein encoded by the *trk* gene. The loss of function of this gene in the germline causes severe neuronal deficiencies in homozygous mutant mice (Smeyne et al. 1994). Similarly, germline loss of the transcription factor encoded by N-*myc* (Stanton et al. 1992) results in severe malformation of the telencephalon (embryonic tissue that gives rise to certain parts of the brain), and knockout mutations in *ret* also eliminate populations of neural crest cells. Thus, despite the diverse functions of the products from the proto-oncogenes *trk,* N-*myc,* and *ret,* all three are critical to normal development of nervous system tissues.

As signaling pathways become defined, the function of specific proto-oncogenes in development has been clarified. A case in point is the participation of *ras*

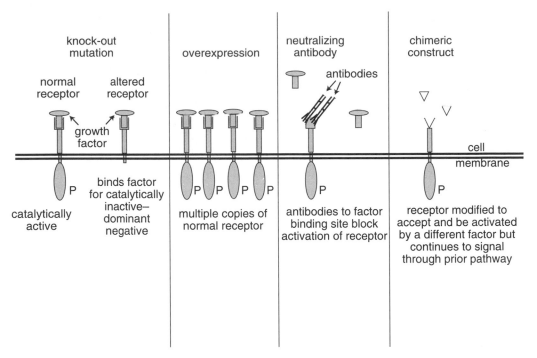

Figure 5–2. Methodologies for investigating gene function. Four approaches are commonly used: knockout mutations to eliminate function, overexpression of gene to distort its activity, neutralizing antibodies to block protein activity, and chimeric constructs to allow control over gene expression.

in the differentiation of neuroectodermal cells (the embryonic tissue that gives rise to nerve cells). A rodent cell line of neuroectodermal origin can be induced in culture to express a neuronal phenotype when stimulated with nerve growth factor (NGF). Microinjection of anti-*ras* antibody blocks this differentiation (Hagag, Halegua, and Viola 1986); a *ras* protein in a constantly signaling or activated form induces neuronal differentiation (Bar-Sagi and Feramisco 1985). Thus, *ras* may participate in the NGF-mediated signaling pathway for the development of neuroectoderm. This also demonstrates another methodology for functional evaluation of gene products: the use of specific neutralizing antibodies to remove the activity of a proto-oncogene product.

Finally, instead of removing gene function, it is possible to evaluate the activity of a gene product by overexpressing or up-regulating its function. In this regard, when normal transforming growth factor-α, a ligand for the epidermal growth factor receptor (EGFR), is overexpressed in a transgenic mouse, it induces hepatocellular carcinomas and mammary and pancreatic abnormalities in the animals (Jhappan et al. 1990). Alternatively, by splicing known regulatory elements with sequences of the catalytic or functional domain of a proto-oncogene, one can tightly control its expression and evaluate the outcome of its regulation. For example, studies of the membrane-bound proto-oncogene *neu* were hampered by the fact that it has no known ligand. Because *neu* shares considerable homology with *erbB*, the gene for the EGFR, chimeric receptor molecules composed of the ligand-binding portion of the EGFR and the intracellular signaling domain of the *neu* protein could be generated (Lehvaslaiho et al. 1989). When introduced into mouse cells, the chimeric receptor localized to the membrane and was activated by treatment with epidermal growth factor, a ligand for the EGFR. This allowed events mediated by the functional domain of *neu* to be monitored within the cells.

5.5 Classification of proto-oncogenes/oncogenes

Proto-oncogenes or oncogenes are classified primarily according to their functional role and position in pathways of signal transduction and subcategorized as growth factors (section 5.5a), receptor (5.5a) or nonreceptor (5.5b) tyrosine kinases, GTP-binding proteins (5.5c), serine/threonine kinases (5.5d, 5.5e), or nuclear proteins (5.5f)/transcription factors (5.5g) (Table 5–2). In this order, they reflect the sequence in which the normal proteins transmit an extracellular signal to the nucleus, resulting in the transcriptional activation of specific responding genes. The common feature among these signal-transducing proteins is their ability to interact directly with proteins that immediately precede or follow them in the sequence and to regulate or be regulated by phosphorylation and/or dephosphorylation reactions as a result of these interactions. Each category then represents a different level in the cascade of phosphorylation events. As described later, aberrant continuous signaling from any of these proteins is often associated with transformation. These categories, however, exclude one very important group of proto-oncogenes: the cell cycle genes. In many cases, the outcome of signal transduction is modulation of cell cycle proteins to either induce or inhibit cell proliferation. It is not surprising, then, that one or more of these proteins may be affected in carcinogenesis. Their involvement in tumor formation is described at length later in this chapter.

5.5a Growth factors and their receptors

Beginning with known growth factors encoded by proto-oncogenes (e.g., platelet-derived growth factor (*sis*), fibroblast growth factor family members *int*-2 (*fgf3*) and *hst* (*fgf4*), and transforming growth factor-α (*tgf*-α), all act through plasma membrane-bound receptors with tyrosine kinase activities and therefore initiate cascades of phosphorylation by activating these receptors. The proto-oncogene receptors are grouped according to sequence homologies and are therefore classified as members of receptor superfamilies for specific growth factors: epidermal growth factor (EGF), proto-oncogene receptors *erb*B-1 and -2 (*neu*); platelet-derived growth factor (PDGF) receptors *fms* and *kit*; fibroblast growth factor (FGF) receptors *flg*, *bek*, and *sam*; nerve growth factor (NGF) receptor *trk*; hepatocyte growth factor (HGF) receptors *met* and *sea*; or insulin receptor *ros*.

The receptors for the ligands consist of three principal domains: an extracellular

Table 5–1. *Classifications of proto-oncogenes/oncogenes with examples*

Type	Location	Function	Tumor association
I. Growth factors	secreted		
sis		PDGF family	sarcoma
hst		FGF family	gastric cancer
int-2		FGF family	mammary cancer
II. Receptor tyrosine	membrane kinases		
erbB		EGF receptor	glioblastoma
erbB-2/neu		EGF receptor family	mammary cancer/ schwannoma
met		HGF receptor	sarcoma and renal carcinoma
III. Nonreceptor tyrosine kinases	cytoplasm or nucleus		
src		mediates integrin signaling	sarcoma
abl		DNA binding and transcription activation	chronic myelogenous leukemia
IV. GTP-binding proteins	membrane		
K-ras		signal junction	colorectal and pancreatic cancer
H-ras		signal junction	mammary and skin cancer
V. Serine/threonine kinases	cytoplasm		
raf		phosphorylates MAPKK proteins in signaling	liver and lung tumors
protein kinase C		signal transduction/ target of promoters	skin cancer
VI. Transcription factors	nucleus		
N-myc		DNA binding	neuroblastoma
jun		DNA binding/transcription activation	osteosarcoma/skin cancer

amino-terminal ligand-binding domain, a hydrophobic transmembrane region, and an intracellular carboxy-terminal tyrosine kinase domain. Receptor activation is characterized by a series of events that alter the configuration of the receptor, allowing its intracellular portion to interact with other membrane-bound or cytoplasmic proteins. For example, when TGF-α or EGF binds the EGF receptor, the receptor's conformation changes, allowing it to interact with another EGF receptor molecule (Yarden and Schlessinger 1987). The resulting dimer promotes phosphorylation (autophosphorylation) on specific intracellular tyrosine residues in the receptor itself (Honegger et al. 1990). These phosphorylation events generate binding sites on the receptor for a group of proteins with an SH2 (*src* homology) domain (Figure 5–3). The SH2 domain recognizes the phosphorylated tyrosine residues on the receptor and allows this group of proteins to interact with the receptors. Adapter proteins Grb2 or CrkII contain SH2 domains that interact with the EGF receptor (Lowenstein et al. 1992, Kizaka-Kondoh et al. 1996).

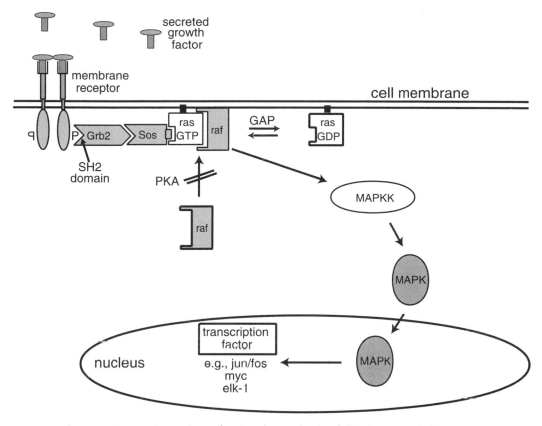

Figure 5–3. A major pathway for signal transduction following growth factor stimulation. The cascade of events begins with ligand/growth factor binding to and dimerization of a specific membrane receptor. This leads to autophosphorylation (P) of the receptor to generate an SH2 domain. The Grb2 protein recognizes this domain, causing the membrane localization of Sos, activated p21^{ras} (GTP bound), and finally Raf. Then Raf phosphorylates a MAPKK, which in turn phosphorylates a MAPK. The phosphorylated MAPK can pass into the nucleus to activate a transcription factor(s). Other signal transduction pathways have been described that may parallel and interact with this pathway; however, this mechanism is critical to many signaling events.

Grb2, and presumably CrkII, can form a ternary complex with the autophosphorylated EGF receptor and a cytosolic protein called Sos, which is recruited to the plasma membrane by Grb2 (Buday and Downward 1993). Sos is a guanine nucleotide exchange factor (replaces a bound inactive nonsignaling GDP molecule with an active GTP molecule) that may play an important function in regulating signal transduction through the next step in the cascade, the *ras* protein (p21^{ras}). Another protein, Shc, can interact with the EGF receptor in the absence of receptor autophosphorylation sites, using instead the receptor's kinase domain for its interactions (Pelicci et al. 1992). As a result, the EGF receptor phosphorylates Shc on tyrosine residues in the Shc protein. These phosphorylated sites may then provide the basis for Grb2 interaction with the Shc protein (Gotoh et al. 1994) and subsequent signaling to p21^{ras}. In both cases, once the Grb2/Sos complex is formed, p21^{ras} becomes activated or capable of signaling.

Other pathways for signal transduction through growth factor–stimulated receptors involve receptor interactions with other proteins that contain SH2 domains, for example, phospholipase C (PLC), Src, GTPase-activating protein (GAP), or phosphatidylinositol 3-kinase (Fantl et al. 1992; Kashishian, Kazlauskas, and Cooper 1992). In each case, autophosphorylation of the receptor is responsible for the interaction, but the resulting proximity then allows the tyrosine kinase activity of the receptor to phosphorylate tyrosine residues in the target proteins.

Signaling from PLC is mediated by the metabolic products of PLC activity and involves members of the protein kinase C (PKC) superfamily in a very complex series of reactions. PLC is catalytically activated by phosphorylation at the membrane. As a result, phosphatidylinositol 4,5-bisphosphate is hydrolyzed to inositol 1,4,5-triphosphate, an intermediate that stimulates the release of sequestered intracellular calcium, and diacylglycerol, which in turn activates protein kinase C (PKC). Interestingly, diacylglycerol and the tumor promoter TPA bind the same regulatory domain of PKC, which increases the enzyme's hydrophobicity and allows it to be recruited to the intracellular side of the membrane (Newton 1995). They also increase the efficiency of PKC's catalytic domain with its serine/threonine kinase activity. PKC then increases the calcium flux from outside the cell (Nishizuka 1992), which further enhances PKC kinase activity. The various phospholipid molecules generated by PLC all tend to prolong the activation of PKC, presumably sustaining its kinase function in certain signal transduction pathways involved in growth control.

5.5b   Nonreceptor tyrosine kinases

An additional group of proto-oncogene products with tyrosine kinase activity is localized to the cytoplasmic side of the plasma membrane. The *src* gene family (*fgr, fyn, hck, lck, lyn, src, tkl,* and *yes*) comprises the majority of the known nonreceptor kinases. Activation of these encoded proteins requires the differential phosphorylation of specific tyrosine residues. Because these family members also contain SH2 domains, they interact with tyrosine-phosphorylated receptor kinases and, indeed, Src protein apparently binds the autophosphorylated EGF receptor (Luttrell et al. 1994). Most recently, however, they have been implicated in signaling from extracellular components, such as the integrins, or from cytoskeletal proteins, notably those in focal adhesion plaques such as talin, vinculin, or paxillin (Liebl and Martin 1992; Kellie et al. 1993, Weng et al. 1993). Under these circumstances, the *src* SH3 domain, which binds to proline-rich peptides such as found in paxillin, may provide the basis for protein interactions (Kaplan et al. 1994), although adhesion plaque proteins like focal adhesion kinase with its autophosphorylation site may also interact with proteins containing SH2 domains such as the *src* family members (Courtneidge 1994). Despite evidence that the *src* group of proteins mediates signaling from a different type of binding ligand (in this case cytoskeletal elements) than do the growth factor receptors, the pathway downstream in the cascade of events is believed to be similar for both receptor and nonreceptor tyrosine kinases.

5.5c   GTP-binding proteins: *ras* activation

Another major group of intracellular membrane-bound proteins includes the *ras* proto-oncogene family members, which bind and metabolize guanine nucleotides. *Ras* activation mediates signal transduction initiated by several growth factors and receptor tyrosine kinases. The *ras* proteins (p21^{ras}) are active and signaling when the protein binds GTP and become inactive upon GTP hydrolysis to GDP by p21^{ras} (Figure 5–3). Once the GTP is metabolized to GDP, it remains bound in the catalytic site of p21^{ras} until the complex of proteins (Grb2/Sos), which is localized at the membrane by the tyrosine kinase receptor, interacts with p21^{ras}, forcing the release of the GDP, allowing entry of another GTP molecule, and converting the p21^{ras} to its active signaling conformation (Chardin et al. 1993; Gale et al. 1993). A second mechanism for modifying p21^{ras} activity involves a group of GTPase-activating proteins (GAPs). GAPs are directly phosphorylated and activated by the receptor tyrosine kinases. They stimulate the inefficient p21^{ras}-associated GTPase activity by more than 100-fold upon interacting with p21^{ras} and thus significantly accelerate the hydrolysis of GTP to GDP. Because p21^{ras}-mediated signals are generated by the conformational change associated with the presence of GTP, GAPs may negatively regulate the activity of the normal p21^{ras} by limiting the duration of signaling. In tumor cells that contain a mutation in *ras*, signaling through p21^{ras} occurs continuously when the GTPase activity is lost due to alterations in its catalytic site or when the ability of p21^{ras} to interact with GAP is inhibited (Downward 1990). In both circumstances, the p21^{ras} remains in its active signaling conformation.

5.5d   Cytoplasmic serine/threonine kinases

The next step beyond p21^{ras} signaling in the cascade of events involving proto-oncogenes is activation of the cytoplasmic serine/threonine kinase Raf. Using purified p21^{ras} and Raf proteins, direct binding of p21^{ras} with Raf has been observed (Zhang et al. 1993; Warne, Viciana, and Downward 1993), and Raf selectively bound the GTP-charged or activated form of p21^{ras} (Figure 5–3). Additionally, Raf inhibited the ability of the GAPs to bind p21^{ras}, which of course would block hydrolysis of p21^{ras}-GTP and presumably stabilize the signaling form of p21^{ras}. A recent report, however, suggests that p21^{ras} is necessary only to localize Raf to the cell membrane. In this study investigators generated a recombinant protein in which Raf was fused to the membrane localization signal of p21^{ras} (i.e., the first 20 amino acids from its carboxy terminus), allowing the recombinant protein to localize to the cellular plasma membrane. The protein was constitutively active and eliminated the cellular requirement for p21^{ras} in mitogen-stimulated signal transduction (Leevers, Paterson, and Marshall 1994), which would suggest that the interaction of p21^{ras} and Raf proteins is only necessary to recruit the Raf to the membrane.

Once activated by its interaction with p21^{ras}, Raf then associates with a family of proteins called mitogen-activated protein kinase kinases (MAPKK or MEKs) (Figure 5–3). Raf can directly phosphorylate these serine/threonine kinases in generating their active or signaling forms in the cascade (Kyriakis et al. 1992). The MAPKKs then, in turn, activate members of the MAPK family, ERK1 or ERK2 (Seger et al. 1992), by phosphorylation of specific tyrosine and threonine residues (Payne et al. 1991). This increases the kinase activity of the MAPKs 1,000-fold. To establish an association between Raf signaling and MAPKK or MAPK activity, several methods have been devised to modulate Raf and monitor its effects downstream. By fusing a steroid-binding domain from a steroid receptor to the catalytic domain of the proto-oncogene *raf*-1, the activity of the Raf-1 protein can be directly regulated using the steroid, because the Raf protein is activated when the steroid interacts with the steroid-binding domain. Using this approach, Samuels et al. (1993) have reported that two cellular enzymes, mitogen-activated protein kinase (MAPK) and MAPK kinase (MAPKK), both become phosphorylated in cells treated with the steroid. Therefore, when expression of the receptor is up-regulated, so is the activity of the protein from the proto-oncogene *raf*-1. Raf, in turn, phosphorylates MAPK and MAPKK, demonstrating that these two kinases receive phosphorylation signals either directly or indirectly from the Raf and are therefore downstream of Raf signaling in the Raf-mediated pathway for signal transduction. When activated, the MAPKs may directly phosphorylate certain nuclear transcription factors such as Fos or Jun (Pulverer et al. 1991) (Figure 5–3), or they may interact with other nuclear kinases, which in turn can phosphorylate transcription factors.

Although the description of this mitogen-stimulated signaling pathway provides the basic details for one of the better defined signal transduction pathways, the actual mechanisms are most certainly more complex than presented. It is known that multiple forms of enzymes exist at each level, so there are, for example, three forms each of the p21^{ras} and Raf proteins (Moodie and Wolfman 1994). Furthermore, there are multiple signaling pathways, which seem to mediate signals for certain classes of mitogens (e.g., growth factors, cytokines, and insulinlike growth factors), although cross-talk occurs between pathways (Cano and Mahadevan 1995). What is known, however, is that several growth factors, including NGF, EGF, and FGF, all signal through the same general mechanism, suggesting this is a common and important regulatory pathway.

5.5e   Negative regulation of *ras* signaling

Whereas p21^{ras} activation stimulates signaling through the MAPK pathway, cAMP down-regulates or inhibits signaling through this same pathway. It accomplishes this by activating the serine-threonine kinase, protein kinase A (PKA), which blocks the transmission of signaling between Raf and p21^{ras} and thus prevents activation of the MAP kinase cascade (Wu et al. 1993; Cook and McCormick 1993) (Figure 5–3). The mechanism appears to involve the ability of PKA to phosphorylate Raf in its N-terminus, producing an N-terminal cap structure that covers its binding site for p21^{ras}. Thus, p21^{ras} cannot convey its signal to Raf, and the pathway is blocked (Marshall 1995). This may explain the long-established effects of cAMP as a negative growth regulatory and differentiation-inducing molecule. Because signal transduction through p21^{ras} has been implicated in tumorigenesis, it may be possible to preempt signaling through p21^{ras} and inhibit abnormal cell growth using molecules that activate PKA. Accordingly, overexpression or cAMP stimulation of PKA has been shown to inhibit the transformed phenotype in ras-transformed cells (Budillon et al. 1995; Bjorkoy et al. 1995).

5.5f   Nuclear signaling

Several links between the phosphorylation cascade in the cytoplasm and signal transduction in the nucleus have been established. As already mentioned, interactions between nuclear transcriptional activation factors and the MAPKs have been reported in vitro, and it appears that an intervening nuclear membrane does not prevent such interactions because both serine/threonine kinases ERK1 and ERK2 are translocated to the nucleus in growth factor–stimulated cells (Lenormand et al. 1993). Additionally, tyrosine kinases also have been localized to the nucleus. The proto-oncogene c-*abl* encodes a tyrosine kinase, which contains a nuclear localization signal and a DNA-binding function (Wang 1994). When translocated to the nucleus, it elevates transcription levels in conjunction with certain transcription factors or perhaps by direct phosphorylation of RNA polymerase II (Baskaran, Dahmus, and Wang 1993). Even *src* family members appear in the nucleus under certain circumstances. The tyrosine kinase STAT (signal transducer and activator of transcription) proteins, which mediate cytokine signaling, become phosphorylated at the cell membrane and are then directly translocated to the nucleus (Silvennoinen et al. 1993). Thus, numerous nuclear proteins are capable of mediating signals to the genome, but the details of their involvement in the various pathways, especially with regard to target specificity, remain to be established.

5.5g   Transcriptional activation

Several transcription factors are encoded by proto-oncogenes (Figure 5–4), including *erbA, ets, fos, jun, myb,* and *myc.* These nuclear protein factors regulate the

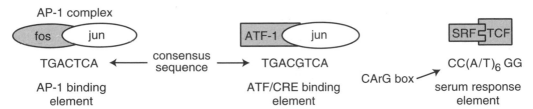

Figure 5–4. Transcription factor recognition sites. Transcription factors generally polymerize to form homo- or heterodimers of factor proteins that recognize specific consensus sequences in the DNA based on the particular complex generated.

transcription or expression of specific genes. This is achieved through the ability of the transcription factor to recognize and interact with specific DNA sequences found in the promoter or enhancer regions of those genes to be transcribed. The AP-1 complex and its components, Fos and Jun, have been extensively studied because of the pivotal role they play in transmitting signals from various growth factors, cytokines, or stress factors. The ability of the AP-1 complex to regulate transcription is mediated at the transcriptional level by controlling *jun* or *fos* expression to increase levels of the complex and at the post-translational level by phosphorylation of the Jun or Fos proteins to enhance AP-1 complex activity, that is, increase its ability to promote transcription. The Fos protein interacts with the Jun protein to generate a heterodimer, which recognizes the consensus sequence TGACTCA (Chiu et al. 1988). This sequence is referred to as the AP-1/TPA-response element (TRE) because the AP-1 complex mediates transcriptional activation by TPA through binding to this sequence. AP-1/TRE and TRE-like sequences are located in promoter regions of several genes, most of which are active during cell proliferation, for example, collagenase, transin, and the Jun protein itself (Gutman and Wasylyk 1991). Following stimulation by certain growth factors or exposure to ultraviolet radiation, a MAPK family member, Jun N-terminal kinase (JNK; Dérijard et al. 1994), phosphorylates the Jun protein specifically on serine residues 63 and 73, enhancing its ability to activate transcription from the AP-1/TRE element (Deng and Karin 1994). The *jun* family members can also dimerize with an activating transcription factor (ATF; also called cAMP-response element binding proteins (CREB) to form a protein complex that recognizes the ATF/CRE consensus sequence TGACGTCA (Hai and Curran 1991). This occurs following cell stimulation with neurotransmitters or polypeptide hormones that activate either protein kinase A (PKA) with cAMP or calmodulin-dependent kinases (Misra et al. 1994) with calcium. This mechanism allows cross-talk among signaling pathways because *jun* synthesis can modulate expression from both the AP-1/TRE and the ATF/CRE elements.

Although the *jun* promoter contains only a modified AP-1/TRE, which is recognized by Jun/ATF heterodimers, the *fos* promoter consists of multiple elements. In addition to the ATF/CRE element, the c-*fos* promoter includes another growth factor–stimulated binding site for transcription factors called the serum response element (SRE) (Treisman 1990). Transcriptional activation is enhanced from the SRE by interaction of the protein serum response factor (SRF) with a DNA sequence called the CArG box (CC(A/T)$_6$GG) in the SRE. An additional protein, ternary complex factor (TCF), is also required to maximize the response to growth factor stimulation of c-*fos* transcription (Shaw, Schroter, and Nardheim 1989) and is recruited to the SRE by the SRF following stimuli that activate MAPKs (e.g., growth factors and ultraviolet radiation).

As just described, the prototypical enhancer region incorporates multiple recognition elements in a gene's promoter. The *fos* enhancer/promoter actually contains an ets, an SRF, and an AP-1-like element, which allows it to self-regulate (Treisman 1994). For c-*met*, which encodes the receptor tyrosine kinase, the promoter region includes a consensus sequence for the transcription factor PEA3 (AGGAA(G/A)), which is induced by growth factors and TPA, and two AP2 sites (CCC(A/C)N(G/C)$_3$) (Gambarotta et al. 1994). Because these transcriptional elements in general can act additively or synergistically in transcriptional activation or repression, signaling from multiple pathways may elicit dramatically different responses in the nucleus. (For a compilation of the various transcription factors, see Faisst and Meyer 1992; for an on-line database, see Ghosh 1993). Furthermore, binding of the transcription factor to its consensus site or phosphorylation are not the only regulatory mechanisms for transcription. The transcription factor *wt1*, the so-called Wilms' tumor suppressor sequence, was initially thought to function only as a suppressor sequence (Call et al. 1990), but further examination has revealed that it can behave as both a repressor and an activator of transcription based on its interactions with other nuclear proteins such as *p53* (Maheswaran et al. 1993). This indicates that certain nuclear proteins can significantly modify the behavior of a transcription factor prior to its interaction with its target site.

5.6   Regulation of DNA synthesis and the cell cycle

Once the signals for growth stimulation have reached the nucleus, they are translated into cellular actions by a complex of proteins that mediate progression of the cell through the various phases of the cell cycle (Figure 5–5): G_0, a quiescent noncycling state; G_1, the preparation for DNA replication; S, DNA synthesis to duplicate the entire genome; G_2, preparation for nuclear and cell division; and M, mitosis. This is a highly regulated process during which the genome is replicated completely and only once for each cycle. The DNA is allocated such that one complete copy resides in each cell progeny following cell division. It is controlled by the stage-specific activation of members of the cyclin-dependent kinase (CDK) family, all serine/threonine kinases, and their interactions with various activating cyclins and suppressing CDK inhibitor (CDKI) proteins. When activated, the CDK proteins phosphorylate other proteins involved in cell cycle regulation. Most

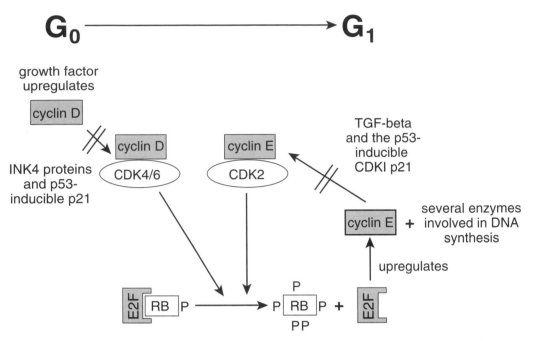

Figure 5–5. Regulation of cell cycle transition from quiescent G_0 to cycling G_1. Cell stimulation with a growth factor causes up-regulation of cyclin D1, which interacts with cyclin-dependent kinase (CDK) 4 or 6. These complexes hyperphosphorylate RB to inactivate it and release the transcription factor E2F. The liberated E2F then up-regulates the expression of cyclin E, which complexes with CDK2 to further inhibit RB activity and allow cells to progress toward S phase. The cyclins are negatively regulated by cyclin-dependent kinase inhibitors (CDKI): the INK4 proteins p15, p16, and p27 and p53- or TGF-β-inducible p21.

notably, the cyclin D–dependent kinases phosphorylate and thus inactivate the retinoblastoma suppressor protein (pRB), preventing pRB from sequestering the transcription factor E2F (Ewen 1994). E2F is then available for up-regulating genes with E2F promoter elements (consensus sequence TTTCGCGC), allowing cells to enter S phase.

The CDK-activating cyclins are categorized according to the cell cycle phase in which they are expressed. Thus, there are G_1 cyclins (D and E) that mediate the G_1 transition to S phase, S cyclins (A) for DNA replication, and G_2 or M cyclins (B) for progression into and through mitosis. Overexpression of certain members of the cyclin family has been associated with carcinogenesis, for example, elevated expression of cyclin D1 has been observed in human esophageal tumors (Jiang et al. 1992), leading to the suggestion that the normal cyclins represent a class of proto-oncogenes. Indeed, cyclin D1 can cooperate with activated *ras* to transform primary rat embryo fibroblasts (Lovec et al. 1994).

In addition to the positive regulation or activation of CDKs by the cyclins, the CDKs are negatively regulated by cyclin-dependent kinase inhibitory proteins

(CDKIs) (Peter and Herskowitz 1994). These factors bind to the CDKs, preventing their phosphorylation by the CDK-activating kinase (CAK), which facilitates cyclin interactions, or they interact directly with cyclin-CDK complexes. The INK4 protein (inhibitors of CDK4) p16 causes a G_1 arrest by specifically binding CDK4 (Serrano, Hannon, and Beach 1993) and blocking its activity. Transforming growth factor-β is a potent antimitogenic factor that also arrests cells in late G_1. It apparently does so by stimulating an interaction between CDK2 and a CDKI protein p27, thus blocking formation of the cyclin E-CDK2 complex (Polyak et al. 1994). Alternatively, it up-regulates another CDKI, p15, which targets CDK4 and CDK6 (Hannon and Beach 1994) and prevents inactivation of pRB. These proteins presumably function as suppressors rather than as proto-oncogenes because loss of function in knockout mice results in an accelerated and/or high incidence of certain tumors (Serrano et al. 1996).

Cell proliferation is initiated by exogenous signaling from a mitogen, which induces the expression of cyclin D family members and moves cells from the quiescent G_0 into early G_1 phase. The cyclin D proteins form a complex with and activate CDK4 and CDK6. Expression of this complex is maximized by late G_1 and persists as long as mitogenic signaling continues. CDK activation occurs when a structurally related CDK-activating kinase (CAK) phosphorylates the CDK on a specific threonine residue. The phosphorylation event facilitates the interaction of the CDK with a member of the cyclin D family, allowing it to interact with targets such as pRB (Pines 1994). Hyperphosphorylation of pRB by the cyclin D-CDK complex prevents pRB from binding the transcription factor E2F (Ewen et al. 1993), which in turn up-regulates itself and cyclin E. Cyclin E then complexes with CDK2, further diminishing the influence of one of its targets pRB and committing the cell to S phase (Neuman et al. 1994, Weinberg 1995). During S phase, cyclin E is rapidly removed by degradation (Won and Reed 1996), releasing CDK2 for interaction with cyclin A, which regulates progression through S phase (Girard et al. 1991). Furthermore, the cyclin A-CDK2 complex inactivates E2F and thus down-regulates the transcription of E2F-responsive genes, including cyclin E (Krek, Xu, and Livingston 1995). Finally, cyclins A and B in combination with CDK1 control entry into mitosis, after which cyclin levels plummet; however, the cycle can be repeated with continued mitogenic stimulation and up-regulation of cyclin D. In the absence of stimulation, cyclin D levels drop while CDKI p27 levels increase, and the cell arrests in G_0 (reviewed in Sherr 1996).

5.7   Signal transduction in general

As details of the various proto-oncogenes have become available, increasing evidence shows a unified or interactive pathway of events that generates, transmits, and interprets signals from extracellular stimuli to the genome. Thus, families of membrane receptors, many of which contain considerable sequence homology in the intracellular domains of the proteins, initiate common pathways of secondary messengers through a number of regulatory enzymes like phospholipase C, protein kinase C, or p21^{ras} or through direct interaction with certain structural proteins that may be anchored in the cell membrane. The signaling appears to involve a flow or cascade of phosphorylation because many of the participants are kinases. It has long been known that regulatory enzymes in metabolic pathways are activated and/or inhibited by phosphorylation reactions, which often alter the conformation of the modified protein to modulate its interactions with substrates, cofactors, or other proteins. In metabolism, phosphatases are invariably involved in reversing the effects of phosphorylation reactions. Now both tyrosine- and serine/threonine-specific phosphatases have been identified as major components of growth signaling pathways and have been shown under some circumstances to act as both positive and negative regulators of signaling pathways (Mourney and Dixon 1994; Tonks and Neel 1996).

The number of signal transduction pathways is rapidly expanding, and certainly more will be described by the time this chapter is published. The emerging picture reveals redundancy of pathways and pathway components. Isoforms of specific signaling enzymes have been described, for example, the protein kinase C family includes more than ten related enzymes or isozymes (Doi et al. 1993), adding another level of complexity to an already complex picture of positive/negative regulatory elements. However, as specific tissues are evaluated for these transducing pathways, it is possible that the active pathways are few and may actually define the differentiation and maintenance of a specific tissue type.

5.8   Mechanisms of oncogene activation

As stated earlier, more than a hundred genes have been identified, which, under certain circumstances, can transform cells when a single gene copy is altered (a dominant transforming gene). Although the majority have been identified by their transforming ability in oncogenic retroviruses, studies of chemically induced and naturally occurring tumors clearly demonstrate that oncogene activation can occur by a variety of mechanisms and in the absence of retroviral promoter sequences. The most frequently observed mechanism involves the incorporation of point mutations causing activation of oncogene proteins. In both naturally occurring human tumors and chemically induced rodent tumors, a *ras* oncogene family member is consistently activated by point mutation in codons 12, 13, or 59 to 61 in certain types of tumors. In humans, K-*ras* mutations in codon 12 occur at high frequency in adenomas and adenocarcinomas of the colon (Forrester et al. 1987) and in pancreatic adenocarcinomas (Almoguera et al. 1988).

It is also possible to induce a high frequency of mutations in *ras* genes of certain rodent tumors using specific carcinogens, suggesting that *ras* activation plays a significant role in carcinogenesis (explained in more detail later). Eighty-three per-cent of mammary tumors, initiated in rats with the alkylating agent N-methyl-N-nitrosourea (MeNU), contained activated H-*ras* sequences, and activation resulted predominantly from GGA to GAA transition mutations in codon 12 (Zarbl et al. 1985). Such a mutation causes a glycine residue to be replaced by glutamic acid in p21*ras*, which alters the conformation of the protein and permits continuous sig-naling through p21*ras*. 7,12-Dimethylbenz[a]anthracene (DMBA) also initiates mammary tumors in rats but does not induce mutations in codon 12 (Zarbl et al. 1985). Instead, codon 61 is affected. This is also the case for DMBA induction of mouse skin tumors in which 90% contained mutations (CAA to CTA) in codon 61 (Quintanilla et al. 1986). X-ray crytallographic analysis and mutagenesis stud-ies of p21*ras* show that mutations in these hot spot regions may impact p21*ras* activity by inhibiting its intrinsic GTPase activity (codons 12, 13, and 61) or by preventing p21*ras* interaction with GAP (codons 12, 59, 61, and 63) (Der, Finkel, and Cooper 1986; Milburn et al. 1990). Either alteration results in the accumula-tion of GTP-bound protein, the activated and signaling form of p21*ras*.

The *neu* oncogene is also frequently activated in rodents by point mutation (Bargmann et al. 1986). The proto-oncogene sequence encodes a receptor tyrosine kinase with considerable homology to the EGF receptor (*erb*B-1). GTG to GAG (valine to glutamic acid) mutations in the putative transmembrane region of the encoded protein were detected in more than 90% of N-ethyl-N-nitrosourea (EtNU)–induced schwannomas, tumors of the peripheral nervous system, in rats but in none of the tumors of the central nervous system elicited in the same proto-col (Perantoni et al. 1987). Such a mutation yields dominant transforming *neu* sequences. These mutations are specifically associated with the carcinogen EtNU because they occur infrequently in spontaneous schwannomas (Perantoni et al. 1994). They also appear very early in the neoplastic process, suggesting they could represent initiating mutations (Nikitin et al. 1991).

Point mutations are not the only mechanism whereby oncogenes acquire their oncogenic potential. Because of genetic sequence differences in *erb*B-2, the human equivalent of *neu*, a similar mutation in the human gene cannot occur in the same codon. However, amplification or elevated expression of the *erb*B-2 gene product p185 is frequently associated with specific human tumors. Notably, a poor progno-sis for women with mammary (Slamon et al. 1987) and ovarian tumors (Berchuk et al. 1990) correlates directly with the extent of overexpression of *erb*B-2 in tumor tissues. Because overexpression of the normal p185*erbB-2* can transform cells in cul-ture (Di Fiore et al. 1987) and provide cells with a selective growth advantage (Nikitin et al. 1991), overexpression of this receptor appears to be a viable mecha-nism for oncogene activation as well.

Chromosomal rearrangements provide another mechanism for oncogene activa-tion and may be a common pathway for induction of leukemias in humans. Chronic myelogenous leukemia (CML) is often associated with the appearance of the Philadelphia chromosome (Nowell and Hungerford 1960), which results from a reciprocal translocation between chromosomes 9 and 22 (Rowley 1973). This rearrangement causes the juxtaposition of the N-terminal exons of *bcr*, a gene with both serine/threonine kinase and GAP domains, from chromosome 22 with exon 2 from the c-*abl* proto-oncogene, which encodes a nuclear tyrosine kinase on chromosome 9 (Kurzrock, Gutterman, and Talpaz 1988). The ability of the *bcr-abl* fusion sequence to cause leukemogenesis has been established using transgenic mice carrying a germline copy of the fusion sequence. Mice with this transforming sequence suffer a fatal acute leukemic crisis shortly after birth (Heisterkamp et al. 1990), clearly implicating this altered fusion oncogene sequence in the etiology of the disease. The mechanism responsible for transformation by the fusion protein is a matter of speculation, although several reasonable models have been described. The fusion proteins retain both the *bcr* and *abl* kinase activities and invariably show enhanced levels of tyrosine kinase activity, which could account for its transforming ability. Alternatively, the ability of c-abl protein to transform cells is regulated by an internal SH3 domain, which suppresses transformation (Muller et al. 1991). Once the SH3 domain is removed, the remaining sequence exhibits transforming activity (Jackson and Baltimore 1989). It has been proposed that the presence of the *bcr* region in the fusion protein alters its conformation in such a manner as to overcome the negative regulatory activity of the SH3 domain. Finally, several other alterations involving proto-oncogenes such as *ras* and *myc* or suppressor genes such as *RB* or *INK4* have been described, and these may contribute significantly to tumorigenesis as well (Melo 1996).

Studies of B cell tumors such as Burkitt's lymphoma have also implicated proto-oncogene rearrangements in tumorigenesis (Croce et al. 1984). Burkitt's lymphomas are characterized by reciprocal translocations of chromosomes 8 and 2 (5%), 8 and 14 (80%), or 8 and 22 (15%). The result is the deregulation of transcription from the c-*myc* locus (Dalla Favera et al. 1982; Croce et al. 1983; Erikson et al. 1983). The transforming potential of this construct has been verified in transgenic mice, which develop B cell tumors (Adams et al. 1985). However, the fact that all B cells are not tumorigenic in these transgenic mice, although all B cells carry the rearrangement, indicates that this rearrangement predisposes to tumor formation but by itself may not be sufficient for tumorigenesis. Indeed, *bcl-2* can cooperate with *myc* under the control of immunoglobulin enhancer elements to rapidly transform B cells in transgenic mice (Strasser et al. 1990).

5.9   Carcinogens and oncogene activation

Although certain proto-oncogenes, notably *ras* gene family members, are readily mutable with carcinogen exposure, do the activating mutations necessarily function in multistage carcinogenesis? Are oncogenes involved in the various stages of tumorigenesis (i.e., initiation, promotion, and progression)? In the rat, transforming mutations in *neu* were detectable in Schwann cells within one week of exposure to EtNU (Nikitin et al. 1991). DMBA-induced skin tumors of mice have a high incidence of H-*ras* mutations in papillomas as well as in carcinomas, and these mutations are detectable early in tumorigenesis as well (Nelson et al. 1992). MeNU, which induces codon 12 mutations of H-*ras* in rat mammary tumors, may act as an initiator because mutations were also present prior to tumor formation (Kumar, Sukumar, and Barbacid 1990). However, in nitrosamine-induced rat kidney tumors, different regions of the same neoplasm were found to contain different K-*ras* mutations (Higinbotham et al. 1996), suggesting that activation occurs after neoplastic conversion in these tumors. In human studies, *ras* mutations arise with increasing frequency as colon tumors progress, but they are observed in adenomas at a significant frequency once the tumors exceed 1 cm, suggesting they represent an event in tumor progression and not initiation (Vogelstein et al. 1988). In human lung, K-*ras* activation was not detected in preneoplastic lesions but occurred in a large proportion of the cancers (Sugio et al. 1994). Additionally, for human prostate tumors, intratumor heterogeneity has been reported for *ras* alterations (Konishi et al. 1995). Although oncogene activation is clearly associated with tumorigenesis in these studies, the exact stage at which it functions seems to be variable.

5.10 Oncogene cooperation

The ability of oncogenes to transform tissues has been tested directly in cultured mammalian cells using purified calcium phosphate–precipitated DNAs to transfect novel genetic sequences. Single activated oncogenes are generally nontransforming for most cells with the notable exception of mouse NIH 3T3 cells, which are particularly sensitive to transfection with certain oncogenes such as activated *ras* or *neu* sequences. However, this immortalized cell line is insensitive to transformation by most oncogenes unless they are introduced in special DNA constructs that provide high levels of expression. In cell preparations from freshly dissociated tissues (primary cultures), individual oncogenes are invariably nontransforming; however, combinations of oncogenes are effective for inducing transformation. For example, when *ras* and *myc* are simultaneously transfected into primary cultures of rat embryo fibroblasts, transformation occurs (Land, Parada, and Weinberg 1983). Individually, *ras* is only weakly transforming, but *myc* can either immortalize cells or markedly reduce cell senescence in culture. Similarly, in reconstituted prostate cultures, cells containing activated *ras* oncogenes were dysplastic, whereas those containing *myc* oncogenes were hyperplastic; however, only cells containing both *ras* and *myc* were neoplastic (Thompson et al. 1990).

These molecular observations support the view that carcinogenesis requires multiple genetic alterations. Through studies of complementation groups of oncogenes, it has been found that growth control is often (but not always) retained when members of the same group are simultaneously expressed; however, members of different signaling circuits or at different levels of signaling are more likely to be transforming in combination. Thus, an activated cytoplasmic tyrosine kinase in combination with a nuclear oncogene often results in transformation of a normal cultured cell (Ruley 1990). Studies of signal transduction pathways provide considerable insight into the basis of this phenomenon. Growth factor induction of DNA synthesis and cell proliferation in cell culture requires signaling from two distinct pathways: one involving *ras* and its nuclear transmitter AP-1 (*jun* and *fos*), and the other involving *src* and its apparent nuclear transmitter *myc* (Barone and Courtneidge 1995). Blocking either pathway prevents the induction of DNA synthesis; therefore, constitutive activation of both pathways would be anticipated for transformation to occur. The studies of tumorigenesis in transgenic mice bearing individual oncogenes are consistent with results of cell culture studies. It is the exception that an individual oncogene can cause rapid tumor formation when introduced in the mouse germline such that it is distributed throughout the animal's tissues (Hunter 1991).

5.11 Normal cells suppress tumor growth

As to why individual oncogenes are incapable of causing transformation, this may be due in part to the fact that normal cells surrounding initiated cells may suppress the outgrowth of the oncogene-containing cell. Thus, v-*src*–infected avian embryonic cells were nontumorigenic in embryos but were morphologically transformed when dissociated and placed in culture (Stoker, Hattier, and Bissell 1990). It appears that the controls placed on the aberrant cells by adjacent normal cells can be abrogated by wounding or treatment with tumor promoters. For example, TPA induces the selective outgrowth of cultured *ras*-transformed rat embryo cells, resulting in their clonal expansion and transformed focus formation in culture (Dotto, Parada, and Weinberg 1985). Again, these findings are consistent with the models of multistage carcinogenesis described in Chapter 4. One might speculate that cells containing an initiating mutation in an oncogene such as *ras* remain quiescent and regulated by adjacent cells until released by immortalization, clonal selection, or death of the surrounding cells. In any of these scenarios, more than one event would be required for tumorigenesis.

5.12 Suppressor genes

Some of the earliest experiments in cell fusions involved the coalescence of tumor and normal cells in cell culture to establish genetic dominance of the neoplastic phenotype. Invariably, the hybrids were less tumorigenic (Harris 1988), and reversion to malignancy occurred with the loss of specific chromosomes in the unstable hybrids (Jonasson, Povey, and Harris 1977). For some tumors, malignancy could be suppressed with the introduction of a normal chromosome to replace lost sequences (Tanaka et al. 1991). These studies suggested that transformation results from a loss of function required for normal cell regulation rather than conversion of a function to a dominant transforming sequence and implicated tumor suppressor genes in transformation.

More than a dozen suppressor genes have now been cloned and characterized, and several more have been localized in the genome. These genes encode proteins that negatively regulate the growth of cells and, just as for the proto-oncogenes, function at a variety of levels in signal transduction and cell cycle regulation. Although studies of experimental models of chemical and viral carcinogenesis have delineated the majority of dominant transforming genes or oncogenes, genetic analysis of specific types of human tumors has led to the discovery of most recessive suppressor genes. Families predisposed to certain types of cancers often exhibit the loss or deletion of one copy of a specific chromosomal region that contains a suppressor gene. This loss occurs in the germline, allowing it to be passed to subsequent generations. Although the absence of one copy (heterozygosity) is often silent, tumorigenesis may occur with the loss of the second copy (homozygosity or loss of heterozygosity). Because the probability of knocking out a single copy of a suppressor sequence is considerably greater than knocking out two copies of the gene, the likelihood of tumor development in families with a germline mutation already present is likewise considerably greater than in the general population.

5.12a   The Rb locus

Retinoblastoma is a highly malignant tumor of the eye that as a genetic disorder is characterized by a loss of heterozygosity in chromosome 13q14. The gene implicated, the *Rb* locus, is either missing or altered (often by point mutations) in nearly every retinoblastoma analyzed whether familial (germline) or sporadic (somatic) (Friend et al. 1987; Yandell et al. 1989). In addition, osteosarcomas (Friend et al. 1986) and small cell lung (Harbour et al. 1988), bladder (Horowitz et al. 1989), breast (Lee et al. 1988), and pancreatic (Ruggeri et al. 1992) carcinomas all have altered pRB proteins to varying extents. Because reintroduction of normal *Rb* sequences into retinoblastoma cells reduces tumor cell malignancy (Huang et al. 1988), alterations in *Rb* appear to be essential for tumorigenesis.

The *Rb* suppressor gene encodes a nuclear protein, p105, which, as described earlier, serves as a negative regulatory factor in cell proliferation. When functioning normally, cells express pRB in a hypophosphorylated or active form at the G_0/G_1 phase. It becomes inactive through the phosphorylation of several serine/threonine residues by a CDK as G_1 progresses (Hinds 1995) and is reactivated by a phosphatase at the conclusion of mitosis (Buchkovich, Duffy, and Harlow 1989). In the hypophosphorylated state, pRB can interact with several proteins. As mentioned, it binds to the transcription factor E2F and thereby represses expression from promoters that contain an E2F-binding motif (Weintraub, Prater, and Dean

1992). In addition, several oncoproteins from viruses bind to pRB. The protein EBNA-5 from the Epstein-Barr virus (Szekely et al. 1993), E7 protein from human papilloma virus (Munger et al. 1989), adenovirus E1A protein (Whyte, Williamson, and Harlow 1989), and SV40 large T antigen (DeCaprio et al. 1988) are all capable of sequestering pRB and thus releasing E2F for transcriptional activation. Mutations in viral proteins that eliminate transforming activity are associated with a loss of ability to interact with pRB. Thus, viruses that can repress *Rb* function may be transforming as a result of this activity because cells would be released from the G_0/G_1 block in the cell cycle. Conceivably, certain nuclear proto-oncogenes when activated might similarly overwhelm the ability of pRB to regulate the cell cycle due to excesses in cell cycle–promoting molecules or altered affinities of effectors for pRB interaction and suppression. This may explain the transforming ability of cyclin D overexpression in cells or activation of the nuclear tyrosine kinase abl, which is normally inhibited by pRB binding to its catalytic domain (Weinberg 1995).

5.12b   *p53* suppressor gene

One of the most frequently altered genes found thus far in human tumors is the *p53* suppressor gene. The loss of function of this nuclear protein is associated with the Li-Fraumeni syndrome (LFS), a familial condition characterized by a high frequency of a diverse group of neoplasms. First identified in families of children with rhabdomyosarcomas, family members also presented with a high incidence of breast and brain cancers, osteosarcomas, and leukemias. These individuals often lose one *p53* allele from chromosome 17 in the germline, and tumors sustain a point mutation in the remaining allele (Malkin et al. 1990; Malkin 1993). Mice harboring a knockout deletion for p53 are grossly normal at birth but eventually develop a variety of tumors with a shortened latency (Donehower et al. 1992). In addition, germline *p53* mutations can cooperate with *Rb* mutations to accelerate tumorigenesis in general and to induce tumorigenesis in tissues for which altered *p53* or *Rb* genes are incapable of inducing individually, for example, pinealoblastomas and pancreatic islet cell tumors (Williams et al. 1994). This may explain why mutations in both genes often occur in human tumors.

The p53 protein behaves like a transcription factor because it recognizes specific DNA target motifs. One particular p21 protein (not to be confused with p21[ras]) with this motif was recently cloned and found to be a potent CDKI, inhibitor of the cyclin-dependent kinases (Xiong et al. 1993). This p21 protein has been reported to interfere directly with DNA synthesis by sequestering proliferating cell nuclear antigen (PCNA), which interacts with and activates DNA polymerase d, the principal DNA replicative enzyme (Waga et al. 1994). Expression of p21 is increased in response to DNA damage, and the cyclin E/CDK2 complex is concurrently inactivated, perhaps precipitating the observed G_1 growth arrest (Dulic et al. 1994). This places *p53* in a pivotal role in the suppression of cell cycle progression and DNA replication. Its frequent inactivation by mutation in a variety of tumor types in humans generally involves its DNA-binding domain, presumably blocking its edit function.

p53 functions as a surveillance factor that monitors the accumulation of DNA damage. Toxic doses of ultraviolet or ionizing radiations normally induce high levels of *p53* expression, and, as a result, cells become growth arrested (Kastan et al. 1991, 1992). This quiescent period allows the cells sufficient time to repair the damaged DNA before alterations are fixed in the genome during replication or to elicit apoptosis (programmed cell death) if the damage is irreparable. For example, normal mouse thymocytes readily undergo apoptosis following exposure to ionizing radiation in culture; however, irradiated thymocytes from mice lacking normal *p53* do not die (Clarke et al. 1993). It has been proposed that *p53* may provide a checkpoint function not only for growth arrest but also for the process of apoptosis (Haffner and Oren 1995).

5.12c   Apoptosis and its role in growth regulation

Apoptosis, or programmed cell death, is the negative regulatory mechanism whereby superfluous tissues are eliminated during differentiation, proper cell numbers are maintained in differentiated tissues, or irreparably damaged cells are eliminated. Simply stated, it is genetically determined cell suicide. Stem cell proliferation ensures continuous production of healthy functional cells; apoptosis regulates the rate at which the cells die, so a balance between cell replacement and cell loss can be maintained. In the developing embryo, this process is crucial to tissue restructuring (e.g., digit formation in the developing limbs). In adults, it accompanies the dynamic process of normal tissue renewal as well as tissue degeneration, as in the postlactating breast. The process even functions in tumor cells, albeit at a reduced rate. As described earlier, it can provide a mechanism for removal of cells under severe stress, such as follows extensive radiation-induced DNA damage or growth factor or extracellular matrix deprivation (Frisch and Francis 1994).

Two major families of proteins have been implicated in the mechanisms of apoptosis as well as the cell cycle regulatory proteins, p53, Rb, and transforming growth factor (TGF)-β. The first, the *bcl-2* family, encodes membrane-bound proteins that can inhibit or activate apoptosis; the second, a group of caspases, cysteine proteases that cleave substrate after specific aspartic acid residues. This includes the ICE (interleukin-1b-converting enzyme) proteins and all activate apoptosis (White 1996). The proto-oncogene *bcl-2* was discovered as part of a translocation event in B cell lymphomas (Cleary, Smith, and Sklar 1986) and was subsequently demonstrated to cause similar tumors in mice when overexpressed (McDonnell et al. 1989). Interestingly, it functioned not by enhancing cell proliferation, but instead by prolonging stem cell survival, that is, inhibiting apoptosis and "immortalizing" the altered cells. Since then, several additional homologous sequences have been identified, some of which behave similarly to *bcl-2,* but others that have the opposite effect and stimulate apoptosis (White 1996). Expression of the *bcl-2* homologue *bax,* one of the presumed activators of apoptosis based on the lymphoid hyperplasia apparent in *bax* knockout mice (Knudson et al. 1995), is up-regulated by p53 (Miyashita and Reed 1995). This may explain, in part, p53-induced apoptosis following cell stress. With stress, p53 is induced and in turn up-regulates both p21, to arrest cells in G_1, and *bax* to prepare the cells for apoptosis in the event the damage cannot be repaired. Alternatively, following growth stimulation, bcl-2 protein can bind bax protein to block p53-mediated apoptosis and might, should the ratios become skewed, result in cell immortalization, a presumed early event in tumorigenesis. As for the caspases, overexpression of ICE induces apoptosis, and its ability to stimulate cell death can be blocked through an interaction with bcl-2 protein (Miura et al. 1993). It has been proposed that the caspases may function in bax protein-mediated apoptosis, although details of possible mechanisms have not been established (White 1996). *Rb* also plays an important role in mediating apoptosis because *Rb* knockouts characteristically show massive cell death in neural tissues (Clarke et al. 1992), and restoration of function inhibits apoptosis (Haas-Kogan 1995). Because many tumor viruses function by sequestering RB protein, the predisposition of *Rb*-deficient cells for apoptosis may provide protection against these pathogens. TGF-β, a soluble secreted factor that binds to membrane-bound receptors, is a potent growth suppressor for many cell types and is capable of inducing apoptosis as well. It functions at several levels to induce cell cycle arrest in G_1, including positive regulatory effects on multiple CDK inhibitors (p15, p21, and p27) and negative effects on the G_1 cyclins and the growth-promoting transcription factor c-*myc* (Alexandrow and Moses 1995). The mechanism for TGF-β stimulation of apoptosis is not well established; however, evidence indicates that both *bax* (Nass et al. 1996) and the caspases (Cain et al. 1996) may be involved.

It has been suggested that the cancer process may result from deficient regulation of apoptosis (Wyllie 1995). That alterations in p53 can impact tumorigenesis via apoptosis has been demonstrated using knockout mutations that inactivate p53. Aggressive choroid plexus tumors with inactive p53 increase rapidly in size in mice, unlike tumors with an active normal form of p53 (Symonds et al. 1994). When these tumors were evaluated for cell proliferation, they exhibited comparable rates; however, the number of apoptotic cells in the slow-growing tumors was markedly higher; so p53-mediated apoptosis can apparently suppress tumor growth.

It has long been known that tumor cells are genetically unstable and this instability precedes tumorigenesis (Hartwell 1992). Accordingly (as discussed in Chapter 3), there has been considerable speculation regarding the existence of a mutator gene, which, once altered itself, could accelerate the rate of endogenous mutations in the genome and thus eventually mutate critical targets in growth control. DNA polymerases are likely candidates and have been observed to be mutated in some human colorectal tumors, which may account for the high mutation rate found in these tumors (Wang et al. 1992). The *p53* gene might also provide such a function. When altered, the cell loses the ability to growth-arrest damaged cells and thus repair DNA mutations prior to their fixation in the genome. The resulting acceler-

Table 5–2. *Cloned suppressor genes with known functions*

Gene	Consequence of gene loss	Function of encoded protein
Rb	retinoblastoma and osteosarcoma	binds and sequesters transcription factor E2F to maintain cells in G_0 of cell cycle
TP53	Li-Fraumeni syndrome multiple tumor types	transcription factor pivotal in progression of cell cycle/surveillance factor for DNA damage
Wt1	Wilms' tumor/nephroblastoma	transcription factor required for renal development
VHL	von Hippel-Lindau syndrome renal cell carcinoma	binds elongin B/C subunits to reduce gene transcription and translocates to nucleus a protein that degrades cell cycle molecules
NF1	von Reclinghausen's disease neurofibromatosis type 1 schwannoma and glioma	GTPase-activating protein (GAP), which regulates signal transduction through $p21^{ras}$
NF2	neurofibromatosis type 2 acoustic nerve tumors and meningiomas	the encoded protein merlin is thought to connect cell membrane proteins with the cytoskeleton
APC	familial adenomatous polyposis colorectal tumors	interacts with β-catenins, proteins involved in a signaling pathway for tissue differentiation
MMR	hereditary nonpolyposis-colorectal cancer	mismatch repair processes

ated accumulation of mutations in a variety of loci could readily expedite tumorigenesis in several tissues, as appears to be the case for Li-Fraumeni syndrome.

5.12d Other defined suppressors

Other suppressors from cancer families (Table 5–2) include the Wilms' tumor 1 locus (*Wt1;* chromosome 11) (Call et al. 1990), which encodes a zinc finger transcription factor expressed during organogenesis from the urogenital ridge. This gene is essential for normal renal development because kidneys do not form in mice with functional knockouts. Wilms' tumor or nephroblastoma accounts for the largest share of solid (nonlymphoid) tumors in children. What makes them especially interesting is the histologic profiles of the tumors; that is, they contain tissue elements typical of a developing kidney. Mutations in *Wt1* are predominantly associated with the familial form of the disease and are rarely observed in sporadic tumors.

Another syndrome, von Hippel-Lindau disease, is characterized by a genetic predisposition to renal and vascular tumors, for example, renal cell carcinomas and hemangioblastomas (Linehan, Lerman, and Zbar 1995). The *VHL* locus has been cloned and implicated as the suppressor sequence responsible for VHL disease. The locus is either lost or mutated in renal cell tumor from essentially all patients with VHL disease (Gnarra et al. 1994). The protein encoded by the *VHL* gene serves as a cofactor in transcription by sequestering subunits of elongin, a protein that activates transcription elongation by RNA polymerase II, and thereby prevents transcription from proceeding (Duan et al. 1995). It also regulates the nuclear translocation of a cullin protein, a family of factors that facilitate the degradation of cell cycle proteins (Pause et al. 1997).

Neurofibromatosis 1 or peripheral von Recklinghausen's disease predisposes to neurofibroma formation in the peripheral nervous system. The responsible locus, NF1 on chromosome 17 (Ballester et al. 1990), encodes a GAP-like protein that negatively regulates signal transduction through p21^{ras}. Other suppressors have been identified (see Table 5–2), and several more loci are currently being characterized at this time. What is clear from studies of the various cancer families is that neoplasia in humans frequently involves the loss of a suppressor function.

5.13   Where pathology meets molecular biology

As described in an earlier chapter, pathologists have long known that cells pass through a series of preneoplastic stages on the way to becoming tumors, and cytogeneticists have shown that specific tumor types contain consistent patterns of chromosomal abnormalities. With the identification and characterization of the dominant (oncogenes) and recessive (suppressors) elements in tumors, we have finally reached the point where we can begin to link certain pathologic changes or chromosomal aberrations with specific molecular alterations. Through exhaustive evaluations of human colon tumors, Vogelstein's laboratory has described a series of events associated with colon tumorigenesis (Fearon and Vogelstein 1990; Vogelstein and Kinzler 1993; Kinzler and Vogelstein 1996). Their studies show that most colorectal tumors whether familial or sporadic suffer a loss of a suppressor sequence called *apc* (adenomatous polyposis coli). This gene encodes a cytosolic protein with β-catenin-binding domains, which mediates signals from the secreted differentiation-inducing Wnt proteins. Furthermore, when wild-type APC replaces a mutated form in colonic epithelial cells, the cells undergo apoptosis, suggesting that APC controls the decision to initiate cell death. Mutations in APC occur early in tumorigenesis and may be essential early events in tumor formation due to the nature of colonic differentiation. The stem cells are maintained in crypts of the colonic villi, and as they differentiate they move away from the crypt, eventually die, and are sloughed in the gut. For transformation to occur, an altered cell would have to be retained as a viable entity. Inhibition of apoptosis would provide an efficient mechanism for achieving this. Once "immortalized," time and probability would then determine neoplastic development in these initiated cells. A second common lesion in colorectal cancer involves a group of genes responsible for a type of DNA repair, namely mismatch repair (*mmr*). Mutations in these genes predispose cells to a high frequency of mutation in the genome in general and can result in the accelerated accumulation of mutations in a variety of genes, including those involved in signal transduction or cell cycle regulation. Thus, the frequently observed alterations in these *mmr* genes may increase the likelihood that the other commonly found mutations in colorectal cancer occur (i.e., K-*ras* activation in the larger adenomas and *p53* inactivation late in the neoplastic process). These studies demonstrate that no single lesion is generally transforming and that neoplasms probably result from an initiating event involving a block in apoptosis followed by a progression event that causes genomic instability such as occurs with mutations in *mmr* and generates the multitude of alterations attributed to malignant cells.

5.14 Summary

Cancer is a multistep genetic process in which a series of mutations involving oncogene activation and/or tumor suppressor loss gradually increase the ability of a cell to proliferate autonomously. Many of the genes implicated in this process have been identified and characterized. They are altered forms of highly regulated, dominantly expressed proto-oncogenes that mediate signal transduction in normal growth control, or they are lost tumor suppressor functions which also serve to regulate growth control but in a negative and recessive manner. These alterations generally arise by somatic mutation, but germline mutations do occur, and, when they do, they predispose the carrier to tumor development. With knowledge of the specific somatic sites, it may be possible to assess the risk for a particular type of cancer through measurements of relevant mutations such as has been reported for the incidence of *p53* mutations in ultraviolet-damaged skin (Nakazawa et al. 1994). Through studies of individuals who develop specific types of cancers, it may be possible to delineate more germline lesions that predispose to cancer. In fact, it is conceivable that as more pathways of tumorigenesis are delineated, we may find that every individual who develops cancer actually belongs to a predisposing group based on some germline mutation. Such a circumstance could lead to early identification of these lesions in predisposed individuals and even to the development of therapies for tumor prevention once we learn to manipulate cellular signaling pathways effectively. It is remarkable that we can now envision such possibilities. Over the past ten years, the study of cancer-related genes has repeatedly provided breakthroughs in delineating the mechanisms of normal growth control. Although many details of the processes remain to be defined, we are well on our way to understanding how growth-inducing signals are transmitted to the nucleus, how the cell progresses from a quiescent to a cycling state, and how these processes are subverted in tumorigenesis.

6

Cancer in nonhuman organisms

Robert G. McKinnell

Workers who deal with cancer problems only in man and his closest relatives among the homeothermic vertebrates may look disdainfully at those who invest some of their energies in studies of neoplasms in creatures such as fish, frogs, snakes, and the numerous species of spineless, often slimy animals that were, in the days of the ancients, lumped together as "vermin." After all, Alexander Pope, a poet of some wisdom, admonished: "The proper study of man is man."

So it may be. But Aristotle, long before modern ecologists, recognized that man is an organism within a greater organism, the earth ecosystem. The ecosystem is not dependent on man, though man is dependent on the ecosystem. Ought not the study of man be extended to include the other animals that coexist with and support him on this motherly planet? They share with man many diseases, including cancers, and it is pertinent to ask: In what ways do neoplasms of animals at various phyletic levels differ from or resemble those of man in relation to etiology, natural history, immune factors, biochemistry, morphology, molecular biology, and the rest? This is the basic question to which comparative oncology addresses itself. . . . [A]s if in reply to Pope, Kipling asked: "What do they know of England who only England know?"

C. J. Dawe 1969

6.1 Introduction

This chapter discusses what some have referred to as a phyletic approach to cancer (Dawe et al. 1981). Descriptions of the distribution and prevalence of malignant tumors in organisms other than humans are considered here for a multiplicity of reasons. A rationale that may not appeal to practical-minded students is a certain intellectual fulfillment in understanding natural phenomena. Which plants have tumors and why? What animals are commonly afflicted with cancer? Learning about phyletic aspects of cancer is an end in itself for some scholars.

There are, however, several very practical reasons for seeking knowledge concerning cancer of plants and animals. For instance, if fish tumors are historically scarce in a particular area but are more recently becoming abundant, then one should seek the cause of the increased rate of neoplasia. Fish live in the water which drains our agricultural and industrial lands, and afflicted fish (or other organisms) may serve as an early warning signal that the level of carcinogens is increasing in that environment. Humans share that environment with fish. Thus, sentinel organisms such as fish serve to protect the environment and focus attention on hazards potentially toxic to humans. The reduction in cancer prevalence in natural populations, when it occurs, comes as a welcome signal that cleanup of a particular site or region has been successful.

Another significant motive for the scrutiny of lower organisms is that their cancers provide information directly useful in understanding human biology. For example, the renal adenocarcinoma of the common leopard frog of North America is caused by a herpesvirus. This frog cancer was the first malignancy associated with a herpesvirus (see later discussion in this chapter). More recently, other animal cancers and human malignancies have been associated with herpesviruses. Whether or not the herpesvirus association in the human cancers proves to be etiologic, certainly the existence of a frog cancer caused by a herpesvirus (and the many other animal cancers caused by viruses) adds credence to that possibility. Moreover, experiments that would be inappropriate in humans relating to viral etiology and cancer treatment can be carried out on frogs. Certainly tumors of lower animals have added immensely to the understanding of cancer in humans.

Another basis for the phyletic approach to cancer relates to the finding of organisms that are relatively, or absolutely, resistant to cancer. For example, great white sharks (described later) seem to be free of cancer. Why? Do these animals live in an environment free of carcinogens? Or do they have a peculiar biology that protects them from cancer-causing agents and renders them free of neoplasia? If the latter is the case, what is that biology and how does it differ from the biology of organisms susceptible to cancer? Is it possible that we could learn from cancer-free animals how to protect humans?

A final rationale for phyletic studies relates to cancer prevention and treatment. Experimental organisms permit novel and unusual approaches to this aspect of cancer research. Consider that an economically important plant cancer can be prevented with an antibiotic substance already in commercial use. Furthermore, a cancer vaccine has kept the price of chicken well within affordable limits for a number of years. And finally, with existing technology, a cat can be protected against a particularly virulent and deadly form of leukemia. These are not pipe dreams. Do not these success stories offer hope to humans?

Prevention, yes, but what about treatment? Many years ago, it was shown that plant tumor cells could be treated to give rise to absolutely normal mitotic descendants and that frog cancer nuclei have the genetic potential to program for a diversity of normal-appearing cell types. The phenomenon of a cancer cell giving rise to normal progeny as illustrated in plant and frog neoplasms is being extended to many other animal tumors, and human studies are underway. Many individuals express hopelessness when they speak of cancer. Some ask, why continue research in a field where progress is so stubborn? Illustrations presented here indicate that progress is being made and at an accelerating rate – and frequently that progress is with plant or animal experimentation. Encouraging results are a driving force that keep experimental biologists at their laboratory benches.

6.2   Plant tumors

Neoplasms, which occur in plants in a diversity of forms, are frequently referred to as "galls." The most common tumor types are the galls caused by a variety of insects and other organisms (Felt 1965; Kalil and Hildebrandt 1981). Insect galls are not malignant. Malignancy in humans and other animals is characterized by anaplasia, invasion, and metastasis (Chapters 1 and 2). Insect galls, while forming the characteristic lumps that permit identification and classification, have well-circumscribed growth potential. Further, their abnormal growth follows specific patterns. Unlike the cells of malignant tumors, normal behavior returns to the cells of insect galls upon removal of the causative insect or its secretions. For the most part, these tumors are not lethal to the plant host. Because they are not invasive, have self-limiting growth potential, and do not cause the death of the plant, insect galls are comparable to benign tumors in humans.

A historical comment is appropriate to add here regarding galls. To some extent, the modern industrial world and its activities have been implicated in the causation of human cancer. Because of that notion it is instructive to consider tumors that have been described in fossil humans (Chapter 7). Paleobotanists have similarly described galls in fossil plants (Whittlake 1981). Those ancient plant tumors are referenced here to indicate the antiquity of neoplasms that predate by a considerable time the onset of the industrial revolution and its ubiquitous pollution.

Tumors with many of the malignant characteristics of animal cancer[1] also occur in plants (White and Braun 1942; Bayer, Kaiser, and Micozzi 1994). Perhaps the most studied of these is the crown gall, which afflicts a diversity of dicotyledonous plants with global distribution. Some of the plants vulnerable to crown gall include apple, almond, walnut and pecan trees, various stone fruits, roses, grapevines, raspberry and blackberry bushes, chrysanthemums, tobacco, and willows. The crown gall presents as large cauliflower-like swellings, usually where lat-

---

[1] Although neoplasms are discussed in this section, the literature of plant cancer is brief. Plants are remarkably resistant to neoplastic change despite their exposure to massive doses of ultraviolet radiation, which is known to cause DNA damage and cancer in animals and humans (see discussion of skin cancer in Chapter 7; also Doonan and Hunt 1996).

eral roots branch from the underground stem but also on stems and leaves. The crown gall was originally thought to be caused by a soil-borne bacterium, *Agrobacterium tumefaciens* (Smith and Townsend 1907; Braun 1943), but later it was shown it was not the bacterium per se that caused the tumor; the disease was found to be mediated by the T_i plasmid, a closed circle of nonchromosomal DNA that varies from 200 to 400 kbp depending on the source (kbp is an abbreviation for kilobase pairs, a measurement of molecular size). A small portion of the plasmid DNA (known as T-DNA with a molecular size of 24 kbp) becomes inserted into the plant cell genome (Chilton et al. 1980; Miranda et al. 1992; Kononov, Bassuner, and Gelvin 1997). The inserted T-DNA contains genes that activate the host plant's synthetic activities, resulting in the formation of the crown gall (Formica 1989; Piper, Beck von Bodman, and Farrand 1993).

Crown galls of tobacco grow well in vitro as chaotic arrangements of tissues and organs. Abnormal shoots and leaves may grow from the neoplastic mass. In an elegant series of experiments, Armin Braun showed that this abnormal tissue could give rise to normally differentiated plants that flowered and set seed. His procedure involved serial grafting of neoplastic tissue to normal hosts (Braun 1951, 1981; Binns, Wood, and Braun 1981). These experiments demonstrated unequivocally that transformed tobacco cells can be manipulated to give rise to the entire spectrum of normally differentiated tissue. Further, and perhaps more importantly, these experiments revealed that the crown gall is not caused by an irreversible alteration of the genome of the neoplastic cell, but rather that the neoplastic state ensues from inappropriate gene activity.

When Braun began his experiments, it was generally believed that cancer cells gave rise only to more cancer cells. Thus, when Braun reported to a scientific audience that his plant cancer cells gave rise to normal mitotic progeny, a Nobel laureate in the audience remarked that although the experiments on plant cell differentiation were interesting, they could have no relevance to cancer. The Nobel laureate stated a longheld dictum: cancer cells can give rise only to more cancer cells. Inasmuch as Braun's cancer cells produced normal cell progeny, it was thought that in no way could they be neoplastic. But because of Braun and others, that dictum is no longer tenable.

Malignant transformation persists after removal of the carcinogenic agent (Chapters 1 and 3). This principle distinguishes temporary perturbations of cell structure and activity from true malignancy. However, the crown gall studies of Braun provided indisputable evidence that the phenotypic expression of a malignant cell may be altered. In this instance, there seems to be a complete remission of the malignant state resulting in normal mitotic progeny. Thus, the principle enunciated earlier should be modified by insertion of the word "ordinarily" in the phrase to read that "malignancy *ordinarily* persists even after the removal of the inducing agent." Another example, the Lucké renal adenocarcinoma of frogs, is discussed later.

Botanists who study the crown gall have contributed to another important area of cancer research. Knowledge of etiology frequently helps in devising rational treatment. Such is true for the crown gall. As indicated earlier, crown galls ensue after the integration of T-DNA into the recipient cell. Thus far, chemical control of the disease seems ineffective. However, about two decades ago, it was reported that *Agrobacterium radiobacter* strain K84 carries a plasmid designated pAgK84 (Kerr 1980; Clare, Kerr, and Jones 1990) which codes for the production of an antibiotic designated Agrosin 84. That antibiotic, a substituted adenine nucleotide, breaks down the opine nopaline necessary for growth of the oncogenic *A. tumefaciens* bacterium. Agrocin 84 then appears to interfere with DNA synthesis of the *A. tumefaciens,* resulting in lethality to the pathogen. It comes as no surprise, therefore, to learn that Agrocin 84 is used commercially for protection of plants vulnerable to crown gall (Pesenti-Barili et al. 1991). Although Agrocin 84 protects plants from infection and *subsequent* malignant transformation, it does not cure already existing crown gall.

At the risk of seeming overly enthusiastic about developments in plant biology, we should stand in awe considering the results reported here. In the case of crown gall, a disease with many similarities to cancer in animals, the cause of the neoplasm is known, malignant cells can be induced to give rise to normal mitotic progeny, and a means of preventing the tumor is already in hand. Biomedical scientists who work with animals and humans should be appropriately impressed.

6.3   Invertebrate animals

As late as 1930 it was asserted by some workers that invertebrates did not, or even for alleged reasons could not, develop tumours; and even today it is sometimes stated that invertebrate tumours are not "really" neoplasms, or are irrelevant to the study of vertebrate cancer. Such statements appear quite ill-founded.

J. Huxley 1958

Are the tumors of invertebrates "true" cancers? That same question may be asked of the plant neoplasms already discussed. Concerning plant tumors, Braun (1972) wrote that although differences in particulars doubtless exist, it is likely that similar cell mechanisms will be found to underlie neoplastic growth in both plants and animals. It follows from this that tumors of invertebrates, properly diagnosed and characterized, may well be a lode awaiting mining by comparative oncologists. Neoplasms, or what appear to be neoplasms, have been reported to occur in mollusks, sipunculids, arthropods, ascidians, and annelids (Scharrer and Lochhead 1950; Dawe 1969; Dawe et al. 1981).

6.3a   *Drosophila melanogaster*

The fruit fly, *Drosophila melanogaster,* is probably the best known species with regard to genetics. If there is a genetic component in cancer, which many believe

(Chapter 4), then studies of tumors in the fruit fly could be of considerable significance in elucidating the connection of genetics to malignancy. Melanotic tumors of *Drosophila* larvae have been widely studied. However, melanotic tumors in fruit flies may not be true malignancies because of their paucity (or absence) of mitosis, their failure to manifest invasion, and their encapsulation by blood cells, coupled with their destruction at metamorphosis. Melanotic tumors are frequently considered to be analogous to granulomas, or pseudotumors, or simply nonmalignant aggregations of cells (Ghelelovitch 1969; Gateff 1981).

However, cancers do exist in *Drosophila*. Malignant neuroblastomas can be recognized in sixteen-hour embryos. In these cases, definitive differentiation of adult optic neuroblasts fails to occur, and the dividing neuroblasts invade and destroy normal portions of the brain. Six independent mutations are known that arrest differentiation of the neuroblasts and give rise to these malignant neuroblastomas. Further, imaginal disc (larval cells that will participate in adult development) tumors of *Drosophila* have been described which have lost the capacity to differentiate and grow in an autonomous and lethal manner. Mutations to specific tumor suppressor genes result in the imaginal disc tumors (Kurzik-Dumke et al. 1992). Malignant plasmatocytes may exist either as mature cells with imperfect control of cell proliferation or as immature plasmatocytes that accumulate in enormous numbers (Gateff 1994; Gateff et al. 1996). These studies provide hope that the "correlation between the primary genetic change and the malignant neoplastic transformation of cells" may be identified and characterized in that organism which is best known genetically (Gateff 1981).

6.3b Other invertebrates

Marine corals were reported to have cancer (Squires 1965), but considering the phylogenetic differences between corals and mammals, the "malignancy" of the marine tumor has been questioned (White 1965). The quandary concerning what is or is not a malignancy plagues much research in nonvertebrate oncology.

Tsang and Brooks (1981) reported transformed cells of the German cockroach, *Blattella germanica,* which had invasive and metastatic potential when injected into appropriate hosts. Old World locusts, *Locusta migratoria,* are vulnerable to mitotically active, less than fully differentiated epithelial tumors that form after severance of sympathetic or gastric nerves. Metastatic colonies of these tumors appear after implantation in host *Locusta* (Matz 1969).

Bivalve molluscans are particularly vulnerable to proliferative benign lesions of connective tissue, perhaps due to encounters with toxic substances in their aqueous environment. However, more malignant conditions are observed that include proliferative blood cell disorders, sometimes referred to as hemic neoplasia (Elston, Kent, and Drum 1988). Oysters and mussels of many species from a number of locations throughout the world manifest the hemic disease that has characteristics thought to be similar to bovine leukemia (Peters 1989).

Biologists studying cancer may someday profit in another way by studying invertebrates. Mercenene is an agent, obtained from marine clams, which was thought to have antineoplastic properties (Schmeer 1969). More recently bryostatin 1, a naturally occurring substance obtained from a bryozoan of the Gulf of Mexico, *Bugula neritina,* has been used as a chemotherapeutic agent. Bryostatin 1 is a potent activator of protein kinase C, has activity against murine lymphomas, leukemias, and melanoma, is a differentiation agent, and is currently undergoing clinical evaluation (Berkow et al. 1993; Stanwell et al. 1994; Stone 1997; Li et al. 1997).

Many lesions have been reported in invertebrates. Some of them may be genuine malignant neoplasms. However, many of these lesions are merely proliferative cell responses to infection or toxic substances, and one must be exceedingly critical when interpreting these "tumorlike" growths.

6.4   Cancer in selected poikilothermic (cold-blooded) vertebrates

Examples of abnormal growth have been encountered in many taxonomic groups of animals and at every level of biological complexity. However, except for the mammals, the study of neoplasia, especially in the lower vertebrates, is to a large extent descriptive, haphazard, and incomplete.

D. G. Scarpelli 1975

Although significant progress in the study of neoplasms in poikilotherms has been made since this comment of Scarpelli was published, it is still true that the oncology of lower vertebrates is, to a large extent, "descriptive, haphazard, and incomplete."

6.4a   Fish

Fish are of greater economic importance than other poikilothermic vertebrates. There are 20,000 or so species that vary enormously in size, habitat, and life expectancy. It is not surprising, therefore, that much is known about tumors of these creatures (Masahito, Ishikawa, and Sugana 1988). Neoplasms of fishes harvested for commercial purposes include all of the major histologic types recorded for mammals, namely, epithelial, mesenchymal, pigment, and nerve cell cancers (Schlumberger and Lucké 1948). The Registry of Tumors of Lower Animals at the Smithsonian Institution reported that in its collection, almost three-fourths of the tumors of chordates were neoplasms of fish (Harshbarger, Charles, and Spero 1981). Most fish neoplasms are found in teleosts (bony fishes).

Section 6.2 discussed fossil galls (plant tumors). Fossil fish are more abundant and better known than fossils of any other vertebrate class. Yet no record of tumors appears in fossil fish. However, living "fossil" fish exist. These contemporary animals have changed little during their immense evolutionary history. In contrast to what is known of their fossil ancestors, the contemporary "fossil" fish are indeed afflicted with tumors (Masahito et al. 1988). One may ask whether the environment of contemporary fish has deteriorated or is it that we simply have not studied (or cannot study) fossil fish adequately?

Although only a few cartilaginous fishes (sharks and their relatives) are vulnerable to malignancies, some shark species seem to be immune to cancer. For example, no tumors have ever been reported from the great white shark, *Carcharodon carcharias*. Perhaps not enough individuals of the species have been examined. However, *C. carcharias* is mentioned here for other reasons, the first of which concerns habitat. The great white shark lives offshore in cleaner water than that found near coastal cities. Moreover, it is not a bottom feeder, in which case it would be expected to encounter foul material from urban wastes. Fish that feed in the open sea, as the great white shark does, have few cancers (especially when compared to their fish relatives who inhabit polluted inland waters). One would surmise, therefore, that environment plays an important role in the epidemiology of tumors found in cartilaginous fish, which may be a message of importance to humans.

The other reason for citing the paucity of shark cancer in this context is that sharks have a skeleton made of cartilage instead of bone. Cartilage is a peculiar fabric when it comes to animal structure. Invasion is an essential component of metastasis (Chapter 2). Cartilage is rarely, if ever, invaded by cancer cells. Further, extracts of shark cartilage are reported to inhibit invasion and the growth of blood vessels that nourish growing cancer cells (Snodgrass and Burke 1976; Luer 1986). It is difficult to prognosticate if shark extracts will ever be important to cancer treatment, but the studies cited suggest that potential therapeutic modalities (Chapter 8) may be derived from exotic sources.

Bony fishes, as stated previously, are more vulnerable to tumors than sharks. Skin tumors, some of which may be malignant, occur in walleye, *Stizostedion vitreum*, taken from Oneida Lake, New York (Bowser et al. 1988). Many mummichogs, *Fundulus heteroclitus*, obtained from a particular site in a river in the state of Virginia, were found to have hepatocellular carcinomas. Almost all of the fish from that site had hepatic lesions, and the site was shown to be contaminated with creosote, a probable carcinogen used as a wood preserver (Vogelbein et al. 1990). Epidemic histiocytic lymphoma occurs in northern pike, *Esox lucius*, from the Åland Islands (Thompson and Kostiala 1990). These fish tumors are examples of organisms that are indicators of possible carcinogens in aquatic environments.

Melanomas have been known for many years in hybrid and wild-caught platyfish of the genus *Xiphophorus*. An observation of potentially great interest is the report of a mutant form that undergoes spontaneous and total remission of melanoma (Schartl and Schartl 1996). Spontaneous remission is known in human cancer (Chapter 1), but it is so rare it is impossible to study it. The fish tumor makes a scientific study of spontaneous remission possible.

Oncogenic viruses include the herpesviruses, which in fishes (Hedrick and Sano

1989) have been associated with the cancers of carp (Sano, Fukuda, and Furukawa 1985) and salmon (Kimura and Yoshimizu 1988). Herpesviruses are associated with an amphibian tumor (see discussion later) and several human cancers (Howley, Ganem, and Kieff 1997). It would be informative to know if a relationship exists between the oncogenic herpesviruses of poikilotherms and the herpesviruses associated with cancers of humans and other organisms. The DNA of a catfish (obviously a poikilotherm) herpesvirus has been sequenced, permitting the comparison of its genome with the DNA of other herpesviruses (Davison 1992).

Of course, oncogenes (Chapter 5) have been isolated and sequenced from fish (Van Beneden et al. 1990). These and other molecular studies permit a characterization of the evolutionary significance of DNA sequences thought to be important in oncogenesis.

The crown gall neoplasm of plants described earlier was shown to give rise to mitotic progeny that differentiated normal cell progeny. Similarly, a pigment cell tumor of goldfish was shown to give rise to cells that differentiated normally (Matsumoto et al. 1993). The genome of a frog tumor can be also manipulated to give rise to a diversity of normal cell types (see next section). One may argue that there is a significant biologic principle relating to cancer when tumor cells from organisms as diverse as plants, goldfish, and frogs have similar competence to produce mitotic descendants that differentiate normally (see also discussion of differentiation in Chapter 1).

6.4b Amphibia

Although amphibian tumors are thought to occur less frequently than fish neoplasms (Harshbarger et al. 1981), amphibians may nevertheless also serve as organisms that monitor the environment for the benefit of humans. For example, a high frequency of tumors was described in tiger salamanders, *Ambystoma tigrinum,* which survived in a sewage lagoon in Texas (Rose and Harshbarger 1977). Other examples include lesions, thought to be malignant melanoma, that have been reported to occur in the newt, *Triturus cristatus,* obtained from several areas in Italy (Zavanella 1974) and a fibrosarcoma of the giant frog of Africa, *Gigantorana goliath* (Frye, Gillespie, and Maruska 1991).

In the United States, cancer of the pancreas is the fourth leading cause of death by a malignancy in the United States. Humans with pancreatic cancer have an extremely short survival time after diagnosis (Parker et al. 1997). Because of the lethality of human pancreatic cancer, it may be reassuring to learn that pancreatic carcinomas of Japanese, Chinese, and Korean pond frog (*Rana nigromaculata* group) hybrids (Masahito et al. 1994, 1995) have also been reported (Figure 6–1). The study of spontaneous pancreatic cancers in frogs may lead to enhanced understanding of the very malignant human equivalent.

The renal adenocarcinoma of the leopard frog, *Rana pipiens* (Figure 6–2), was originally described by Lucké (1934). Scarpelli (1975), when writing about tumors of lower animals, said the Lucké tumor deserves "special mention" because of its

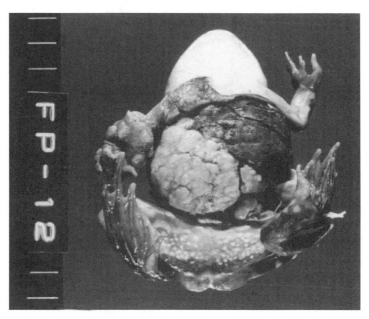

Figure 6–1. A pancreatic carcinoma in a hybrid frog. (From Masahito, Prince, Nishioka, M., Ueda, H., Kato, Y., Yamazaki, I., Nomura, K., Sugano, H., and Kitagawa, T. 1994. Frequent development of pancreatic carcinomas in the *Rana nigromaculata* group. *Proceedings 8th Int Conf Int Soc Diff,* pp. 183–186, Hiroshima, Japan: International Society of Differentiation, Inc.)

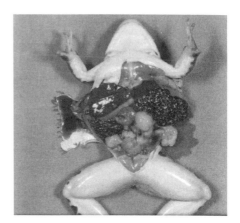

Figure 6–2. Sexually mature female northern leopard frog, *Rana pipiens,* with Lucké renal adenocarcinoma. (Photograph from author's files.)

major contributions to knowledge concerning the biology of cancer. The frog tumor was the first neoplasm associated with a herpesvirus (Lucké 1938), and that virus was subsequently shown to be the etiologic agent of the tumor (Naegele, Granoff, and Darlington 1974). The Lucké tumor herpesvirus (formerly designated "LTHV" but now redesignated "RaHV-1"; Roizman et al. 1995) has been observed in electron micrographs (McKinnell and Cunningham 1982) (Figure 6–3), and the viral DNA has been purified and sized by field inversion gel electrophoresis

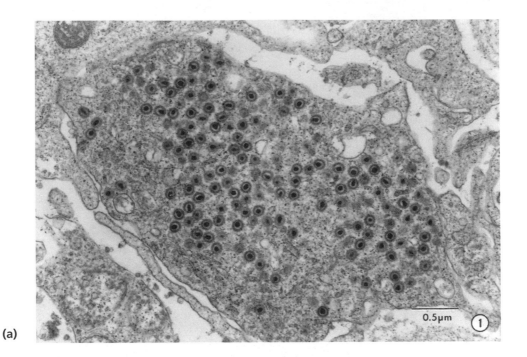

(a)

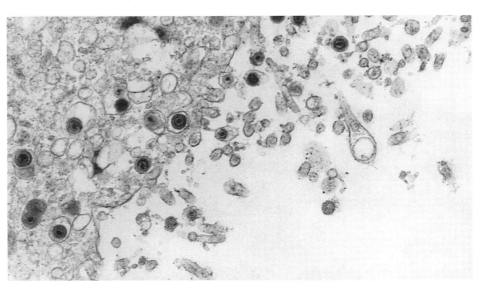

(b)

Figure 6–3. Electron microscope illustrations. (a) RaHV-1 viruses in the cytoplasm of a Lucké renal adenocarcinoma. (From Sauerbier, W., Rollins-Smith, L.A., Carlson, D.L., Williams, C.S. Williams III, J.W., and McKinnell, R.G. 1995. Sizing of the Lucké tumor herpesvirus genome by field inversion gel electrophoresis and restriction analysis. *Herpetopathologia* 2: 137–143.) (b) RaHV-1 viruses of a spontaneous renal carcinoma that had been maintained at 4° C in the laboratory for several months. The viruses are found in both tumor cell cytoplasm and the extracellular environment. The tumor was obtained from a frog in the author's laboratory; the electron microscope illustration is courtesy of Carol S. Williams of Tuskegee University.

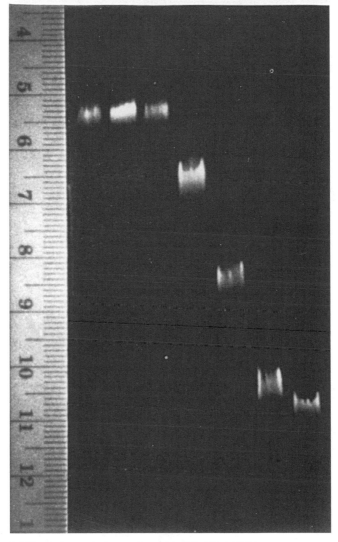

Figure 6–4. Field inversion gel electrophoresis of RaHV-1 DNA on an agarose gel (lane 2). Lanes 1 and 3: DNA from chromosome 1 of *Saccharomyces cerevisiae* with a molecular size of 220,000 bp. The *S. cerevisiae* DNA flank and comigrate with the RaHV-1 DNA, giving a size estimate of RaHV-1 DNA of 220,000 bp. Lanes 4, 5, 6, and 7 are other size markers. (From Sauerbier, W., Rollins-Smith, L.A., Carlson, D.L., Williams, C.S., Williams III, J.W., and McKinnell, R.G. 1995. Sizing of the Lucké tumor herpesvirus genome by field inversion gel electrophoresis and restriction analysis. *Herpetopathologia* 2:137–143.)

(McKinnell et al. 1991, 1993). Genomic size of RaHV-1 was estimated to be approximately 220 kbp (Figure 6–4) (Sauerbier et al. 1995). More recently, a map has been prepared showing the order of BamHI restriction enzyme fragments with a more recent estimate of genomic size of 217.14 kbp (Andrew Davison, personal communication). The genomic calculations, produced by different procedures, are similar and provide confidence that the size estimates are reliable and essentially correct (McKinnell and Carlson 1997).

Herpesviruses have been associated with other animal and human cancers (Howley et al. 1997). It has been suggested that a herpesvirus is likely involved in the etiology of Kaposi's sarcoma (Kedes et al. 1996). Knowledge concerning the oncogenicity of herpesviruses began with frogs. Perhaps knowledge of virus/cancer cell interaction in frogs may someday provide insight into herpesviruses and their relationship to human cancer cells.

The presence of viral DNA sequences in normal as well as tumor tissue and cell lines has been sought by amplification of a restriction enzyme fragment of that DNA with the polymerase chain reaction (Carlson et al. 1994a, 1994b, 1995; McKinnell et al. 1995; Williams et al. 1996).

Nuclear transplantation was developed in R. pipiens as a means of characterizing the developmental potential of somatic cell genomes (Briggs and King 1952; DiBerardino 1997; DiBerardino and McKinnell 1997). Successful nuclear transplantation experiments mandate a normal or near-normal complement of chromosomes. For that reason, chromosomes of the Lucké tumor were studied, and most cells were found to be euploid or nearly euploid (DiBerardino, King, and McKinnell 1963; Williams et al. 1993; Carlson et al. 1995; see also Chapter 4). A study of the capacity of the genome of the Lucké renal adenocarcinoma to promote development was undertaken (King and McKinnell 1960). Instead of forming a small population of cancer cells, tumor nuclei transplanted into enucleated ova (ova or eggs from which the maternal genes have been removed) formed early embryos (Figure 6–5) that were abnormal primarily because they failed to reach maturity (McKinnell, Deggins, and Labat 1969). A significant enhancement of the differentiation and growth of a number of tissue types was reported in grafts of tumor nuclear transplant tissue (Figure 6–6) (Lust et al. 1991; McKinnell et al. 1993; McKinnell 1994). The investigators believed that death of the nuclear transplant animal might be due to improper activation of one or a few genes of the embryo but that other genes necessary for normal development were present and could be expressed if the tissue was placed in an optimal environment. Grafting fragments of the tumor nuclear transplant animal to a normal host was thought to enhance the microenvironment of the tissue fragment. In other words, the limited development of nuclear transplant embryos was enhanced by grafting fragments of the tissue to normal hosts. These nuclear transplant studies reveal clearly that the genome of the Lucké renal adenocarcinoma has the competence to direct the differentiation of a number of cell types in addition to neoplastic cells. In a sense, these frog renal adenocarcinoma studies and the studies of goldfish pigment cell tumors complement the studies of normally differentiated plant cells obtained

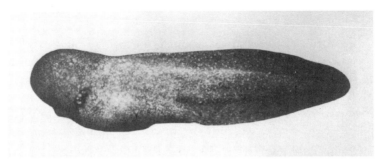

Figure 6–5. Swimming tadpole produced from the insertion of a Lucké renal adenocarcinoma nucleus into an activated and enucleated ovum of the northern leopard frog, *Rana pipiens*. The tadpole swims, has a beating heart, with well-formed body and tail. Abnormal primarily because it fails to feed, the tadpole dies shortly after reaching the swimming stage. (From McKinnell, R.G. 1973. Nuclear transplantation. In *Seventh National Cancer Conference Proceedings,* pp. 65–72. Philadelphia: Lippincott-Raven publishers.)

Figure 6–6. Histology of an eye primordium, obtained from a tumor nuclear transplant tadpole similar to that shown in Figure 6–5, allografted to the tail of a normal tadpole. The grafted eye rudiment differentiates a normal-appearing lens and retina. (Courtesy of Dr. J. M. Lust.)

from the crown gall tumor. Because normally differentiated mitotic progeny have been obtained from both plant and animal tumors, it would be expected that other tumor types could be manipulated to give rise to normally differentiated cells and tissue: they do (McKinnell 1989a, 1989b; see differentiation of malignant stem cells, Chapter 1).

6.4c   Reptiles

Laboratory animals tend to be small, inexpensive, and easy to manage. Perhaps because reptiles are *not* commonly used as laboratory animals and therefore may be scrutinized for cancer less often than other animals, relatively few cancers have been reported. Nevertheless, neoplasms have been reported in the Boa constrictor (*Boa constrictor*), the Komodo dragon (*Varanus komodoensis*), the California king snake (*Lampropeltis getulus*), the marsh terrapin (*Pelomedusa subrufa*), the rattlesnake (*Crotalus horridus*), the rock python (*Python molurus*), and the box turtle (*Terrapene carolina*). Malignancies of reptiles are cataloged in Harshbarger et al. (1981) and Zwart and Harshbarger (1991).

6.5   Cancer in selected ectothermic (warm-blooded) vertebrates

6.5a   Birds

Two bird cancers are discussed in this section. The first bird neoplasm to be considered, Marek's disease of chickens (Calnek 1992), is worthy of note for several reasons. This lymphoproliferative neoplasm was of considerable importance to farmers because of the financial loss associated with the disease (estimated to be about $200 million per year in the United States in the late 1960s). The economic importance of the neoplasm led to a search for its causative agent, which turned out to be the Marek's disease herpesvirus(es) (formerly MDHV; now known as GaHV-2 and GaHV-3; Roizman et al. 1995) (Rong-Fu et al. 1993). Perhaps most striking is the fact that a vaccine was developed for the cancer. The vaccine is prepared from a herpesvirus of turkeys, which is highly effective in reducing mortality in inoculated chicks with a resultant economic gain (Payne 1994). The efficacy of the chicken vaccine led one biologist to exclaim with gusto that the vaccine was "worth crowing about!"

Another bird neoplasm, originally described in barred Plymouth Rock chickens, is named Rous sarcoma for the person who discovered the tumor and the agent that causes it (Rous 1911). Peyton Rous of the Rockefeller Institute showed that the sarcoma of chickens could be transmitted by means of a cell-free filtrate. The studies of Rous were difficult to ignore because they were carefully done, but they were not in harmony with the then current beliefs concerning the origin of "spontaneous" tumors. Rous clearly showed that something (a virus) was in his cell-free preparations, and it was this "something" that caused the sarcoma of chickens. How could a malignancy be spontaneous if the malignancy had a virus etiology? Even after a half century, Rous was thought to have misled cancer researchers by insisting that cancer could be caused by a virus. However, at age 87, Rous received the Nobel Prize for his seminal observation concerning viral causation of some cancers – resulting from landmark experiments made on a tumor in chickens.

Figure 6–7. Mouse with a spontaneous (i.e., not induced by humans) mammary carcinoma. (From Gross, L. 1970. *Oncogenic Viruses.* Oxford: Pergamon Press.)

6.5b Mammals

It has been stated that the development of a multicellular body results in vulnerability to cancer. There probably is much truth to that statement, which applies to mammalian species as it does to all other creatures. Thus, a complete review of malignancy in mammals would require a review of all mammals. Because that is not possible here, brief mention is made of certain neoplasms of special interest.

A marsupial, the Brazilian opossum *Monodelphis domestica,* develops melanoma on exposure to ultraviolet radiation, the only species of mammals other than humans to do so (Robinson et al. 1994). Interest in this animal model of melanoma is obviously enhanced by the increased prevalence of this lethal form of skin cancer in the human population (see melanoma in fish here and human melanoma in Chapter 7).

Mice are better known with respect to their genetics than any other mammal. Furthermore, the creatures are small, well adapted to laboratory husbandry, and have been studied extensively. For example, the mouse mammary carcinoma (Figure 6–7) was shown to be transmitted via an agent (a virus) in milk (Bittner 1937), marking another historical landmark in viral oncology. More recently, the mouse mammary carcinoma has been used in experimental studies of metastasis in which the production of collagenase was shown to be an important factor (Tarin, Hoyt, and Evans 1982; Tarin 1992).

Teratocarcinomas, a kind of "germ cell tumor" (see appendix), occur spontaneously in certain strains of mice (Stevens 1973), or they may occur as the result of grafting embryos to adult structures (Stevens 1970). These mouse tumors are comprised of many kinds of cells and tissues including nervous system, cartilage, bone, skin, striated muscle, and respiratory and gut mucosa (Stevens 1981). The arrangement of these tissues seemingly without regard to each other was described

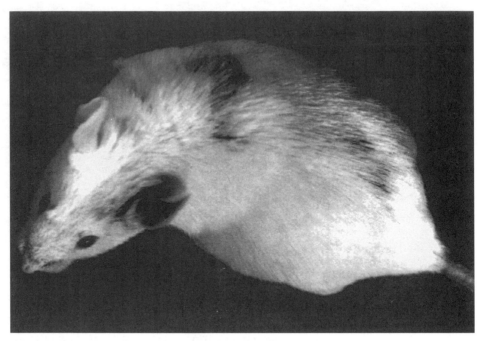

Figure 6–8. Albino mouse with pigmented hair at anterior base of left ear and several locations on the right flank and thorax. The normal pigmented hair developed from embryonal carcinoma cells of agouti (pigmented) genotype. This is a now classic example of normal cell differentiation ensuing from malignant cell mitotic progeny. (From Brinster, R. 1993. Stem cells and transgenic mice in the study of development. *Int J Dev Biol* 37:89–99.)

as an "approximation to chaos" (Needham 1942). Mouse teratocarcinoma cells have been used in a great diversity of experiments demonstrating differentiation competency (Figure 6–8) (see discussion of the induction of differentiation in the mitotic progeny of plant, fish, and frog tumors presented earlier in this chapter, also Chapter 1 and Pierce, Dixon, and Verney 1960; Pierce 1970; Damjanov 1993; see also Tienari et al. 1995; Jiang et al. 1995). Similarly, mouse myeloid leukemic cells have the potential to differentiate (Jimenez and Yunis 1987).

Cancer occurs spontaneously in dogs, which have about twice the prevalence of cancer as their owners. It has been stated that canine malignancies have been underutilized as a resource in the study of cancer therapy (Hahn et al. 1994). Tumors of dogs, which occur commonly and share significant pathologic characteristics with their human counterparts, include lymphoma, mammary neoplasia, melanoma, and lung cancer. Although rodents are small and relatively easy to rear in the laboratory, dogs have a longer life expectancy, share food and lodging with their owners, and when they become afflicted with neoplastic disease, they often are treated. Studying the response of a dog to chemotherapy, which may lead to

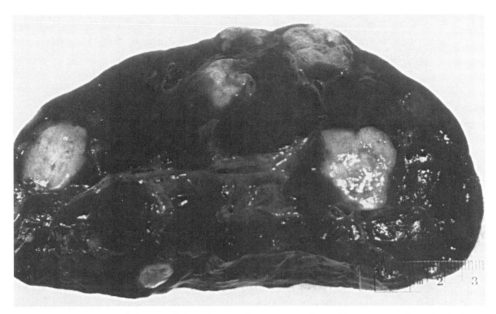

Figure 6–9. Multiple white nodules of a renal carcinoma found in a mature female California sea lion, *Zalophus californianus*. The malignancy was anaplastic and widely metastatic. (Reprinted with permission from Howard, E.B., Britt, Jr., J.O., and Simpson, J.G. 1983. Neoplasms in marine mammals. In E.B. Howard, ed., *Pathobiology of Marine Mammal Diseases,* Vol. II, p. 146. Copyright © 1983 CRC Press, Boca Raton, FL.)

better treatment options for other dogs and humans, may be the last friendly gesture of a devoted companion.

Cats are vulnerable to feline leukemia, which results from infection with the feline leukemia virus (Jarrett et al. 1964). Actually, the viral infection causes a number of conditions in cats in addition to leukemia, among which are lymphosarcoma and immune deficiency syndrome. A virus antigen, known as FeLV-gp70, has been developed as a vaccine to provide protection for cats against the feline leukemia virus (Lewis et al. 1988). If the development of a vaccine against a chicken cancer "is worth crowing about," then one might observe that a protective vaccine for felines should be referred to as "the cat's meow."

Metastatic transitional cell carcinoma of the urinary bladder, as well as other tumors, have been reported from the great white beluga whale (*Delphinapterus leucas*). The tumors were noted at autopsy of stranded carcasses (Martineau et al. 1985; De Guise, Lagace, and Beland 1994). Figure 6–9 illustrates a kidney carcinoma of a California sea lion, *Zalophus californianus.* Metastatic mammary carcinoma was reported in a horse (Munson 1987), and a diversity of neoplasms are known from cattle and other domestic animals (Naghshineh, Hagdoost, and Mokhber-Dezfuli 1991; Rostami et al. 1994).

It is impractical, as stated earlier, to provide a catalog of mammalian cancer. For more information on the malignancies of mammals, refer to other sources (e.g., Moulton 1978; Turusov and Mohr 1994).

6.6 Summary

The rationale for a study of cancers in organisms other than humans is explained nicely in a paragraph taken from the preface written for a symposium volume by Clyde Dawe et al. (1981). This chapter began with Dawe and it ends with Dawe – only because he phrased it so well.

By retracing the steps followed in the phylogenetic development of regulatory mechanisms, are we not likely to find understanding of these mechanisms, just as developmental biologists expect to reach such understanding by retracing ontogenetic development? Regrettably, it does not necessarily follow that with understanding will come control of neoplasia. An escape from this uncomfortable thought lies in the advice of Michael Faraday: "But try anyway; no man knows what is possible."

7

Epidemiology

Robert G. McKinnell

The most desirable way of eliminating the impact of cancer in humans is by prevention.

L. W. Wattenberg 1985

The ultimate goal in the control of any disease is prevention, and so it is with cancer.

Y. Hayashi et al. 1986

The evidence suggests that about one third of the 500,000 cancer deaths that occur in the United States each year is due to dietary factors. Another third is due to cigarette smoking. Therefore, for the large majority of Americans who do not smoke cigarettes, dietary choices and physical activity become the most important modifiable determinants of cancer risk.

American Cancer Society 1996 Advisory Committee on
Diet, Nutrition, and Cancer Prevention

7.1 Introduction

Epidemiology is the study of the distribution and prevalence of disease. Risk factors identified in epidemiologic studies can be used to prevent or control neoplasia. These studies include an examination of the effects of occupational exposure to carcinogens as well as lifestyle activities that tend to make certain individuals prone to cancer. A practical example, already widely recognized, relates to lung cancer and smoking. Lung cancer does not randomly afflict individuals within a population. Rather, it appears disproportionately in people who smoke cigarettes. As smoking has increased, so, too, have the incidence and deaths due to lung cancer (Figures 7–1 to 7–3). Identifying tobacco as a risk factor for lung cancer has implicit with that knowledge the potential for prevention of much of the disease. It follows that lung cancer prevalence will be reduced if cigarette consumption is

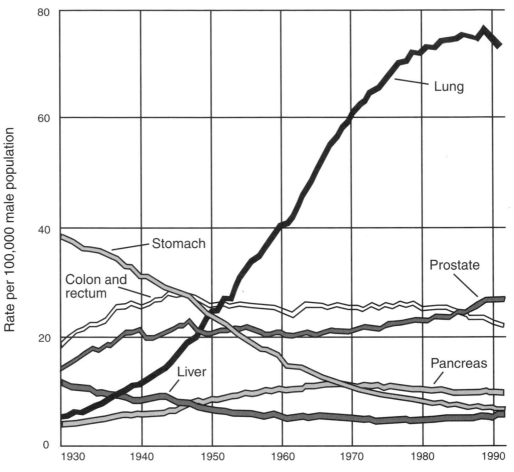

Figure 7–1. Age-adjusted cancer death rates for males of the United States. Rates per 100,000 are indicated on the ordinate, and years are indicated on the abscissa. (From *Cancer Facts and Figures – 1997,* American Cancer Society.)

decreased. This, in brief, illustrates some of the problems and issues that are of concern to individuals who study human cancer epidemiology.

Epidemiologists are aware that cancer strikes neither capriciously nor randomly. The risk for specific cancers varies with different populations, and maps have been constructed that identify populations at varying risk for specific malignancies (Figures 7– 4 and 7–5). The differences in cancer prevalence suggest that perhaps 75% to 80% of cancers in the United States are due to factors in the environment (Doll and Peto 1981). "Environment" for the purposes of this chapter includes the personal environment of diet, exercise, and other factors such as exposure to the sun and tobacco use. Breast cancer is 7 times more frequent among Hawaiian women than among non-Jewish women of Israel. Prostate cancer in African American males of Atlanta is 70 times more common than in males of Tianjin, China (Frau-

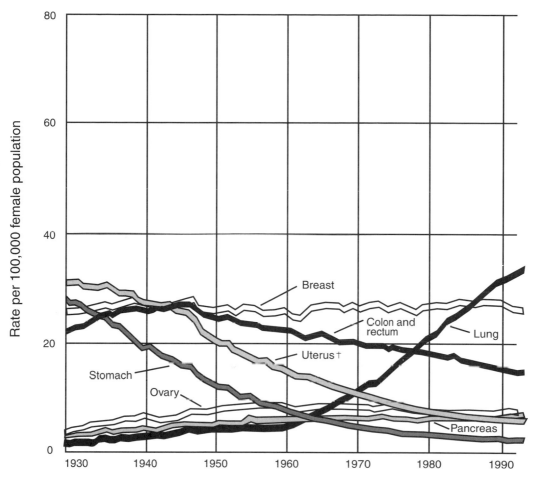

Figure 7–2. Age-adjusted cancer death rates for females of the United States. Rates per 100,000 are indicated on the ordinate, and years are indicated on the abscissa. Uterine cancer death rates are for cervical cancer and cancer of the corpus combined. (From *Cancer Facts and Figures – 1997*, American Cancer Society.)

meni et al. 1993). Perhaps of greater impact are the studies of immigrants who rapidly assume the cancer rates of their adopted country. For example, gastric cancer is common in Japan. Japanese who immigrate to the United States undergo a marked dietary change and experience a decrease in stomach cancer rate. The offspring of the immigrants have an even greater decrease in stomach cancer mortality. In contrast, Japanese who move to California experience an increased colon cancer incidence presumably associated with dietary change (Dunn 1975; Wogan 1986). For these reasons, the environment is considered to be a major factor in cancer vulnerability.

About 1,228,600 new cases and approximately 560,000 deaths from cancer in the United States will occur in the year 1998 (Landis et al. 1998). More people will die of heart disease. Thus, to the uninitiated, heart disease is more important

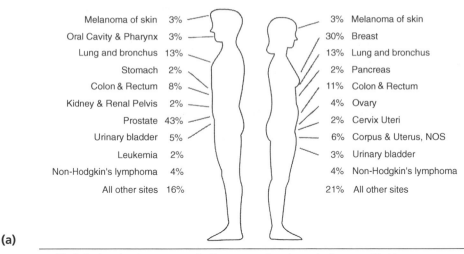

(a)

*Excludes basal and squamous cell skin cancer and carcinoma in situ except bladder.

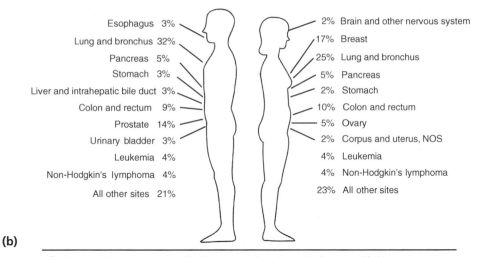

(b)

*Excludes basal and squamous cell skin cancer and carcinoma in situ except bladder.

Figure 7–3. (a) Estimated new cancer cases and (b) estimated cancer deaths in 10 leading sites for male and females of the United States, 1997. (From Parker, S.L., et al. 1997. Cancer statistics, 1997. *CA Cancer J Clin* 47:12.)

than cancer. A disproportionate number of older people die of heart disease. Cancer, in contrast, may afflict the very young (as well as older people, of course). Hence, for individuals under the age of 65, the number of years of human life lost due to disease is greater for cancer than for heart disease. Compare years of life lost by a young child with leukemia with years of life lost by an elderly person who dies of a heart attack.

The reduced prevalence, or absence, of a number of diseases other than cancer

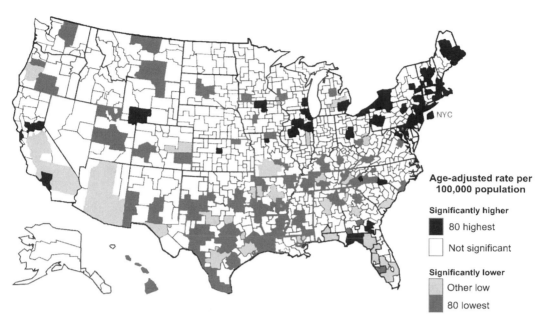

Figure 7–4. Breast cancer death rates of white females, 1988 to 1992. The map indicates that southern states have lower rates of breast cancer than the urban northeast, Chicago and vicinity, and metropolitan areas of California. (From Pickle, L.W., et al. 1996. *Atlas of United States Mortality,* p. 69. Hyattsville, MD: National Center for Health Statistics.)

speaks eloquently of the efficacy of control by prevention. Examples of diseases particularly appropriate in this context include pellagra, smallpox, scurvy, cholera, typhoid fever, and typhus. Many other examples could be cited, of course, but these are striking both because of their lethality *and* their present rarity. These diseases are currently so infrequent that few Americans of college age have had any experience with them. The absence or very low prevalence of these and a number of other serious illnesses results from their *prevention*. Public health measures such as vaccinations and attention to the purity of the food and water supply have been extraordinarily effective in disease control. Perhaps because of their complex and poorly understood etiologies (see carcinogenesis, Chapter 3), cancers are almost alone as a group of diseases for which preventive measures have received scant attention. This neglect to the area of prevention is rapidly being remedied, and there is now much research into cancer prevention. There is heightened anticipation that such a preventive effort will be rewarded with reduced cancer frequency.

An impetus to this research is the stated goal of the National Cancer Institute to reduce cancer deaths by 50% by the year 2000 (Butrum, Clifford, and Lanza 1988). It is unlikely that cancer cures by chemotherapy (Chapter 8) will improve sufficiently in the next few years to reduce the cancer death rate significantly. Although mortality due to stomach and uterine cancers is decreasing, lung cancer

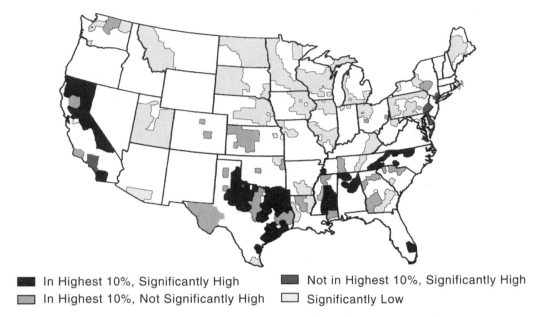

In Highest 10%, Significantly High   Not in Highest 10%, Significantly High
In Highest 10%, Not Significantly High   Significantly Low

Figure 7–5. Melanoma cancer death rates of white males, 1970 to 1980, is not random. Highest rates of this skin cancer occur in southern states and wide areas of California. (From Pickle, L.W., et al. 1987. *Atlas of U.S. Cancer Mortality Among Whites,* p. 83. Bethesda, MD: Public Health Service, National Institutes of Health.)

mortality is increasing in white females, and melanoma incidence is increasing among white males (Trichopoulos et al. 1997; Landis et al. 1998). If not through cure, then by what means will a decrease in cancer mortality occur by the year 2000? Prevention has the possibility of leading to much of the sought for reduction in the cancer death rate.

7.2   Cancer in fossil humans: A brief digression concerning paleopathology

It is sometimes suggested that cancer is caused by the pollution of food, air, and water due to modern technology. Some argue that we could avoid cancer altogether if we would only go back to a more "simple time." Cancer detected in fossil humans informs us, however, that not all cancer is attributable to carcinogens produced by chemical industries and power plants of the late twentieth century. Such a discussion belongs in a chapter on epidemiology because it brings a perspective to the considerations of what causes cancer and who is vulnerable.

Multiple myeloma is a malignant disease with an ancient past. It is a disease of bone marrow cells that has the competence to invade and destroy calcified bone.

When it does so, the osteolytic lesions leave a lasting record in the bones of those unfortunate individuals who manifest the disease. Holes in the skull and other bones are formed with sharply defined borders that have been described as appearing "punched out." Of course, after several thousands of years most people leave nothing of their mortal remains. However, in exceptional cases, fossil bones are discovered. Because multiple myeloma may leave its mark in bones, researchers seek evidence of this neoplasm in fossil human bones. Human remains with evidence of multiple myeloma have been described from such diverse locations as Kérpustza, Hungary; Ipswich, England; Indian Knoll, Kentucky; and Kane Mound, Missouri. These paleopathologic specimens vary in age from 3,000 to 5,000 years (Morse, Dailey, and Bunn 1976). They are witness to the antiquity of human cancer in general and to multiple myeloma in particular.

Other malignancies have been reported in fossils, including nasopharyngeal carcinoma in a mummy, osteosarcoma in Celtic warriors (Ortner and Putschar 1981), and metastatic melanoma. Let the skeletal remains described in these few paragraphs remove the naive notion that cancer is only a disease of contemporary society.

7.3   Epidemiology of selected human cancers

Clearly, only a few cancers can be considered here. They were judged to merit consideration for differing reasons. Lung cancer is increasing among women and has passed breast cancer as a cause of death. Skin cancer is the most common malignancy in the United States and deaths from one form, malignant melanoma, are increasing faster than any other lethal cancer. The reason for inclusion of other cancers in this section will become obvious upon reading.

7.3a   Lung cancer

"Bronchial carcinoma became the most common cause of death from cancer in the world in the early 1980s" (Doll 1992). The neoplasm has been described as "pandemic" in most of the world (Higginson, Muir, and Munoz 1992). "Tobacco smoking is the single most important cause of lung cancer and, in fact, of all human cancer considered as a group" (Trichopoulos et al. 1997). Most lung cancer deaths are preventable if people do not smoke cigarettes. Prior to World War I, lung cancer was rare. Deaths due to lung cancer soared among males in the United States after World War I (Figure 7–1), and, more recently, death rates have increased in American women (Figure 7–2). Lung cancer is the leading cause of cancer death in both males and females in the United States (Figure 7–3). Consumption of tobacco in the form of cigarettes surged during and after World War II. Probably 95% of lung cancer in men is attributable to smoking, and the

93,100 American men who will die of the disease in 1998 comprise about 32% of all male cancer deaths (Landis et al. 1998). An additional 67,000 American women will die of lung cancer in 1998, making a total of 160,100 largely unnecessary American deaths. About two-thirds of Chinese males smoke; they also smoke at an increasingly younger age, and despite their knowledge of the hazards of tobacco consumption, they evidence little interest or desire to quit (Gong et al. 1995). China, the world's leading tobacco grower, clearly has an incipient health problem of colossal proportions.

Smoking is estimated to either cause or contribute to the deaths of 400,000 Americans per year (McGinnis and Foege 1993). That figure includes, in addition to cancer, heart disease and stroke-caused deaths attributed to smoking.

Lung cancer causes the loss of 1 million years of life each year (i.e., if the loss of life measured in years of each smoker who dies of lung cancer per year were added together, the toll would be about 1 million years of life lost – approximately 6 years of life lost by every person who dies of lung cancer in the United States). The overall cancer death rate would be declining except for the annual increase in the lung cancer death rate.

The magnitude of the drop in cancer mortality that would occur if Americans (and all other citizens of the world) ceased smoking could not be matched by any treatment of surgery, radiation, or chemotherapy currently available, being developed, or even imagined by cancer research workers.

The concern for tobacco as it affects health is not new. In 1761, John Hill of London admonished against the use of snuff, reporting that it caused cancers of the nose. This warning was followed by the report of carcinoma of the lips "where men indulge in pipe smoking" by Sammuel Thomas von Soemmering of Mainz in 1795 (Shimkin 1977).

Prior to the present, lung cancer was uncommon. In fact, Adler (1912) stated that "primary neoplasms of the lung are among the rarest forms of the disease." However, lung cancer became increasingly common after World War I, and a number of observers correctly noted a probable relationship between cigarette smoking and the increased rate of lung cancer (Tylecote 1927; Mertens 1930; McNally 1932); worthy of special note were Ochsner and DeBakey (1939, 1940, 1941). Epidemiologic studies followed (Wynder and Graham 1950; Doll and Bradford-Hill 1950) and showed that the lung cancer death rate is dose dependent; that is, it is directly related to the number of cigarettes smoked, and lung cancer incidence would be reduced by 80% to 90% if smoking could be eliminated. These and other major studies were reviewed by the Advisory Committee to the Surgeon General (1964), which reported that "the risk of developing lung cancer increases with duration of smoking and the number of cigarettes smoked per day, and is diminished by discontinuing smoking. In comparison with nonsmokers, average male smokers have approximately a 9- to 10-fold increased risk of developing lung cancer and heavy smokers at least a 20-fold increased risk."

Furthermore, cigarette smoking greatly increases risk when combined with other hazards. For example, asbestos workers who smoke are far more vulnerable to lung cancer than their nonsmoking coworkers or smokers who do not work with asbestos (Selikoff, Seidman, and Hammond 1980).

The association between involuntary exposure to tobacco smoke, that is, the passive breathing of other people's smoke, and lung cancer has been shown in most studies to be positive and dose dependent (Lofroth 1989; Trichopoulos et al. 1997). Smokers are known to excrete mutagens in their urine (Yamasaki and Ames 1977). Nonsmokers exposed to heavy smoke (e.g., bartenders) are reported to have elevated levels of carcinogens in their blood (Maclure et al. 1989). The urine of nonsmokers who are exposed to environmental tobacco smoke has been found to be mutagenic (as is the urine of smokers) (Sora et al. 1985).

7.3b   Breast cancer

Breast cancer in women is second only to lung cancer as a cause of death by malignancy in the United States (Figure 7–3). Although breast cancer is the leading cause of cancer death in women aged 15 to 54, lung cancer becomes the primary cancer killer from age 55 through 74 and still exceeds breast cancer in women 75 and older. It is estimated that 43,900 women in 1998 will die of breast cancer in the United States. The lung cancer toll will be 67,000. Although breast cancer was formerly the principal cause of death by cancer in women, its current position as second to lung cancer is *not* a result of decreased prevalence of the malignancy. Rather, it is the sharply *increased* mortality due to lung cancer. Between two and three times more women are afflicted with breast cancer than with lung cancer. However, it is the *lethality* of lung cancer that accounts for its greater death rate (Landis et al. 1998).

Men will suffer from about 1,600 of the 180,300 new breast cancer cases expected in 1998 (Landis et al. 1998). American women of Japanese and Chinese ancestry as well as American Indian women have less than half the breast cancer mortality of their white and African American compatriots.

By no means does the United States lead in breast cancer mortality. Women of Ireland, Denmark, the Netherlands, the United Kingdom, and Israel (in descending order; 1992–1995 data) all suffer from a higher breast cancer mortality than U.S. women. In contrast, China has significantly less breast cancer; its mortality rate is about one-fifth that of the United Kingdom (Landis et al. 1998).

A long history concerns this particular malignancy. Bernardino Ramazzini noted almost three centuries ago that breast cancer was more common among nuns than among other women (Shimkin 1977). His observations were correct, but not because the religious life renders nuns more vulnerable than other women to breast cancer. A "protective" role of marriage was suggested. However, it is not marriage per se that is protective. Rather, not having children (nulliparity) is associated with increased prevalence of breast cancer among women. The risk factor of nulliparity, or of having children after 35 years of age, is as real now as it was when

Ramazzini first wrote about the peculiar vulnerability of nuns to breast cancer. Most of the increased risk associated with nulliparity is for postmenopausal breast cancer.

As stated earlier, epidemiologists search for risk factors in the hope that they may give insight into the causes of cancer and provide means of reducing cancer prevalence. In this discussion, *cause* is used in the sense of that which produces a consequence. A risk factor has the potential of inducing a consequence (in this instance, cancer). *Cause* is not used here in the sense of a change in a nucleotide sequence or altered transcription. Risk factors are unlikely ever to reveal mechanisms of fundamental changes in cell activity. An individual at increased risk will not necessarily succumb to cancer. Rather, the risk factor simply identifies individuals who are at higher risk than others who are not in that category. Consider breast cancer and the nuns studied by Ramazzini. Although they had (and still have) a higher prevalence of breast cancer than women who are not nuns, remember that the great majority of nuns die of conditions other than breast cancer. And so it is with the other risk factors listed next.

Reproductive history is an important risk factor, which in turn implicates hormonal factors in the genesis of breast cancer. Age at the birth of the first child is a risk factor. Women who have children prior to age 20 have a reduced risk of breast cancer when compared to women who have children after age 35. Associated with reproduction is lactation and breast feeding. Obviously, women who do not have children do not breast-feed. A moderate reduction of risk with breast feeding for extended time has been documented (McTiernan and Thomas 1986). Early menarche (the first menstrual period) and late menopause are also factors that increase risk for breast cancer. There may be an increased risk for premenopausal breast cancer with long-term use of oral contraceptives (Henderson, Bernstein, and Ross 1997; Trichopoulos et al. 1997).

On a worldwide basis, fat in the diet is believed to be positively related to breast cancer (Figure 7–6). However, within the United States, little evidence supports the role of dietary fat in promoting breast cancer. The availability of fats, and other nutrients, leads to increased height during the growing years, and the increased height (and weight) is correlated positively to breast cancer risk. Further, an energy-rich diet may lead to obesity, which is similarly correlated positively to breast cancer after menopause. Dietary fat as it relates to breast and other cancers is discussed at greater length in the sections on diet, nutrition, and cancer.

Tobacco use, although extraordinarily important in the genesis of lung cancer (section 7.3a), probably has little effect on breast cancer prevalence. Alcohol in the diet is related to breast cancer (section 7.4i), although some have speculated that individuals who drink frequently also smoke, and there may be a synergistic effect of the two.

Geographic pathology is concerned with mapping distributions of cancer prevalence. Breast cancer is far from random in its geographic distribution. Women of the northeast United States, Chicago and vicinity, and the metropoli-

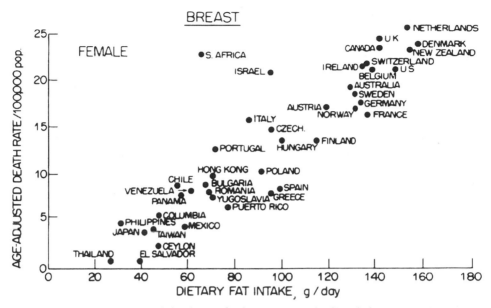

Figure 7–6. The age-adjusted death rate for breast cancer (ordinate) shows a positive correlation with the average amount of fat consumed per day per person (abscissa). (From Carroll, K.K., and Khor, H.T. 1975. Dietary fat in relation to tumorigenesis. *Prog Biochem Pharmacol* 10:331. Reprinted by permission of Karger, Basel.)

tan areas of California are more vulnerable to breast cancer than women who live in the South, Midwest, and West (except for parts of California previously mentioned) (Figure 7–4) (Pickle et al. 1996). Affluent white women in the United States have a higher prevalence of breast cancer than women of lower socioeconomic status. The higher prevalence of breast cancer in the North may well be related to the historically greater affluence of the North compared with other parts of the United States.

Age is a risk factor. Breast cancer is uncommon before age 30 but rises rapidly until age 40, after which time it continues to increase in prevalence but at a slower rate. In the United States, the likelihood of developing breast cancer is less than 1 in 200 prior to 40 years of age, but that likelihood increases to 1 in 15 between the ages of 60 and 79. Breast cancer comprises over 30% of the malignancies in women between the ages of 35 and 54 compared to less than 17% in women 15 through 34 years of age (Landis et al. 1998).

Heredity, however, probably is the most important breast cancer risk (Chapter 4). A woman with a mother or sister with a history of breast cancer may be at increased risk. For example, a woman whose mother or sister developed breast cancer before age 45 has a 3.85 times greater risk than women of the general population (Houlston et al. 1992).

Epidemiologic studies suggest the major determinant for breast cancer is the cumulative exposure of breast tissue to estrogens and progesterone. This exposure is mediated by reproductive history (nulliparous versus one or more pregnancies), lactation, duration of fertility (the time from menarche to menopause), and dietary factors (Higginson et al. 1992). But note that the majority of women in the highest risk categories do *not* develop breast cancer (Madigan et al. 1995), and the risk of dying of breast cancer is one-third the risk of developing that awesome malignancy. Surely students of epidemiology will, in time, seize on the nonrandomness of one or more risk factors for breast cancer of age, reproductive history, economic status, geographic origin, family history, and other considerations to discover factors that can be manipulated to lead to the ultimate control of this feared malignancy. In the meantime, perhaps the best advice to reduce the risk of breast cancer is to "limit intake of alcohol beverages, eat a diet rich in fruits and vegetables, be physically active, and avoid obesity" (American Cancer Society 1996).

7.3c   Skin cancer

Skin cancer is common and increasing. Of the three skin cancers, the prevalence of basal cell carcinoma greatly exceeds that of the less common squamous cell carcinoma and malignant melanoma. In the U.S. white population, about 75% to 80% of skin cancer is basal cell carcinoma; squamous cell carcinoma accounts for 20% to 25%. Nonmelanoma skin cancers are the most common cancers among whites. They are the most treatable forms of skin cancer with a high cure rate of 95% (probably due to the fact that they are easily noticed and early treatment is generally sought). Nonmelanoma skin cancers account for about 700,000 new cases per year (Safai 1997).

About 8,500 deaths per year are due to all skin cancers. Malignant melanoma accounts for 7,300 of those skin cancer deaths, making it by far the most lethal of the skin cancers (Landis et al. 1998).

Basal cell carcinoma (Figure 7–7) is more common in men than in women. Occurring usually on the upper part of the face, it was formerly associated with aging. With increased skin exposure to ultraviolet radiation (see discussion later), it now occurs in younger individuals. It is slow growing and extraordinarily late in its metastatic behavior (the metastatic potential has been described as "practically nil"). Nevertheless, the malignancy has the competence to invade and destroy normal tissues. Tumors that persist for a long time may form lesions known as "rodent ulcer."

Sporadic basal cell carcinomas occur in small numbers on the exposed skin of older patients. A genetic form of basal cell carcinoma occurs in large numbers in teenage individuals. This latter form is thought to be due to a mutated tumor suppressor gene located on chromosome 9 (Johnson et al. 1996).

Squamous cell carcinoma occurs in men more often than in women and arises primarily in skin exposed to the sun in a dose-dependent manner. Squamous cell carcinoma is a cancer of light-complexioned individuals.

Nonmelanoma skin cancer frequency varies enormously throughout the world.

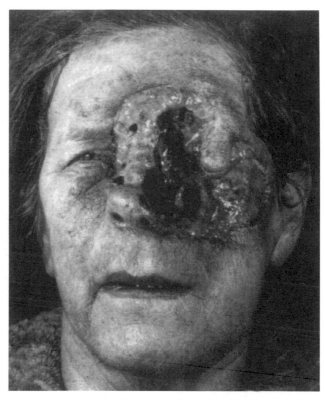

Figure 7–7. Basal cell carcinoma in a female patient. (From Walter, J.B. 1982. *An Introduction to the Principles of Disease,* p. 245. Philadelphia: Saunders.)

For instance, male citizens of Bombay, India, experience less than 1 percent of the skin cancer that afflicts the white males of Australia and Tasmania (compare 1.5 per 100,000 in the former country to 167 per 100,000 in the latter). Obviously, the indigenous people of Bombay are dark skinned, and the whites of Australia and Tasmania are descendants of British and other northern European migrants. Nonwhite Americans have less basal cell and squamous cell skin cancer than their fellow white Americans.

Malignant melanoma is a lethal cancer compared with the high survival rate of individuals who have basal cell carcinoma and squamous cell carcinoma. Malignant melanoma arises from pigment cells in the skin, and the prognosis is based, in part, on the thickness of the tumor and whether it has transgressed the epithelial basement membrane at biopsy. Almost all patients with melanoma less than 1 mm thick survive compared with less than 50% survival among those with tumors over 4.6 mm thick (see letter in Introduction). Alternative expressions of survival with lesions of varying thickness still convey the same message. For example, another author states that patients with melanomas less than 0.75 mm thick have a 97% to 99% survival rate compared with less than 50% survival rate if the lesion

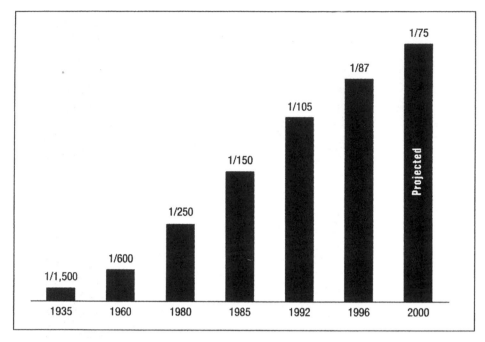

Figure 7–8. The lifetime risk of developing malignant melanoma has risen from 1 in 500 in 1935 to 1 in 87 by 1996. It is projected to increase to 1 in 75 by the year 2000, an increase of over 600% in two-thirds of a century. (From Rigel, D.S. 1996. Malignant melanoma: Perspectives on incidence and its effects on awareness, diagnosis, and treatment. *CA Cancer J Clin* 46:196. © American Cancer Society.)

is more than 3 mm. Regardless of which set of figures are used, the message is the same: survival is inversely related to lesion thickness (Rigel 1996).

Malignant melanoma was once uncommon, but it is now increasing rapidly (Figure 7–8) with a death toll in the United States of 7,300 expected in 1998 (Landis et al. 1998). It is increasing in the world faster than any other cancer. This is also true for men in the United States. Only lung cancer is increasing more rapidly in American women. About one-third of the victims of melanoma are women; the other two-thirds are male. Although susceptibility increases with age, a significant proportion of relatively young individuals die of malignant melanoma (years of life lost due to this malignancy are proportionally greater than years of life lost due to a cancer that afflicts primarily older people). As in basal cell and squamous cell skin cancers, solar exposure is believed to be of great importance in the induction of melanoma. Because solar radiation is more intense in southern latitudes than in the north, melanoma mortality is found predominantly in the south-central and southeastern states as well as parts of sunny California. There is no measurable ultraviolet radiation (due to overcast skies and the low angle of the sun with the horizon) during much of Minnesota's winter. The salutary effects of the sky and sun angle is apparent by scrutiny of the map of melanoma rates (Figure 7–5) (Pickle et al.

1987). The take home message is that lethality (especially of melanoma) can be largely eliminated by protection from sun exposure.

Lifestyle during the Great Depression precluded frequent sunbathing expeditions and prolonged exposure to the sun for most people. A retrospective scrutiny of recreational attire over the past half century reveals that more skin is exposed during the time spent in the outdoors now than in previous years. Combine these two considerations, that is, more time for recreation and clothing that exposes more skin, with a cultural value placed on a "healthy" tan in white people, and it is not surprising that the lifetime risk for malignant melanoma has increased from 1 in 1,500 in 1935 to about 1 in 100 (a 15-fold increase in about a half century) (Friedman et al. 1991), and melanoma is *projected* to occur in about 1 in 75 by the year 2000 (Rigel 1996). The projected prevalence is an increase of 20-fold since 1935.

Although malignant melanoma is the least frequent type of skin cancer, it is of considerable concern because of its rapidly increasing incidence (Figure 7–8). However, there are at least two salutary aspects of this potentially fatal cancer. The first is an enhanced survival of malignant melanoma patients (the five-year survival rate has doubled since the 1940s). The second aspect is that most people can easily minimize their exposure to ultraviolet radiation (exposure to the sun has been referred to as a "modifiable" activity). Despite the enhanced survival rate, there is a linear increase in deaths from malignant melanoma due to the exponential increase in incidence (Rigel, Kopf, and Friedman 1987; Rigel 1996) (Figure 7–8).

An ominous fear further adds to the already immense concern for ultraviolet exposure: the possibility of increased ultraviolet radiation associated with a diminished ozone layer. The ozone layer of the atmosphere provides protection from skin-damaging ultraviolet, and considerable concern has been expressed about loss of ozone protection with resultant increase in skin cancer (Kerr and McElroy 1993; see also Herman et al. 1996).

A long history relates exposure to the sun to skin damage and skin cancer (Hyde 1906; Findlay 1928; Roffo 1935). Light-complexioned people who burn easily, tan poorly, and freckle, such as those of Irish, Scot, and English origins, and who are regularly exposed to the sun (note letter to Jill from Amy in Introduction) are much more vulnerable to skin cancer than those who have darkly pigmented skins. Furthermore, the exposed portions of the body are more vulnerable than the portions of the body protected by clothing. For example, women who wore bikini bathing suits, or who swam nude, were 12.97 times more likely to develop malignant melanoma in the trunk region than women who wore one-piece bathing suits with high backs (Holman, Armstrong, and Heenan 1986). Fishermen, farmers, ranchers, and other outdoor workers are frequently afflicted with skin cancers on the backs of their hands, their necks, and their faces. Clothing is an important consideration for protection from exposure to ultraviolet radiation. Hats with brims are protective, and the broader the brim, the more protection. Dark, closely woven cloth protects more than white loosely woven material (a simple test for protection is to see how dense a shadow the cloth produces in bright sun; the more dense the shadow, the more protection the cloth provides from ultraviolet radiation). Commercially available sunscreens are available to provide further protection (Marks 1996).

Because the intensity of sunshine varies with latitude, southern peoples are more vulnerable than those from the north (Figure 7–5). The fact that skin cancer among African Americans is relatively rare gives rise to the notion that melanin pigment protects against ultraviolet damage and subsequent skin cancer. This hypothesis is substantially strengthened by the observation that albino blacks, who lack skin pigmentation, are particularly vulnerable to skin cancer (Luande, Henschke, and Mohammed 1985; Witkop et al. 1989).

Studies of animals and of the epidemiology of skin cancer have revealed it is the ultraviolet portion of solar radiation that causes skin cancer. Mice, rats, and opossums develop skin cancer when exposed to ultraviolet radiation (Findlay 1928; Roffo 1935; Kripke 1986; Robinson et al. 1994). Ultraviolet light consists of longer wave lengths of 320 to 400 nm known as ultraviolet A (UV-A) and shorter wave lengths of 280 to 320 nm known as ultraviolet B (UV-B). DNA absorbs light in the UV-B wavelengths, and it is these wavelengths that have been associated with skin cancer (Setlow 1974). Ozone in the earth's atmosphere, 6 to 30 miles above the earth's surface, filters out some of the UV-B, providing partial protection to exposed human populations. Ultraviolet C (UV-C) is completely absorbed by the ozone layer and seems to be of little or no concern to human health. The ozone layer may have been damaged by the release of chlorofluorocarbons, substances formerly used widely as refrigerants, insulating foams, and solvents, with a resulting increase in UV-B encountered on the surface of the earth (Kerr and McElroy 1993; Herman et al. 1996). Other ozone-depleting substances include the pesticide methyl bromide and halons used in fire extinguishers. The process of ozone depletion is continuing because it may take a decade or more after release of a potentially damaging substance for ozone diminution to occur.

Ultraviolet radiation is mutagenic. It can break DNA strands and induce the formation of cyclobutane pyrimidine dimers between adjacent pyrimidines (Setlow 1974; Haseltine 1983). It is this DNA damage that presumably results in carcinogenesis. Fortunately, most of the damage is repaired by cellular enzymes, which probably accounts for a lower cancer prevalence and mortality than might otherwise be expected with exposure to UV-B. Individuals who do not have the proper enzymes for DNA repair are particularly vulnerable to skin cancer (Cleaver and Kraemer 1989).

Moreover, studies with mice have shown that the immune response is diminished by ultraviolet irradiation (Kripke 1990; Kripke et al. 1992; Kripke and Yarosh 1994). Although mouse skin cancer cells are highly antigenic, the host immune response fails to impede their growth in UV-irradiated mice. The tumor does not persist when grafted to unirradiated syngeneic compatriots of the mouse with the UV-induced tumor. These studies have demonstrated clearly that the immune response is damaged in UV-irradiated mice and the damage has a direct effect on the capacity of the induced tumor to grow (Kripke 1986; Hostetler, Romerdahl, and Kripke 1989).

What is the message of this information for lifestyle management? If indeed much or most of skin cancer is UV induced, then a significant portion of the more than 700,000 new cases could be prevented by a deliberate effort to limit exposure to sunshine. The best known salubrious effect of ultraviolet irradiation to the skin is the production of vitamin D_3 (cholecalciferol) by the photoconversion of 7-dehy-drocholesterol. But even this ultraviolet-mediated reaction is not required for those who receive dietary vitamin D. Thus, because of aging effects and the potential for skin cancer, it is wise to avoid excessive exposure to UV radiation.[1]

Until recently, the tanned body has been deemed by many to be more attractive than a body of lighter hue. Inasmuch as UV-B is believed to be the culprit in skin damage, one might ask if exposure to UV-A would be safe and still result in the desired bronze color for light-skinned descendants of northern Europeans. For a variety of reasons that include damage to dermal elastic tissue, possible lens injury, and the failure of UV-A tanning to protect from UV-B damage (and even the production of UV-B by UV-A lamps), most health-care providers warn individuals, especially people with light complexions, from partaking of cosmetic tanning (Diffey 1987; Morison 1988; Bruyneel-Rapp, Dorsey, and Guin 1988). If these considerations fail to convince you, it may be food for thought to note that UV-A causes tumors in hairless mice (Diffey 1987).

7.3d   Prostate cancer

The prostate gland is important in reproduction because it provides most of the fluid that carries sperm to the exterior. The prostate is located at the neck of the bladder and surrounds the urethra, which is a passageway for urine. Cancer of the prostate is the most common deep-seated (internal) human malignancy in the United States, accounting for an estimated 43% of new cancer cases in males (Figure 7–3) (Parker et al. 1997). The actual rate may be higher because many elderly men die of stroke, heart disease, lung or colon cancer, or any of a number of other diseases, and with them dies a malignancy that had not yet manifested itself as a life-threatening disease. The small, latent, and undiagnosed cancers are nevertheless real malignancies. Prostate cancers have the potential in elderly men of exhibiting the lethal malignant propensities of other cancers. American men are becoming more health conscious: they smoke less, exercise more, and eat fruits,

---

[1] Light-skinned people of northern Europe who migrate to latitudes closer to the equator become vulnerable to skin cancer. What happens to darkly pigmented individuals who migrate in the reverse direction, who move to the northern United States or northern Europe? The dark-skinned people receive less UV radiation than they would have at lower latitudes. The reduction of protective vitamin D_3 related to reduced sunshine exposure has been postulated as causing the greater prevalence of breast and prostate cancer found in nonwhite citizens (Studzinski and Moore 1995). The message here is that UV radiation may cause cancer in some people (whites) while *possibly* preventing it in others (blacks).

vegetables, and whole grain cereals. Perhaps as a result of their enhanced lifestyle, more and more men are living to an age at which they may manifest the malignant aspects of their prostate cancers. Thus, a gradual increase in prostate cancer mortality is apparent (Figure 7–1).

Historically, significantly less money was spent on this malignancy of men than was spent on many other cancers and diseases. For example, more than 41,800 men in the United States will die from prostate cancer in 1997. With a recent research budget of $28 million, about $700 per prostate cancer death is available (compare that figure with the $2,700 spent on research per breast cancer death and the $15,000 spent on research for each AIDS death, although obviously research budgets will vary from year to year; the figures given are from the mid-1990s and will probably remain relatively similar for several years). A possible reason for fewer dollars dedicated to prostate research compared with breast cancer research is that the years of life lost due to prostate cancer in an elderly patient are less than the years of life lost in a young woman who dies of breast cancer. Because of the tragic implications of breast cancer in relatively young women, especially a young mother with children, it is not difficult to understand the greater funding of the less common breast cancer.

Prostate cancer will be diagnosed in about a third of a million American males each year during the last decade of the century, although there is evidence of a declining rate in the late 1990s (Haas and Sakr 1997). It is suggested that an even greater number would be discovered with increased use of biopsy (Doll 1992). African American males have the highest frequency of prostate cancer in the world, about twice the frequency of white American males. African American males of Atlanta, Georgia, have an incidence of prostate cancer 14 times greater than Japanese males. Little information is available at the present time to provide convincing evidence of the role of diet, occupation, or other identified factors in the genesis of this disease. Surely the epidemiologic differences in prostate cancer prevalence, especially the extraordinarily high rate of prostate cancer in African American men compared with the very low mortality of the disease in Japanese men, will offer insight into the cause(s) of this very common malignancy and ultimately into its prevention.

An intriguing but unproven hypothesis concerns the reduction of ultraviolet radiation that darkly pigmented individuals sustain when living at the latitudes of the United States and western Europe compared with more tropical areas. Less ultraviolet radiation results in reduced vitamin D_3, which may cause increased vulnerability to prostate cancer (Studzinski and Moore 1995).

Because the causes of prostate cancer are, at best, poorly understood (Aronson et al. 1996; see Sinha 1995 and the entire issue of *Microscopy Research and Technique*), how to prevent this most common of deep-seated malignancies is similarly unknown. Recently, however, a diet rich in the antioxidant lycopene has been suggested as protective against prostate cancer (Giovannucci et al. 1995) (see discussion of lycopene in the section on non-nutrient compounds in food, section 7.4h).

The American Cancer Society (1996) recommends that men limit consumption of foods with saturated fats (red meats and dairy products) to reduce risk for this most common cancer of males.

Although the etiology of prostate cancer remains poorly known (a gene that increases risk for prostate cancer when mutated is described in Chapter 4), Charles Brenton Huggins received the Nobel Prize for his studies of this cancer. Huggins and Hodges (1941) showed that most prostate cancer cells are *not* autonomous and these androgen-dependent cells of the prostate undergo cell death (apoptosis) after androgen ablation. Castration (orchiectomy) of prostate cancer patients resulted in an immediate reduction in the elevated serum acid and alkaline phosphatases of patients and a regression of the tumor with a concomitant reduction of pain. Unfortunately, not all prostate cells respond to androgen ablation, and those cells that do not will eventually increase in number with an expanded tumor mass. The new cells are not androgen sensitive. The absence of prostate cancer among men without testes and individuals with bilateral testicular atrophy supports the view that prostate cancer cells respond to their endocrinologic environment and are not autonomous.

7.3e Colorectal cancer

A problem for epidemiologists is why the much longer small intestine (about 20 feet in length in an adult with 90% of the absorptive surface) is almost immune to cancer, whereas the shorter (5 feet in length with only 10% of the absorptive surface) large intestine is so vulnerable. Cancer of the colon and rectum is the third leading cause of cancer death in the United States in both men and women (Figure 7–3). In 1998, almost 57,000 American women and men will die of cancer of the large intestine and rectum compared with the 600 women and 600 men who will die of cancer of the small intestine (Landis et al. 1998). Of course, the answer to the query about differences between small intestine and large intestine cancer is not known. The rhetorical question is posed because it provides insight into the "why" of epidemiology. If data were not collected concerning how many people die of what kind of cancer, it would be unlikely if anyone would ever know that cancer of the small gut has a distinctly different frequency than does cancer of the large gut. Because both the small and large intestines are comprised of gut epithelium, smooth muscle, glands, and connective tissue, one can seek an answer to the puzzle. For example, one may examine the bacterial flora as it changes along the length of the gut. Small intestine bacteria are found in lower number and are metabolically less effective in activating procarcinogens than the bacteria of the large intestine. It has been postulated that transit through the small intestine is faster than transit through the large, and inasmuch as water is removed from the contents of the large intestine, the gut lining of the large intestine would be exposed for longer periods to a more concentrated concoction of noxious substances (Coit 1997; see also section 7.6a). Further, the epidemiologic information

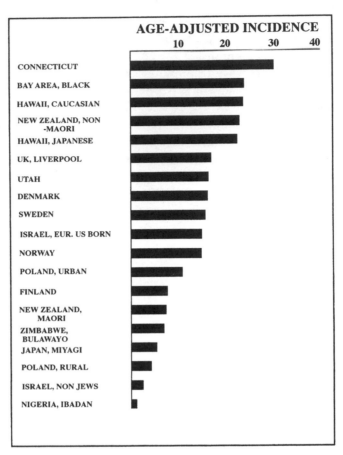

Figure 7–9. Cancer prevalence varies in different populations throughout the world. Illustrated here is age-adjusted colon cancer incidence by geographic area. (From Greenwald, P., Lanza, E., and Eddy, G.A. 1987. Dietary fiber in the reduction of colon cancer risk. *J Am Diet Assoc* 87:1179. Copyright The American Dietetic Association. Reprinted by permission of the Journal of the American Dietetic Association.)

provides an intellectual framework from which solutions (altered diets, for example) can be devised.

As with most cancer, the rate of colorectal cancer differs in different parts of the world. The citizens of Connecticut have 10 times as much colorectal cancer as do the people of Bombay, India (Higginson et al. 1992). Men of the Czech republic are at least 15 times more likely to die of cancer of the colon and rectum than men of Albania (Landis et al. 1998). In general, people who live in the less developed countries of Africa and Asia have a lower frequency of this cancer than people who live in more highly developed North America and northern Europe (Figure 7–9) (Trichopoulos et al. 1997). Similarly, people who live in the northeast part of the United States are more vulnerable to cancer of the large intestine than other Americans (Pickle et al. 1987). These data suggest that affluent people (perhaps because they can afford to buy meats marbled with fat and other fat-rich foods) are at greater risk than poor people.

Is there any way that a healthy person can reduce risk for colon cancer? Of course, no one knows how to prevent cancer in all people, but emerging evidence indicates that diet modification has the potential for risk reduction for colon cancer. Evidence is particularly strong for a diet containing five or more servings of fruits and vegetables each day. Associated with plant food intake should be reduced fat intake, especially from red meat (Doll 1992; American Cancer Society 1996; Trichopoulos et al. 1997). Diet and cancer is discussed at greater length later.

Factors other than diet are believed to influence colorectal cancer, including physical activity, avoidance of obesity (discussed later in this chapter), and smoking. Women who are current smokers or former smokers are at increased risk for cancer of the colon and rectum (Newcomb, Storer, and Marcus 1995). Ulcerative colitis and the hereditary condition of polyps known as familial adenomatous polyposis (FAP) (Chapter 4) both are associated with increased risk. Although the latter syndrome (FAP) is rare, most affected individuals will develop cancer of the large intestine if left untreated. A number of other genetic factors are implicated in colon cancer susceptibility (Chapter 4).

7.3f Cervical cancer

The uterus, a pear-shaped hollow organ about 3 inches long, 2 inches wide, and 1 inch thick, consists of an expanded upper portion known as the corpus and a more narrow lower portion, the cervix, that extends into the vagina. The cervix provides an opening to the interior of the uterus. The stratified squamous epithelium of the vagina changes abruptly to the columnar epithelium of the uterine interior at the cervix. Most cervical cancer is squamous cell in origin.

Although the death rate of uterine cancer (cervix and corpus combined) has been declining since well before World War II (Figure 7–2), nevertheless about 13,700 new cases of cervical cancer with about 4,900 deaths are expected from that malignancy in the United States in 1998 (Landis et al. 1998). Cervical cancer is found infrequently in women younger than 20, among celibate females (e.g., nuns), and among those who have had only one sexual partner who in turn has had only one sexual partner. It is uncommon among Jewish women. Worldwide, cervical cancer is the second most common cancer in women and the *most common* cancer in developing countries. Mexico is almost last in overall cancer deaths (per 100,000 people), but it is first in cervical cancer deaths (Landis et al. 1998). The cancer is frequently found in individuals who have experienced sexually transmitted infections such as herpes simplex type II, cytomegalovirus (another herpesvirus), and human papilloma virus (Herrero et al. 1990).

The association between these two herpesviruses and cervical cancer is not thought to be causal. Rather, the herpes infections may simply suggest a lifestyle that results in greater susceptibility to whatever the causal agent may be. Emerging evidence indicates that infection with one or more of the 77 human papilloma viruses (HPVs) may be etiologically related to cervical cancer (zur Hausen 1987). The changes in cell structure designated "cervical dysplasia" (seen in pap smears and biopsies) are believed due to HPV infection; HPV-infected individuals often show a significant increase in the number of dysplastic cells in a pap smear. The dysplastic cells contain HPV DNA and virus-specific antigens. Cervical dysplasia is in turn believed related to carcinoma in situ, which is ultimately linked with invasive squamous carcinoma of the cervix (Howley, Ganem, and Kieff 1997). The close association of the DNA-containing human papilloma viruses with cervical cancer, and the abundance of cervical cancer worldwide, led a German virologist to state at a scientific meeting that DNA viruses are now second only to tobacco smoke as a cause of human cancer.

If epidemiologic studies do indeed confirm one or more human papilloma viruses as the causal agent for cervical carcinoma, there will be abundant opportunity to devise ways to prevent the spread of the virus, which in turn should reduce the frequency of the cancer.

7.3g   Hodgkin's disease

Hodgkin's disease is of particular interest because it is one of the few curable cancers. It was invariably fatal when first described in 1832 but can now be successfully treated in most cases. Mortality from the neoplasm has been decreasing for the past 30 years with 1,400 expected deaths in the United States in 1998 (Landis et al. 1998). The disease, a primary malignancy of lymph nodes, has characteristic multinucleated giant Reed-Sternberg cells that are diagnostic of this neoplasm.

Hodgkin's disease increases in incidence until about 25 years of age, then decreases until age 45, after which it increases again (Doll 1992). The bimodal incidence suggests there is more than one kind of Hodgkin's disease. This disease is more common among the educated than uneducated, the urban dweller than rural person, and among siblings of a patient with the cancer. Americans and Italians have among the highest prevalence of Hodgkin's disease; Chinese and Japanese have the lowest (Higginson et al. 1992). A relationship of Hodgkin's disease with infection of the Epstein-Barr herpesvirus has been suggested. The viral relationship needs further study but is of interest because some of the giant Reed-Sternberg cells contain viral DNA. An interesting sidelight relating to the Epstein-Barr virus is that it causes infectious mononucleosis, and this nonmalignant disease may also have Reed-Sternberg cells or cells that are so similar they cannot be distinguished from Reed-Sternberg. The comparison of Hodgkin's disease to infectious mononucleosis is not inappropriate because there is concern that some Hodgkin's disease may not be malignant; that is, it may not be a "true" malignancy, but may resemble more closely a nonmalignant granuloma.

7.4 Occupational cancers

A classic in epidemiology is the description of cancer of the scrotum by Percivall Pott, in 1775. The plight of young boys, purposely kept small to permit them to fit into and clean tiny chimneys that were hot and sooty, is a compelling story that over 200 years later elicits compassion in the hearts of readers (Pott 1775). The description of brutal treatment of these boys brings to mind the writings of Charles Dickens in the following century. Pott was probably the first to associate a particular malignancy (the scrotal cancer) with a specific occupation (chimney sweeping). He believed it was the coal soot caught in the skin folds of the scrotum that caused the cancer. Later, coal tar was indeed shown to cause experimental skin cancer in rabbits (Yamagiwa and Ichikawa 1918). Coal tar is a mixture of substances. Kennaway (1924) showed that coal tar contained polycyclic aromatic hydrocarbons, particularly benzo-(a)-pyrene, and it was these substances that conveyed a carcinogenic potential to soot.

Since those early days, there has been increased interest in occupational cancer. No one knows how much cancer is due to occupation, but Landrigan (1996) suggests that about 10% of cancer deaths are due to occupational exposure. About 564,800 U.S. citizens are expected to die from cancer in 1998 (Landis et al. 1998). Ten percent of that figure, 56,000, is twice the number of American military personnel who were killed during the *entire* Korean War (1950–1953). Almost all occupational cancers can be prevented. Thus, it is a rich area to be explored in the quest for reduced cancer deaths.

Other substances and the cancers they cause include (but are not limited to) benzene and leukemia, asbestos and lung cancer, and vinyl chloride and angiosarcoma of the liver (Landrigan 1996). Listings of human carcinogens, identified by occupational exposure and judged to cause cancer, are found in Stellman and Stellman (1996), Trichopoulos et al. (1997), and Chapter 3. It is gratifying to learn that "in the United States and other developed countries, exposure to established occupational carcinogens has been substantially reduced" (Trichopoulos et al. 1997). However, it is disquieting to note that fewer than 1,000 of the 50,000 chemicals regularly used in commerce have been examined for their cancer-causing potential (Stellman and Stellman 1996). Occupational cancer remains a rich area for exploration.

7.5 AIDS-related cancers

7.5a Kaposi's sarcoma (KS)

An individual with the human immunodeficiency virus (HIV) infection has an increased risk for KS of about 40,000. Although Kaposi's sarcoma is the most frequent neoplasm afflicting HIV/AIDS patients, it has a declining prevalence with about a third of patients developing the neoplasm. Risk factors for KS, other than

HIV infection, include male homosexual lifestyle, cytomegalovirus disease, infection with other sexually transmitted diseases, inhaled nitrates, and the number of sexual partners in HIV-endemic metropolitan areas. The neoplasm occurs less commonly among HIV-infected individuals who are drug users, hemophiliacs, and women and children. Kaposi's sarcoma also arises in some transplant patients who are maintained on immunosuppressive drugs as well as elderly men of Mediterranean or eastern European origin (Ensoli, Barillari, and Gallo 1991). Young men of central Africa are vulnerable to a particularly malignant form of the cancer.

The neoplasm, of vascular or lymphatic endothelial cell origin, causes pink, red, or blue spots on the skin of the trunk, arms, head, and neck, and in the mouth (Safai, Diaz, and Schwartz 1992; Miles, Mitsuyasu, and Aboulafia 1997). The disease also spreads to the gut, lymph nodes, and other internal organs. Although KS has been associated with HIV infections for a number of years, a herpesvirus (HHV-8) is now being studied as a possible etiologic agent for this malignancy (Chang et al. 1994; Kedes et al. 1996).

7.5b   Other AIDS-related malignancies

Non-Hodgkin's lymphoma (NHL) is the second most common malignancy of AIDS patients with a rate about 60 times greater than in the general population. Persons with AIDS also have increased vulnerability of primary central nervous system lymphoma (PCL). The average age of non-HIV infected patients with PCL is about 60; in contrast, the average age of AIDS patients with PCL is 35 (Safai et al. 1992). Immunosuppressed women are particularly at risk for infection with human papilloma viruses (HPV), and thus at increased risk for cervical cancer (section 7.3f) (Miles et al. 1997). Of course, immunocompromised patients who are *not* AIDS patients also suffer from an increased risk for Kaposi's sarcoma and non-Hodgkin's lymphoma. Because of this, there is emerging interest in differences in the epidemiology of these tumors among AIDS patients and other immunocompromised patients as a means of studying pathogenesis of the neoplasms.

7.6   Diet, nutrition, and cancer

For a number of years, there has been strong but indirect evidence that common neoplasms could be made less prevalent if suitable alterations were made to the diet (Doll and Peto 1981; Watson and Mufti 1996; Trichopoulos et al. 1997). It is a commonly accepted notion that diet can affect cancer prevalence, either by avoiding substances in food that enhance risk or by selecting foods thought to protect against cancer (Newberne and Conner 1988). Unfortunately, because of many problems relating to the identification of what is an optimal diet for cancer prevention, little definitive information is available about which foods to select or avoid. Even less is known about how some nutrients in these foods effect cancer protection. Nevertheless, major health organizations concur concerning dietary recommendations. See Table 7–1 for the recommendations of the American Cancer Society (1996) (at the end of this section). These recommendations are virtually identical to those of the National Cancer Institute, the National Research Council, the U.S. Department of Agriculture/Department of Health and Human Services Guidelines, and the American Heart Association. They are also harmonious with those made in Japan and by the European Organization for Cooperation in Cancer Prevention (Palmer 1986; Weinhouse 1986; Butrum et al. 1988; Bal and Foerster 1991). Although this uniformity of opinion (of several agencies) does not in itself establish truth, it does suggest that a number of thoughtful individuals who have examined the evidence have arrived at the same conclusions. Although diets potentially optimal for reducing cancer risk have been identified and recommended (Chen and Meguid 1986), only a minority of Americans deign to consume these foods; thus, progress toward attaining compliance with dietary guidelines appears to be minimal (Bal and Foerster 1991). For example, of 11,658 adults studied, only 16% consumed high-fiber breads and cereals, and only 21% ate fruits and vegetables high in vitamin A (Patterson and Block 1988). No change for the better has been noted since the 1988 study.

Sections 7.6a through 7.6j consider several major dietary components that have been linked with cancer risk reduction as well as substances which occur in trace or limited quantities. Concerning the latter, two vitamins (A and C) are thought by some to be important in cancer risk reduction. There may be others. In addition to vitamins, a large number of nonvitamin, non-nutrient dietary components that occur in small quantities in certain foods may be important in protection against cancer. The existence of this array of protective substances adds credence to the admonition of dieticians to not rely on vitamin pills in lieu of a balanced diet for good nutrition. Several kinds of non-nutrient, nonvitamin, protective dietary substances not yet available in pills are discussed here.

As stated, the information that follows is concerned primarily with cancer risk reduction by consumption of foods thought to protect. The other side of the coin is the avoidance of harmful substances. Alcohol is considered in this context. Other substances are thought by some to be potentially harmful, but many of these are of little day-to-day concern to most residents of the United States. Included in this category are aflatoxins produced by *Aspergillus* molds. A direct relationship has been found between the level of food contamination by aflatoxins and human liver cancer in parts of Africa and East Asia. Esophageal cancer in several provinces of the People's Republic of China seems to be related to the consumption of pickled or moldy food. Furthermore, in the United States concern has been expressed about food additives, nitrites as preservatives, and pesticide contamination of food. Despite that concern, at present no epidemiologic evidence supports the view that such substances are dangerous to human health (Pitot 1989).

7.6a   Dietary fiber and colorectal cancer

Dennis P. Burkitt, an English surgeon who spent considerable time in Africa, noted that individuals on a so-called native diet did not suffer, or suffered less frequently, from a number of diseases common in his home country. These maladies included hemorrhoids, diverticular disease, benign polyps, appendicitis, and ulcerative colitis. Most significantly, for the purposes of this chapter, was his observation that cancer of the colon and rectum occurred less frequently in Africa than in Burkitt's native England or the United States (Figure 7–9). Note the remarkable differences in cancer rates in the incidence of large bowel neoplasms in Connecticut[2] (32.3 per 100,000) compared with India's incidence (3.5 per 100,000). The Connecticut rate seems even higher when compared with Dakar, Senegal (in West Africa), where the incidence is 0.6 per 100,000 (Weisburger and Wynder 1987). (Note: crude cancer rates may be misleading for comparison of countries where survival past 50 years of age is relatively rare compared to countries such as Sweden, Norway, Belgium, the United Kingdom, and the United States where survival past 75 is common. Because of this, the age structure of a population is included in the computation of the *age-adjusted rate* [Higginson et al. 1992] for cancer prevalence or mortality that is used throughout this book.)

The cluster of digestive tract diseases correlating with bowel cancer suggested a common or related etiology. Burkitt noted that Africans consumed a diet with far more fiber than was found in contemporary English (and American) food. He and others postulated that fiber in the diet provides protection against colorectal cancer (Burkitt 1971, 1975; Burkitt, Walker, and Painter 1974; Kune, Kune, and Watson 1987; Rosen, Nystrom, and Wall 1988; Vogel and McPherson 1989). A review of many epidemiologic studies suggests an inverse relationship between total dietary fiber intake and colon cancer (in most of the studies) (Greenwald, Lanza, and Eddy 1987; Trock, Lanza, and Greenwald 1990; Shankar and Lanza 1991). Despite the fact that a majority of studies attribute a protective role to fiber, caution should be used in the interpretation of the studies because many people obtain fiber from fresh fruits and vegetables. There may be other protective substances besides fiber in the fruits and vegetables (discussed subsequently). Alternatively, several agents may work together to cause colorectal cancer, and fiber and other substances may perform a risk-modifying role. Clearly, the role of fiber in colorectal cancer is intriguing and needs more study (Jensen 1989).

The imperfect correlation of fiber-rich foods and reduced cancer risk may relate to the diversity of substances classified as "fiber" and thus to differing foods with dissimilar physiologic effects. If dietary fiber is protective, it would be useful to know what it is in these various nondigestible, nonstarch polysaccharides (which include cellulose, hemicelluloses, pectins, gums, and mucilages) and lignins (Slavin 1987) that protects. Fiber has a water-holding capacity that contributes to fecal bulk and weight. Bulky feces mandate a short "transit time" because feces of large

---

[2] The prevalence data cited here is old. The 1997 estimated colon and rectum cancer mortality rate in Connecticut is 690 per 100,000 (Parker et al. 1997), which indicates an increase in recent years.

volume enhance intestinal peristalsis and passage (i.e., transit) through the digestive tract. This in turn increases defecation frequency (Shankar and Lanza 1991). The differences in transit time can be rather remarkable. For example, transit in an African on a native diet takes only a third of the time for that of students in English boarding schools on a diet of refined foods. How might rapid transit and a large volume and weight of feces be related to colorectal cancer? Burkitt postulated that slow transit permits time for fecal bacteria to convert intestinal procarcinogens to active carcinogens while a reduced fecal volume results in greater concentration of the ultimate carcinogen. Slow moving, reduced volume and weight feces permit concentrated fecal carcinogens to have a long time to affect the intestinal mucosa (Burkitt 1971). The chemical change of substances without cancer-causing capabilities (procarcinogens) to carcinogenic substances (ultimate carcinogens) is referred to as "metabolic activation" (see Chapter 3).

It has been suspected since 1933 that mutagens may in fact occur in feces. Researchers showed that the bile acids, deoxycholic and cholic acids (steroids derived from cholesterol), could be converted to the potent carcinogen 3-methylcholanthracene (Wieland and Dane 1933; Fieser and Newman 1935). Hill (1974) showed that fecal bacteria convert bile salts to carcinogens in vivo, and Burkitt (1975) observed that populations with a high incidence of colorectal cancer have elevated levels of fecal steroids.

Other substances or conditions may be involved in the etiology of colorectal neoplasia. Factors that may contribute to cancer risk include fecapentaenes (highly mutagenic hydrocarbon ethers produced by fecal bacteria) (Curren et al. 1987), inadequate vitamin D and dietary calcium (Garland et al. 1985), fat intake (Carroll and Khor 1975; Armstrong and Doll 1975), as well as intake of potential carcinogens produced by cooking meat over an open fire or frying at high temperature (Weisburger and Wynder 1987) and elevated fecal pH (Thornton 1981). (The significance of high pH in the colon is controversial: a recent study concerning alkaline pH and its relationship to carcinogens in the gut records no difference in gut pH of patients with colorectal cancer and normal controls; Pye et al. 1990.) The risk of colon cancer in the United States is significant. Cancer of the large bowel is third in mortality from neoplastic diseases, following only deaths due to lung cancer and breast or prostate cancer. An estimated 56,500 men and women will die of cancer of the colon and rectum in the United States in 1998 (Landis et al. 1998).

7.6b Correlations between food substances and cancer prevalence: Significance

Before proceeding to other dietary factors that may relate to cancer, consider the following. Epidemiologists have provided much of the insight into the putative role of fiber (and other substances) in protection against cancer. It was noted earlier that colorectal cancer prevalence and dietary fiber consumption differ between African villagers and English boarding school students. This suggests a protective role of dietary fiber for decreased colorectal cancer risk. However, other differences between the populations exist, including affluence, possible racial vulnerability, other dietary factors, occupation, education, geography, and recreational activities.

Fiber is itself a complex mixture of substances difficult to quantitate. Furthermore, fiber-containing vegetables and grains contain many substances other than fiber that may act alone or in concert with fiber to reduce risk. Finally, individuals who eat a large quantity and diversity of fruits and vegetables either may not desire or be unable to afford substances related to the etiology of cancer. Fast-food hamburgers are less available in a rural country than in metropolitan England or the United States. These cautions are inserted to warn you that, thus far, we know very little concerning what it is in a particular food substance that may (or may not) provide protection against cancer.

Consider the remarkable correlation between breast cancer mortality and total dietary fat (Figure 7–6). The relationship seems compelling. However, the "per capita intake" of a food item, in this instance fat, is calculated by adding the amount of the food item produced to the amount imported less the amounts exported, fed to animals, put to nonfood use, and lost in storage. This quantity is divided by the total population to provide an estimate of the per capita intake. The estimate does not take into account the quantities produced by individuals in gardens or on private farms, waste, and different patterns of consumption by subgroups differing in age and ethnic and economic background (National Research Council 1982). Hence, it is prudent to be circumspect when ascribing a cause/effect relationship of a substance with cancer in populations with differing cancer prevalences.

7.6c   Dietary fat

Fat in the diet of mice was implicated in the development of experimental mouse mammary tumors by Albert Tannenbaum (1942) (also see Tannenbaum and Silverstone 1953). Further, reduced caloric intake in rats and mice resulted in reduced cancer. For example, leukemia in mice was reduced from 50% to 4% by restriction of food intake (Gross 1988). There may be a similar relationship to food intake and cancer in humans because individuals who are 25% or more overweight suffer a two-thirds increase in cancer incidence (Doll and Peto 1981).

Epidemiologic studies in humans present a remarkably positive correlation between total dietary fat intake and age-adjusted death rate for breast cancer in various countries (Figure 7–6) (Carroll, Gammal, and Plunkett 1968; Carroll 1986, 1991). The positive correlation of dietary fat intake and breast cancer has also been reported in other (Lea 1966; Wynder 1969; Richardson, Gerber, and Cenee 1991) but not all (Willett et al. 1987; Rohan, McMichael, and Baghurst 1988) studies. Similar to the case of fiber and colon cancer (considered previously), controversy surrounds the role of fat in tumorigenesis and whether it is total dietary calories, obesity, and/or other factors that predispose some populations to breast and other cancers (Kolata 1987; Berrino, Panico, and Muti 1989).

Fat itself may not be the cause of cancer; rather, it may act as a promoter (Chapter 3). Promotion by dietary fat occurs in mammary cancer induced by chemicals, hormones, and ionizing radiation in experimental animals. Although polyunsaturated fats seem to be better promoters than saturated fats, the mechanism of promotion is not known. In this context, note that case control and cohort studies do not provide strong support for a causal relationship between dietary fat and breast cancer. Some have argued, however, that difficulty in estimating past fat intake, small differences in people living within a common culture, and genetic components in disease vulnerability are methodologic problems. As such, the epidemiologic evidence is not invalidated, or so it is conjectured (Carroll 1991).

Dietary fat has been correlated with cancers other than those of the breast: these include colonic, pancreatic, rectal, prostatic, and ovarian cancers, and other malignancies (Carroll and Khor 1975; Palmer and Bakshi 1983; Vogel and McPherson 1989; Graham et al. 1988).

The American Cancer Society 1996 Guidelines and the National Cancer Institute's Guidelines (Butrum et al. 1988; Bal and Foerster 1991) recommend that people limit their intake of high-fat foods. Skepticism has been voiced concerning the competence of medical advisers to effect a reduction in fat intake or other dietary changes in their patients. Although it may take considerable effort, emerging evidence indicates that compliance to dietary recommendations is feasible, and, thus, there is room for optimism as diet management becomes more important in risk reduction (Boyar et al. 1988; Post-White et al. 1989).

7.6d Vitamin A (beta-carotene)

Vitamin A and its related compounds, beta-carotene and the retinoids, have been studied with respect to their effect on cell differentiation and cancer risk reduction (Pastorino et al. 1991; Stahelin et al. 1991). Beta-carotene is a plant pigment converted by the gut and liver to vitamin A (retinol). This vitamin is essential for the normal growth and differentiation of epithelia and bone and important in normal reproduction and vision. Foods rich in vitamin A or beta-carotene include milk, liver, green leafy vegetables, and yellow fruits and vegetables. Retinoic acid (vitamin A acid) has been used extensively to modulate the differentiated state of neoplasms in vitro (Chapter 1) (see also Pierce and Speers 1988; McKinnell 1989a, 1989b; Janick-Buckner, Barua, and Olson 1991). Consumption of foods rich in vitamin A and its related compounds has been associated with risk reduction for cancer of the lung, larynx, bladder, esophagus, stomach, colon and rectum, prostate (National Research Council 1982), and ovary (Miller 1989; Slattery et al. 1989). However, vitamin A supplement in excess of recommended levels is toxic, especially to pregnant women. Further, supplements of vitamin A are not known to reduce cancer risk (American Cancer Society 1996).

How beta-carotene or its vitamin A derivative effects protection is not known. Beta-carotene may act as a free radical scavenger and singlet oxygen quencher to prevent DNA damage (Stahelin et al. 1991). Also, ornithine decarboxylase, which is associated with tumor promotion in mice, is inhibited by retinoids (Boutwell 1982), as are the transforming growth factors, TGFs (Todaro, DeLarco, and Sporn 1978). Moreover, a vitamin A–induced enhancement of immune function may suppress tumor growth (Watson 1984). Other roles for vitamin A and its related compounds will undoubtedly be identified.

Although the mechanism of vitamin A may not be understood, there is sufficient evidence to conclude that eating a diversity of foods rich in beta-carotene or vitamin A is associated with a reduced cancer risk (Harris et al. 1991).

7.6e   Vitamin C (ascorbic acid)

The consumption of fruits and vegetables rich in vitamin C has been correlated with reduced cancer risk in a number of epidemiologic studies. Malignancies that seem to be inhibited most by elevated vitamin C intake are oral, larynx, esophagus, stomach, pancreas, lung, and breast neoplasms (National Research Council 1982; Fontham et al. 1988; Block 1991). Controversy surrounds the role of vitamin C in cancer risk diminution. For instance, people who have a high intake of vitamin C in their diet also consume generous quantities of fiber, beta-carotene, and other vitamins. Is the reduced cancer risk due to the other nutrients, an interaction of ascorbic acid with the other substances, or to vitamin C itself? Note that vitamin C taken as a dietary supplement has little or no effect on cancer risk (American Cancer Society 1996).

The reduced cancer risk obtained from *foods* rich in vitamin C may be significant. For instance, reduced fat intake coupled with elevated vitamin C consumption is cumulative and has the potential for reducing breast cancer risk by as much as 25% (Block 1991). If vitamin C protects, how does it do so? Oxidation and free radicals are associated with oncogenesis. Ascorbic acid acts as an antioxidant and a free radical scavenger. Vitamin C also has efficacy in inhibiting the formation of nitroso carcinogens from precursors and has a number of other complex effects on enzyme and nonenzyme mediated reactions. How these reactions relate to reduced cancer risk is far from understood (Henson, Block, and Levine 1991).

7.6f   Vitamin E

Cancer risk reduction has not been associated with vitamin E (American Cancer Society 1996).

7.6g   Selenium

A role for selenium in cancer prevention has been suspected for several decades as a result of epidemiologic studies (Salonen 1986; Ringstad et al. 1988; Miller 1989; Clark et al. 1996). Selenium, a nonmetallic element, is an essential component of the enzyme glutathione peroxidase, which functions in the antioxidant defense system by removing hydrogen peroxide. The enzyme helps prevent hydroxy radical formation, which, in turn, may inhibit the mutagenic action of some carcinogens (Hocman 1988; Neve 1991).

The amount of selenium in meat and vegetables varies with the level of that element in animal food, water, and soil. Meat ordinarily contains more selenium than vegetables. Because of the generous use of meat in their diet, most Americans are not likely to suffer a deficiency of selenium. Selenium is added to fertilizer in Finland because of its putative value as an agent that may reduce cancer prevalence. Note, however, that an excess of selenium is toxic and because of the narrow margin between safe and harmful levels, dietary supplements have not been recommended (Brown 1990; American Cancer Society 1996).

7.6h   Non-nutrient compounds in food that may protect against cancer

Many chemically diverse, non-nutrient substances in food may protect against cancer. These substances have been studied intensively in animal experiments because of their potential role in the chemoprevention of cancer in humans.

The consumption of cruciferous vegetables (e.g., cabbage, cauliflower, Brussel sprouts, kale, and turnips) is associated with a reduced incidence of cancer in human populations. These vegetables include sulfur-containing dithiothiones among their many organic non-nutrient constituents. Dithiothiones inhibit tumors in mice caused by chemically diverse carcinogens (Bueding, Ansher, and Dolan 1985; Wattenberg and Bueding 1986).

More recently, other organic sulfur-containing compounds of cruciferous vegetables have been studied. One such compound, benzyl isothiocyanate, has the happy propensity of inhibiting 7–12-dimethylbenz(α)anthracene-induced mammary cancer in rats and stomach and pulmonary adenomas in mice. This non-nutrient dietary substance also inhibits stomach cancer in benzo(α)pyrene-fed mice (Wattenberg 1985, 1990).

Species of *Allium* (e.g., garlic, leeks, and shallots) contain organic sulfur compounds. One of these, diallyl disulfide, may have a cancer-preventing competence. However, the American Cancer Society believes not enough evidence supports a specific role for this substance in cancer prevention (American Cancer Society 1996). Monoterpenes are formed in citrus fruit oil (e.g., D-limonene) and caraway seed oil (e.g., D-carvone). These also have cancer prevention capabilities in animal studies (Wattenberg and Coccia 1991).

Many chemical carcinogens are not direct acting but require metabolic activation for the compound to have cancer-causing competence (Chapter 3; Miller 1978). The non-nutrient dietary chemicals discussed in this section are believed to obtain their cancer-blocking action by inhibiting the metabolic activation of carcinogens (Wattenberg 1990).

Another mechanism for protection by non-nutrient compounds may be as oxygen scavengers (see also discussion of Vitamins A and C). Prostate cancer (see 7.3d) is the most common internal malignancy in Americans. No certain method of diminishing vulnerability to this nearly ubiquitous neoplasm of older men is known. However, tomatoes and tomato-based products (except tomato juice) seem to afford some protection against prostate cancer. The protective effect is believed due to the antioxidant lycopene, almost all of which in the diet is obtained from tomato products. Lycopene is a carotenoid that occurs in ripe tomatoes and gives them their characteristic red color. The biologic availability of lycopene seems to be due to cooking tomatoes with oil. Hence, spaghetti sauce and pizza are excellent sources of lycopene, whereas tomato juice, containing no oil, affords less protection (Giovannucci et al. 1995).

7.6i   Alcohol

Alcohol is probably the only chemical thought to cause cancer in humans that has not been shown to cause cancer in animals (Tuyns 1990). If dietary recommendations were made exclusively from animal studies, then there would be no section on alcohol. Furthermore, for many years evidence has suggested that individuals who consume modest quantities of alcohol have a greater life expectancy than those who eschew alcohol consumption entirely (Pearl 1922; Marmot et al. 1981; Colsher and Wallace 1989). Why then does the National Cancer Institute recommend that alcoholic beverages be consumed "in moderation *if at all*" (Butrum et al. 1988) and the American Cancer Society (Table 7–1) urges a *limited* consumption of alcohol?

The reason for the recommendations on alcohol ensues from studies that report an increased risk for cancers of the upper digestive tract with alcohol consumption (Garro and Lieber 1990; Tuyns 1990; Hebert and Kabat 1991). Particularly alarming is the reported synergistic relationship of tobacco and alcohol in which alcohol-consuming smokers are 15 times more likely to develop oral cancer than their nonsmoking, nondrinking compatriots (Butrum et al. 1988). Moreover, emerging evidence indicates that drinking is correlated with risk for breast cancer. Consumption of 3 to 4 glasses of wine per day (or an equivalent quantity of alcohol as beer or spirits) constitutes a highly significant and consistent risk for breast cancer (Howe et al. 1991). This statistical relationship is supported by the observation of low breast cancer risk among religious groups that do not use alcohol (Tuyns 1990). Other studies suggest a relationship between alcohol consumption and several other cancers, including liver and rectal neoplasms.

As with so many aspects of cancer biology, we have no simple explanation of how alcohol may cause its reputed pathologic effect. It has been postulated that alcohol stimulates the cytochrome P-450 microsomal enzyme system, which is involved in procarcinogen activation, that alcoholic beverages may contain carcinogen contaminants (some beverages have been reported to contain nitrosamines), and that alcohol induces changes in hormone profiles resulting in cancer. Alcohol may also have direct cytotoxic effects. Cytotoxicity results in increased mitotic activity (increased cell division) to replace killed cells, resulting in more

Table 7–1. *American Cancer Society (1996) Guidelines*

- **Choose most of the foods you eat from plant sources.**
  Eat five or more servings of fruits and vegetables each day.
  Eat other foods from plant sources, such as breads, cereals, grain products, rice, pasta, or beans
      several times each day.

- **Limit your intake of high-fat foods, particularly from animal sources.**
  Choose foods low in fat.
  Limit consumption of meats, especially high-fat meats.

- **Be physically active: Achieve and maintain a healthy weight.**
  Be at least moderately active for 30 minutes or more on most days of the week.
  Stay within your healthy weight range.

- **Limit consumption of alcoholic beverages, if you drink at all.**

cells as targets for carcinogenic initiation (Chapter 3). Be that as it may, the multiplicity of purported mechanisms suggests much is yet to be learned. It is perhaps wise to consider the judgment of the International Agency for Research on Cancer (IARC), which stated in 1988 that "alcoholic beverages are carcinogenic to humans" (Tuyns 1990).

7.6j   American Cancer Society (1996) *Guidelines on Diet, Nutrition, and Cancer Prevention*

It is not yet possible to ascertain the precise role of diet in cancer risk. Nevertheless, many studies (some of which were reviewed here) suggest a consistent relationship between substances in the diet and vulnerability to certain cancers. On the basis of this information, the Americn Cancer Society has established guidelines (Table 7–1) thought to be prudent and likely to reduce cancer risk.

7.7   Exercise as it relates to cancer

The American Cancer Society (Table 7–1) recommends physical activity. Most of the following supports that point of view. A number of years ago, a fibrosarcoma was transplanted to several male mice. Some were forced to exercise; the controls were not. Transplanted tumors were substantially smaller in the exercised group (Rusch and Kline 1944). Similar results were obtained with a transplanted rat tumor. In some cases, complete tumor regression was reported among exercised animals (Hoffman et al. 1962). More recently, chemical carcinogen-induced colon cancer in rats was reduced by exercise (36% in the exercised animals versus 75% in the sedentary animals) (Kritchevsky 1990). However, contrary to the expectation based on these studies, it was reported that exercise *enhanced* chemically induced rat mammary tumors (Thompson et al. 1989). These contrasting studies reveal a

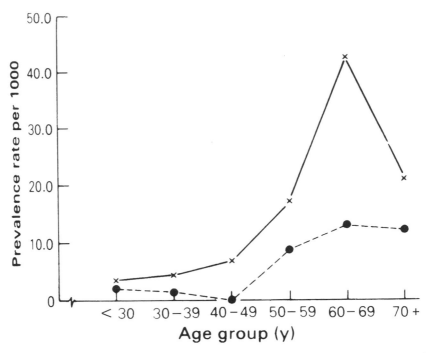

Figure 7–10. Women who were athletes in college (data indicated with solid black circles) have been reported to have a lower prevalence of cancers of the reproductive system (uterus, ovary, and vagina) than nonathletes (data points indicated with x's). (From Frisch, R.E., et al. 1985. Lower prevalence of breast cancer and cancers of the reproductive system among former college athletes compared to non-athletes. *Br J Cancer* 52:887. Reprinted by permission of The Macmillan Press Ltd.)

complex relationship between specific tumors and the multiplicity of physiologic effects induced by varying kinds of exercise. Nevertheless, despite the complexity that is yet to be unraveled, it is not premature to ask if a relationship can be detected between cancer and exercise in human populations.

Certainly exercise seems to have an influence on age at death. Nearly 17,000 Harvard alumni were studied, and not surprisingly regular walking, stair climbing, and sports participation were inversely related to mortality (Paffenbarger et al. 1986). Exercise and a decrease in mortality are related primarily to cardiovascular fitness. However, even though it may take 20 to 40 years for cancer to appear because of a particular lifestyle, studies have detected an altered risk associated with exercise. For example, colon cancer was 60% to 80% higher in sedentary workers of Los Angeles County compared with occupations requiring greater physical activity (Garabrant et al. 1984). Colon cancer among 1.1 million Swedish men was similarly lower in active men (Gerhardsson et al. 1986). Agricultural, forestry, and sawmill workers were at reduced risk for colon cancer when compared with sedentary workers (Fredriksson et al. 1989), and breast cancer and neoplasms of the female reproductive system (Figure 7–10) were less common in former athletes compared with nonathletes (Frisch et al. 1985, 1987). These studies present a compelling argument that exercise may have a protective role in risk abatement of certain cancers, namely, colon, breast, and female reproductive organ cancers (Vena et al. 1987).

How exercise may exert this putative protective effect is not known. Peristalsis is stimulated with exercise, which may result in reduced transit time, thought by Burkitt and others to be important in the genesis of colon cancer (section 7.6a). Perhaps physical activity leads to a healthier diet (also discussed previously) because of appetite stimulation. Furthermore, active people may be less obese, more interested in healthful living, and shun activities that could contribute to certain cancers (Eichner 1987). Whatever the mechanism, accumulated evidence suggests a beneficial effect of physical activity in risk reduction for cancer (Table 7–1; American Cancer Society 1996).

As with most aspects of cancer biology, there are some exceptions to the generalization that exercise is good for the body. Harvard athletes born between 1860 and 1889 had a greater death rate due to neoplasms than did nonathletes (Polednak 1976). Longshoremen who engage in vigorous physical activity, although protected from coronary heart disease, were not similarly protected from cancer (Paffenbarger, Hyde, and Wing 1987). One could postulate that the athletes and longshoremen had less optimal diets than their compatriots (did they eat more fatty meat?) or engaged in other activities that negated the effects of physical activity. Be that as it may, these studies remain a curious exception to the generalization that exercise is essential for good health.

7.8 Tobacco as a lifestyle hazard

Lung cancer caused by tobacco smoke is discussed in section 7.3a. There are other effects of tobacco consumption. It has been known for a number of years that cigarette smoking is also associated with cancer of the larynx, oral cavity, and esophagus. More recently, cancers of the lower end of the digestive tract also have been associated with smoking. Women who smoke, or who have smoked in the past, are at an increased risk for cancer of the colon and rectum (Newcomb, Storer, and Marcus 1995). Tobacco use in other forms such as snuff, chewing tobacco, cigars, and pipes is similarly hazardous. In an article on the role of the primary care physician in tobacco use prevention and cessation, Bal, Lloyd, and Manley (1995) stated, "Doctors better than anyone know what smoking does to people. They know the *pain, debilitation, and premature death* [italics added] experienced by at least half of all smokers, and, most importantly, they are in a position to make a difference."

Of course, not all tobacco is consumed by burning. Tobacco may also be chewed or used as snuff. Smokeless tobacco comes in two categories: moist, which is frequently flavored and placed in the mouth between the cheek and gums; and dry, which is finely powdered and drawn forcibly into the nostrils with inhalation (hence the Dutch-derived term "snuff"). Although smokeless tobacco is less widely consumed than the burned varieties, it is of growing concern because of the recent increase in its use by school-age children. It is estimated that 17% of boys in the United States in grades 3 to 12 partake of smokeless tobacco at least one time per week. In Louisiana, the figure is estimated to be 26% (Squier 1988). Smokeless tobacco is strongly related to cancer of the cheek, gums, and pharynx (Winn 1988). Thus, tobacco, whether incinerated or used in some other manner, constitutes a greater cancer risk than any other known factor.

7.9   Other lifestyle hazards

Probably all human activity has some impact on cancer prevalence. Discussed in the section on breast cancer is the observation that nuns are at a higher risk than other women. With present knowledge, we recognize it is not the religious community that results in the greater risk for breast cancer but rather not having children.

Similarly, it is well known that Kaposi's sarcoma (KS) is particularly prevalent among homosexual men. KS is the most common AIDS-associated malignancy. However, it is not being gay per se that causes Kaposi's sarcoma. The cancer occurs in individuals with immune dysfunction, seen among acquired immunodeficiency syndrome (AIDS) patients, gay or not. Present opinion indicates specific viruses, not lifestyle, as the cause of Kaposi's sarcoma (section 7.3h).

7.10   Summary

Considered in this chapter are factors that affect cancer risk. It is the judgment of a number of scientists as well as the former secretary of health and human services, Margaret M. Heckler, that modification of lifestyle is an enormously rich lode to be mined for cancer risk reduction. Heckler (1984) stated, "Too few Americans realize the simple truth that cancer is usually caused by the way we live, and its risks can be reduced by the choices we make." Heckler estimated that 80% of cancers are linked to "lifestyle and environmental factors." We are not entirely at the mercy of our environment. In fact, we have the capacity to modify, control, or influence vulnerability to several important cancers.

One question is how many cancers can be prevented. The answer is not known. Heckler was correct in noting that about 30% of all cancer deaths are due to lung cancer (this is as true in 1998 as it was in 1984; see Landis et al. 1998), and she and her advisers suggested that perhaps as many as 35% are due to diet (Doll and Peto 1981; see also American Cancer Society 1996). Other environmental factors account for a much smaller percentage of cancer deaths. Heckler did not consider the effects of exercise, but we do (section 7.5) and so does the American Cancer Society (1996). What percentage of these deaths could be prevented with education remains unknown.

Perhaps more important than speculation about numbers is the reported teaching success for diet modification and efforts to entice smokers to quit. These preliminary successes should serve to motivate educators as to the feasibility of risk reduction and stimulate even greater efforts. Instruction by health professionals is essential. Over 70% of the U.S. population sees a physician at least annually, which presents an exceptionally important opportunity to educate at least 38 million of the 54 million smokers (Gritz 1988; also see Bal et al. 1995). Books on cancer hopefully will also provide needed information about cancer prevention and risk reduction – and that is a goal of this chapter.

8

Cancer treatment

Surgery, radiotherapy, cytotoxic chemotherapy, and their combination

Ralph E. Parchment

8.1   Introduction

One philosophy for treating cancer involves the eradication of all cancer cells from the body, or at least a sufficient number that the time required to regrow to relapse is longer than a patient's life expectancy. The clinical strategies for eradicating cancer cells are the subject of this chapter. A second philosophy entails the alteration of the nature of the cells in a cancer, so that disease progression is slowed or halted entirely and quality of life improved even though cancer cells remain in the body. The clinical approaches for accomplishing this are the subject of Chapter 9, Biotherapies.

Specific cancers and their treatments are used in this chapter and the next only when they clearly illustrate the principles of therapy. Cancers of interest to you might not be mentioned because they do not add scientifically to the topic. For detailed information about treatment of specific cancers, see *Cancer Treatment* (Haskell 1990), *Cancer: Principles and Practice of Oncology* (de Vita, Hellman, and Rosenberg 1997), *Cancer Medicine* (Holland, Frei, Bast, Kufe, Morton, and Weichselbaum 1993), and *Medical Oncology: Basic Principles and Clinical Management of Cancer* (Calabresi and Schein 1993). Likewise, this chapter is not intended to be a compendium of drug targets and mechanisms. Rather, a few drugs are discussed to illustrate the pharmacologic principles of cancer treatment. For detailed information about anticancer drugs and their use, refer to *Cancer Chemotherapy*

(Chabner and Collins 1992) and the *Cancer Chemotherapy Handbook* (Dorr and Von Hoff 1994).

8.2 Eradicating cancer cells with conventional modalities: Surgery, radiotherapy, and chemotherapy

Three established strategies are used to eradicate cancer cells in a patient's body. Physical removal of the tumor mass (resection) is the foundation of surgery; radiotherapy and chemotherapy use exposure to toxic ionizing radiation and to cytotoxic chemicals, respectively, to destroy cancer cells without having to find and remove them. Each of these strategies is called a "modality," and the term "conventional" refers to therapy accepted as the best available standard treatment. Because of their different strategies, each modality is associated with specific risks and side effects. This chapter briefly describes the principles of surgery, radiotherapy, and cytotoxic chemotherapy, and it provides a scientific understanding of the indications, successes, limitations, and toxicities of these modalities.

8.2a Surgery as a cancer treatment

Surgical procedures are used to physically remove malignant tissue. If surgery is the only treatment, the assumptions are made that the location of all malignant tissue is known, it can all be removed, and undetectable micrometastatic disease is not left behind. Surgery is the most effective modality for treating localized disease, so prompt diagnosis from early warning signs of cancer results in a higher probability of surgical cure. Adjacent healthy tissue is also usually removed to provide a surgical "margin" between the diseased and healthy tissues. This area is used by the pathologist to assess whether the tumor is invading the adjacent tissue or not. Local lymph nodes may also be removed (dissection) to estimate a probability of systemic disease (Chapter 1).

Because diagnosis of solid tumors occurs at approximately the same size in each patient, the toxicity (morbidity) and the risk of death (mortality) of tumor resections are determined in large part by the location of the tumor, the percentage of healthy tissue removed during the operation, and the amount of lost normal tissue that can be replaced by the body. When the resected tumor is small compared to the size of the involved tissue, or in a noncritical site, surgical morbidity is lower. For example, surgical removal of a localized squamous carcinoma of the skin has low morbidity and mortality because a small proportion of skin is removed, the residual tissue can regrow, and the tumor is easily accessible. Resection of a brain stem tumor of the same size is a higher risk procedure in a tissue that cannot regrow, and morbidity can be much higher. Cancer surgery carries the risk of surgery in general: dangers of anesthesia, loss of hemostasis, and infection. As with all major surgeries, risks increase with age: for example, the mortality rate of surgery for colorectal cancer is 15% higher in octogenarians than those less than 70 years of age (Haskell, Selch, and Ramning 1990a). However, the experience of the surgical team with the procedure can also be an important factor (Garnick 1994).

Even very large tumors can be successfully resected. However, larger tumors are more likely to be associated with tumor extensions into critical or fragile normal tissue, greater risk of lethal hemorrhage during resection, and suppressed immune function that increases the risk of infection. For example, the mortality rate of optimal cytoreductive surgery for ovarian cancer is 3% (Berek and Hacker 1990), but it rises to 16% in patients with advanced ovarian cancer when the operation includes relief of bowel obstruction, exploratory surgery, or extensive tumor resections (Pecorelli 1994).

In some cases, inaccessible anatomic sites or extensive intermingling of tumor and critical normal tissues results in an "unresectable tumor." In other cases, clinical experience indicates that the cancer has very likely spread to other sites in the body by the time of surgery; even though not detectable at the time, this metastatic disease will cause a relapse in spite of successful surgery. In both cases, a nonsurgical treatment is also required.

8.2b   Radiation therapy as a cancer treatment

Radiation therapy ("radiotherapy") for cancer originated in the finding that x-rays sterilize rams by killing the proliferating germ cells in the testes that maintain spermatogenesis (Regaud and Ferroux 1927; Regaud 1930). The historical understanding of cancer as a disease of overly rapid cellular proliferation made it logical to treat cancer patients with x-rays, and the initial tumor responses encouraged the development of this treatment (Coutard 1932). During radiotherapy, malignant cells are exposed to ionizing radiation from either an external or implanted radiation source, and the resulting damage causes the death of the cell when it tries to divide, which leads to a gradual reduction in tumor mass – hence the descriptive name cytotoxic therapy or cytoreductive therapy. Radiotherapy is regional therapy; the radiation is focused like a beam of light on the treated area called a radiation field, but cancer cells that reside outside of the irradiated area will not be damaged.

Radiation is most toxic to proliferating cells, and higher doses are required to kill cells that are capable of proliferation but are not actively dividing (quiescent cells) at the time of exposure. Mammalian cells are most sensitive to radiation-induced damage in the late G_2 and M phases of the cell cycle. A radiation dose that kills a cell in G_2/M will usually not kill the same cell in the G_1 or S phases, or in a quiescence state, because there is time for the cellular damage to be repaired before progressing through G_2/M (Kaplan 1981; Parker 1990; Weichselbaum, Hallahan, and Chen 1993). Cellular damage produced by radiotherapy is an indirect result of ionization of chemicals in the cell to very reactive compounds. Oxygen is the predominant electron capturer in cells, and cytotoxicity is due primarily to damage caused by oxygen free radicals like hydrogen peroxide (H_2O_2), superoxide anion (O_2-), and hydroxy radicals (OH·) (Kaplan 1981; Parker 1990; Weichselbaum et al. 1993). In fact, in the absence of oxygen or at very low concentrations of oxygen (hypoxia), mammalian cells become two- to four-fold more resistant to radiation toxicity (Kaplan 1981; Parker 1990; Batzdorf, Black, and Selch 1990; Weichselbaum et al. 1993). Therefore, poor blood perfusion of the tumor or a poorly developed microvasculature in the tumor, both of which cause poor oxygenation, leads to decreased effectiveness of radiotherapy. Chemicals that can substitute for molecular oxygen have been developed for use with radiotherapy of poorly oxygenated tumors. However, it may not be possible to deliver these blood-borne radiosensitizers to the tumor if oxygen itself cannot be delivered.

Because malignant cells and their normal counterparts use the same basic mechanisms for cell division, it is not surprising that radiotherapy kills tumor cells as well as rapidly dividing normal cells within renewing tissues. Thus, radiation treatments cause cytoreduction not only in tumor tissue but also in healthy, normal tissues (toxicity) that lie within the radiation field. Death of rapidly dividing normal cells in renewing tissues causes the acute side effects (toxicities rapid in onset) of radiotherapy, and cells that divide as fast or faster than cancer cells are highly sensitive: hair follicles, oral-gastrointestinal epithelium including the oral cavity, and hematopoietic tissue (blood cell–producing cells) in the bone marrow. Death of the dividing cells in these normal tissues produces the side effects that people associate with cancer therapy: infection and hemorrhage (bone marrow toxicity), diarrhea (gut mucosa toxicity), mucositis (oral mucosa toxicity), and hair loss (hair follicle toxicity). For example, abdominal radiation for metastatic ovarian cancer also exposes the intestine, causing nausea, vomiting, or anorexia in 75% of the patients (Berek and Hacker 1990). In contrast, most of the bones that contain the hematopoietic marrow lie outside of this radiation field, so decreased peripheral blood counts of neutrophils (neutropenia) or platelets (thrombocytopenia) occur in only 10% of patients treated in this way (Berek and Hacker 1990).

The dose of radiotherapy administered to the treatment field can also be limited by chronic (irreversible) side effects in nonproliferative tissues or in slowly renewing tissues like liver and kidney. Clinicians view severe chronic toxicities as more serious side effects than acute toxicities, which can be reversed if the patient can be kept alive until recovery. Chronic toxicities are quite insidious in that many months or years may pass before they manifest clinically, more time is required for recovery, and they may be irreversible. There is also a general lack of effective clinical management of patients with these side effects, and recovery may take weeks to months to reverse toxicity to tissues with protracted tissue renewal. Damage to nonproliferative, terminally differentiated tissues like peripheral nerve and heart may be irreversible. These chronic toxicities most often limit the total cumulative dose that can be administered. A relatively slow rate of repair of sublethal damage in these tissues may cause the delayed nature of these toxicities that depend on cumulative total dose because repair from the prior exposures has not been made by the time of the subsequent exposure. The tissues susceptible to chronic toxicity are shielded during radiotherapy whenever possible. For example, shielding is used during abdominal radiotherapy for ovarian cancer to limit the total radiation dose to the kidney and liver.

Usually one healthy organ or tissue in the treatment field is more sensitive to radiation damage than the others, and the patient cannot be exposed to higher levels of radiation than this dose-limiting tissue can safely tolerate (Grever and Grieshaber 1993). The radiation dose might be limited by the severity of an acute toxicity to a rapidly renewing tissue or by the severity of an irreversible toxicity that appears several months after treatment is completed. As a general rule, the dose given per unit time, called dose intensity, is limited by acute toxicity usually to rapidly renewing tissues, whereas cumulative dose is limited by irreversible toxicity or delayed toxicity to nonrenewing tissues. Dose-limiting toxicity (DLT) in the dose-limiting tissue limits the radiation exposure a patient can safely receive. Too high a dose poses risks of protracted or permanent organ damage, significant morbidity, and risk of death. The maximum tolerated dose (MTD) is defined as the dose that does not produce life-threatening toxicity in the dose-limiting tissue in most patients. Therefore, this is the most commonly used dose for treatment.

For radiotherapy, the term "efficacy" denotes the effectiveness of a therapy. The term "therapeutic index" is the ratio between the MTD and the efficacious dose, in other words how much differential sensitivity exists between the target tissue (the cancer) and the dose-limiting tissue. For cytotoxic therapy, the therapeutic index is usually low, which means the radiation dose that causes tumor regression and that which causes DLT usually differ by very little, so radiotherapy has little margin for error in dose calculation.

8.2c   Cytotoxic chemotherapy as a cancer treatment

Cancer chemotherapy evolved along a course quite similar to radiotherapy: out of the observed effects of chemical toxicants on rapidly proliferating tissues of the body. The bone marrow suppression and lymph node atrophy that followed combat exposure to nitrogen and sulfur mustard gases during World War I suggested a trial of these toxicants against leukemia and lymphoma, cancers of the bone marrow and lymphoid tissues (Gilman and Philips 1946; Goodman et al. 1946). The clinical effectiveness of these compounds led to an effort to discover cytotoxic drugs for each type of cancer (Noble, Beer, and Cutts 1958), an effort that is still an active area of investigation today and the foundation of the cancer pharmaceutical industry. Even toxicity to actively dividing bacteria that contaminate electrode baths led to one of the most active solid tumor drugs called cisplatin (Rosenberg et al. 1965, 1969). Like radiotherapy, cytotoxic chemotherapy is designed to kill proliferating cells, so it is another type of cytoreductive therapy. Cytotoxic chemotherapy works best against cells actively progressing through the cell cycle, and these drugs are generally less effective against the same cells in a quiescent state.

Like radiotherapy, chemotherapy eradicates dividing cells in renewing tissues, and the anticancer drugs act as cytoreductants in both tumor tissue (efficacy) and in the gastrointestinal mucosa, bone marrow, hair follicles, and germ cells (toxicity). The most sensitive normal tissue in which life-threatening toxicity occurs is again called the dose-limiting tissue, and the dose just below that which causes life-threatening toxicity is called the maximum tolerated dose. Usually the therapeutic index of chemotherapy (MTD to efficacious dose ratio) is quite small, and patients are dosed with the drugs to toxic levels in order to achieve maximum clinical benefit. Chemotherapeutics are administered by injection directly into the bloodstream or are absorbed into the blood after oral dosing where they circulate in the bloodstream. In contrast to surgery and radiotherapy, which are local and regional therapies, respectively, chemotherapy is a form of systemic therapy because the dose is distributed throughout the body. Although this means toxicity will be more extensive, systemic therapy is the only conventional modality that potentially can treat every malignant cell of a metastatic cancer, even in micrometastatic disease with unknown locations of tumor foci. However, tumor foci with poor perfusion or underdeveloped capillaries may be exposed to suboptimal drug levels because "drug delivery" is poor.

A great variety of cytotoxic drugs has been developed, and all cell cycle phases can be targeted (Table 8–1). For many of these drugs to be cytotoxic, cells must be exposed to the drug during the sensitive phase of the cell cycle, primarily because the drug's molecular target is only expressed, or is only required by the cell, during this particular phase of the cycle. Such drugs are therefore called cell cycle phase–specific drugs. To achieve exposure of every target cell during this specific phase, prolonged exposures to these drugs are required, and doses are usually divided over multiple days (e.g., daily for 5 days every 3 weeks). Improper scheduling of drug dosing usually leads to a dramatic loss of effectiveness; hence, these drugs are also called "schedule-dependent drugs" (Dorr and Fritz 1980; Berenson and Gale 1990).

Relatively few schedule-dependent drugs are toxic to cells in G_0 or G_1. Although it is usually considered an inhibitor of mitosis, taxol is most toxic to cells making the $G_0 \rightarrow G_1$ transition (Donaldson, Goolsby, and Wahl 1994). During the $G_1 \rightarrow S$ transition to early S phase, camptothecins inhibit topoisomerase I, an enzyme that relieves tension in DNA (Slichenmyer et al. 1994).

There are many S phase–specific drugs and many biochemical targets for cytotoxic drugs during this phase. Inhibition of thymidine biosynthesis with nucleoside analogues is a favorite strategy because thymidine is the only nucleotide specific to DNA. Inhibition of DNA synthesis can be accomplished with 5-fluorouracil or floxuridine, which inhibit thymidylate synthase, and methotrexate, which inhibits dihydrofolate reductase production of reduced folate required by thymidylate synthase (Heidelberger 1973; Allegra 1990). The combination of methotrexate and 5-fluorouracil is sometimes used to increase the effect (Grem 1990). Tumors that rely heavily on the "salvage pathway" that phosphorylates thymine and bypasses the

Table 8–1. *Some characteristics of commonly used antineoplastic drugs*

Drugs that specifically inhibit enzymes or biochemical processes required only at one point in the cell cycle are usually schedule dependent whereas those that exhibit multiple cytotoxic mechanisms are generally schedule independent. See bibliographic citations for detailed information about these drugs. Drugs that affect cells at several points in the cell cycle are active against several types of cancer (indicated by *). Nucleophilic targets include both macromolecules and small biochemicals like glutathione. The G_1 phase remains underutilized as a target.

Phase of the cell cycle	Name of drug product	Common name	Molecular target
$G_0 \rightarrow G_1$	Taxol*	paclitaxel	microtubules
G_1	none		
$G_1 \rightarrow S$ transition	Hycamtin, Camptosar	topotecan, CPT-11	topoisomerase I
S phase	Cytosar, Cladribine	ara-C, 2-chlorodeoxyadenosine	DNA synthesis
$S \rightarrow G_2$ transition	Etoposide	VP-16	topoisomerase II
G_2	Blenoxane	bleomycin	???
M	Oncovin, Taxol*	vincristine, paclitaxel	microtubules
nonspecific	Platinol*	cisplatinum	nucleophiles
nonspecific	Adriamycin*	doxorubicin	topoisomerase II, DNA
nonspecific	Cytoxan*	cyclophosphamide	nucleophiles

need for the de novo pathway would be resistant to these drugs (Rabinowitz and Wilhite 1969). Some S phase–specific nucleoside analogues act as competitive inhibitors of nucleoside kinases, and some of these have additional mechanisms, such as terminating DNA chain elongation. Examples of such nucleoside analogues are ara-C, 6-mercaptopurine, and 2-chlorodeoxyadenosine, which are highly active against many leukemias (Elion and Hitchings 1965; Skipper, Schabel, and Wilcox 1967; Calabresi and Parks 1980; Carson et al. 1980; Chabner 1990a; McCormack and Johns 1990). During the $S \rightarrow G_2$ transition, etoposide (VP-16) inhibits topoisomerase II, which contributes to organization of DNA topology for mitosis (Bender, Hamel, and Hande 1990).

Few drugs target the G_2 phase of the cell cycle. Bleomycin causes fragmentation of DNA into small pieces and inhibits $G_2 \rightarrow M$ progression (Chabner 1990b). Because exposed cells accumulate at the G_2/M boundary, bleomycin has been used to synchronize tumor cells for subsequent exposure to other cytotoxic therapies (Calabresi and Parks 1980).

The M phase–specific cytotoxics disrupt microtubule function. Vincristine and vinblastine, extracted from the periwinkle plant, depolymerize microtubules (Noble et al. 1958; Johnson, Wright, and Svoboda 1960; Johnson, Armstrong, and Gorman 1963). Taxol, extracted from bark and needles of yew trees, hyperstabilizes the microtubules, making them rigid and static rather than dynamic (Wani, Taylor, and Wall 1971; Fuchs and Johnson 1978; Bender et al. 1990).

A second class of drugs causes the death of cells that have been exposed during any phase of the cell cycle. Because cytotoxicity does not require drug exposure during a sensitive phase of the cell cycle, they are called schedule-independent drugs, and they are administered to patients by a variety of schedules from bolus to long IV infusions. Several hypotheses have been proposed to explain cytotoxicity, including ATP depletion (Smulson et al. 1977), asynchronous progression of cell cycle mechanisms (Li and Deshaies 1993), and interference with mitochondrial DNA function (Lin et al. 1994). Several of these drugs are mainstays of modern cancer chemotherapy: doxorubicin (Adriamycin), the nitrosoureas, cisplatin, cyclophosphamide, melphalan (L-PAM), and chlorambucil. Cyclophosphamide is a prodrug that decomposes into cytotoxic species following metabolic bioactivation by a type of enzyme called cytochrome P-450 (Colvin and Chabner 1990). These drugs cause many of the same side effects as the schedule-dependent drugs: bone marrow suppression, mucositis, and hair loss. Some of them also cause cumulative dose-dependent toxicities to slowly or nonproliferating normal tissues like kidney, liver, nervous system, and heart. For example, doxorubicin is limited to a cumulative total dose of about 550 mg per m^2 of body surface area because of the risk of congestive heart failure (Myers and Chabner 1990).

Why are these drugs schedule independent even though they are preferentially toxic to dividing cells? Cisplatin, nitrosoureas, chlorambucil, and melphalan are alkylating agents that covalently modify proteins and nucleic acids and cause long-lasting chemical modifications within the cell (Calabresi and Welch 1962; Johnston, McCaleb, and Montgomery 1963). Cytotoxicity results when exposed cells attempt cell division before repairing the critical damage. Second, schedule independency could be due to multiple drug targets and a high probability that at least one of these critical targets is expressed during every phase of the cell cycle. For example, cisplatin is most toxic to cells in early G_1, but it is cytotoxic throughout the cell cycle (Rosenberg et al. 1965, 1969; Donaldson, Goolsby, and Wahl 1994). Doxorubicin interacts directly with topoisomerase II and the plasma membrane, intercalates into DNA, and generates free radicals (Lane et al. 1987; Myers and Chabner 1990). Multiple mechanisms may explain why these drugs are so frequently used clinically.

Another important research question is why the dose-limiting toxicities of drugs differ when they are all cytotoxic to dividing cells. Most drugs cause bone marrow toxicity, causing a decrease in neutrophils or platelets at 8 to 14 days after exposure (myelosuppression). However, a few drugs are not myelosuppressive at standard doses: floxuridine, vincristine, bleomycin, streptozotocin, and cisplatin (Dorr and Fritz 1980). Because of their lack of bone marrow toxicity, they are often included in combination chemotherapy protocols to achieve an additional antitumor effect without additional suppression of white blood cell and platelet counts. Another interesting example is the comparative toxicology of two vincas: the dose-limiting toxicity of vincristine is neurotoxicity, but the closely related analogue vinblastine is myelosuppressive (Dorr and Fritz 1980). The alkylating agent cisplatin exhibits kidney toxicity but rarely suppresses platelet levels even though its analogue, carboplatin, causes depression in platelet counts but less nephrotoxicity (Reed and Kohn 1990).

8.3   Cytoreduction theory and cancer "cure"

8.3a   Absolute versus fractional cytoreduction

Cancer usually cannot be diagnosed until the tumor burden in the patient reaches at least 10^{10} malignant cells, and a tumor burden of 10^{13} is approximately the number of cells that are lethal. Surgery will reduce the tumor burden by the amount of cells localized within the macroscopic disease. The tumor burden does not affect the outcome of surgery because even very large tumors can be removed. A patient with a tumor burden of 10^{12} will be cured if all of these cells are contained within the tumors (patient 1, Figure 8–1). However, any tumor burden in unresectable lesions, including micrometastatic disease, will not be removed and will regrow the lost tissue at a rate dependent on the doubling time of the tumor (patients 2 and 3, Figure 8–1). The patient with the faster growing tumor will relapse and die sooner than the patient with the slower growing tumor, even though the surgeries were equally successful.

In contrast, cytoreduction from radio- or chemotherapy depends on the number of cells, and each course (cycle) of radiotherapy or chemotherapy kills a proportion or fraction of the dividing cells rather than a constant number. There is a significant difference in the cytoreduction achieved over time between therapy that kills a constant proportion of cells and one that kills a constant number of cells per cycle (Figure 8–2). It is conventional to quantify the cytoreduction in units of logs of cell kill, where each log-cell kill means a 10-fold reduction in cell number (one logarithm reduction). For example, a 2-log cell kill means that 99% of the cells are destroyed (100 reduced to 1). Although a 99% kill rate seems excellent, such a 2-log reduction leaves 10,000 surviving cells in a tumor of 1,000,000 cells and 10^{10} survivors in a patient with a typical tumor burden of 10^{12} malignant cells (Figure 8–2). Therefore, multiple exposures to the cytotoxic substance are required to eradicate the tumor burden completely, which is why patients must receive multiple cycles (courses) of cytotoxic therapy to have a chance at cure (Figure 8–2).

The concept of renewing tissues as mosaics of discrete and separate populations or phenotypes in symbiotic relationship, so-called unit characters during tumor progression (see Chapter 1), explains why a constant fraction of the tumor cell population is killed by cytotoxic therapy with each cycle. If phenotype A always occupies 90% of the malignant tissue at homeostasis, then no matter how many tumor cells are present, a dose of therapy that selectively kills cells of phenotype A will result in one log-cell kill. Other biologic patterns beside mosaics may determine sensitivity to therapy.

When the 10^{12} malignant cells are localized, surgery is obviously preferred over cytotoxic therapy. This tumor burden would take seven courses of 2-log cell kill per cycle of toxic therapy to eradicate. However, what if 1,000 malignant cells have

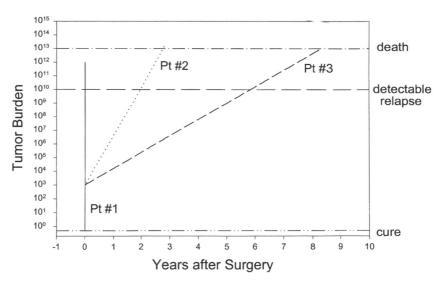

Figure 8–1. The treatment of cancer with surgery alone. Three patients each with tumor burden of 1×10^{12} were treated with surgery alone. In patient 1, all of the tumor burden was localized within the resected tissue, and the tumor burden remaining after surgery was on average <1 cell, so cure resulted. In both patient 2 and 3, 1,000 malignant cells had spread outside the resected area by the time of surgery. However, patient 2 suffered relapse from disseminated disease 2 years after surgery and death at 3 years from lethal tumor burden. Patient 3 did not reach these same end points until 6 and 8.5 years, respectively, even though surgery was equally effective in the two patients, because the residual tumor multiplied more slowly in patient 3 than 2.

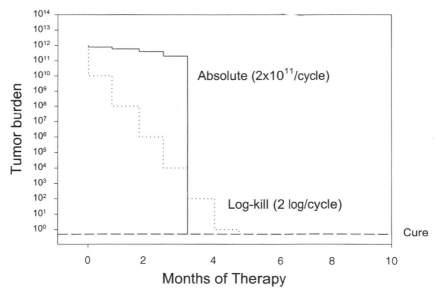

Figure 8–2. The different nature of cytoreduction between absolute cell kill and log-cell kill per cycle of treatment. Modern cytotoxic therapies such as radiotherapy or chemotherapy reduce tumor burden by a fractional amount each cycle, rather than an absolute amount. The proportion of malignant cells killed per cycle is quantified in orders of magnitude as "log-cell kill." In this example, an initial tumor burden of 10^{12} cells is reduced to <1 cell by 7 cycles of therapy that achieves 2-log cell kill per cycle (99% reduction) or by 5 cycles of therapy that achieves a 2×10^{11} absolute cell kill per cycle.

disseminated outside the surgical margins before resection? In this case, the patient will be declared "cancer free" after surgery (without clinically detectable disease), but these residual 1,000 cells will eventually regrow and cause a relapse. The time to relapse will depend on the tumor's growth rate (Patient 2 vs. 3, Figure 8–1); tumors with the fastest growth rates will relapse after the shortest period of time.

To eradicate every malignant cell from the body of a patient with metastatic disease obviously requires the combined use of surgery, radiotherapy, and chemotherapy. Combining these modalities into an integrated treatment strategy is called multimodality therapy, and in most cases it offers a better chance of cure than any single modality. Obviously, cytotoxic therapies must be used if the tumor is unresectable.

8.3b   What is a cancer cure anyway?

The log-cell kill principles of curative cytotoxic therapy were discovered in animal models of chemotherapy (Schabel 1969, 1975; Skipper et al. 1970; Shackney, McCormack, and Cuchural 1978; Norton 1979, 1990; Norton and Simon 1986). These principles show that cure occurs only when the total log-cell kill from all courses has eliminated the last malignant cell because a single malignant cell has the capacity to regrow into a symptomatic tumor and kill the host (Skipper et al. 1970). Thus, the goal of cytoreductive therapy is often stated as reducing the surviving number of malignant cells to less than one. This principle that cure only occurs after eradication of the last remaining malignant cell is often called the Skipper hypothesis to honor its discoverer.

Although patients also conceptualize a cancer cure in the same way, in fact what they usually mean is that they want to live the rest of their lives without any symptoms of the cancer. They want to live the rest of their lives "cancer free." This goal is quite different from the Skipper hypothesis and does not necessarily require eradication of every last cancer cell. Surgical resection can leave behind some micrometastatic disease that will regrow at a rate determined by the tumor doubling time. For cancers that usually grow at slow rates (many solid tumors), complete surgical resection that provides about 2.5 years of disease-free survival will result in a "cure" of a patient who lives only 2 more years after the operation and dies of causes unrelated to the cancer, but will "fail" a patient who lives longer, experiences a relapse around 2.5 years after, and eventually dies of a lethal tumor burden after about 10 years (Figure 8–3). Chemotherapy or radiation therapy might be indicated in the latter patient but not in the former, because it makes a difference in survival only in the patient with longer life expectancy but causes only toxicity without benefit in the other patient (Figure 8–3). It is often quite difficult to distinguish these two types of patients, however, so cytotoxic therapy is generally given to everyone in a specific prognostic group.

A major difference between the animal models and the human patient has to be considered when extrapolating the principles of cancer cure to the clinic: most

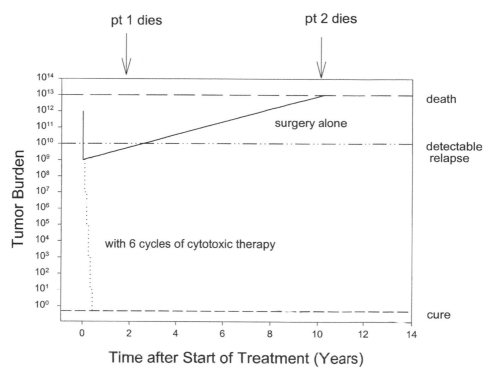

Figure 8–3. What is the meaning of cancer cure? Two patients have successful surgery for cancer that reduces the tumor burden from 10^{12} to 10^{9} (not clinically detectable). The residual tumor burden multiplies at the same rate in both patients such that relapse is detectable at 2.5 years and death from cancer will occur at 10 years. However, patient 1 dies from causes unrelated to the cancer at 2 years, prior to relapse, and would undoubtedly be thought of as a "cure" to patient and his or her family. Patient 2 will be a cancer fatality because he or she lived long enough to be one. This example underscores the importance of understanding the term "curing cancer" in the context of the patient's life expectancy. Instead of trying to eradicate all malignant cells from the body, the patient desires only to never be bothered again by the malignancy for the remainder of his or her life. Although 6 cycles of therapy would cure both patients, only patient 2 would benefit.

human cancers regrow much more slowly than the tumors used in the animal models. Patient life expectancy from other causes and conditions is a variable that simply cannot be factored into the animal models. Although the log-cell kill principles and the Skipper hypothesis definitely apply to clinical oncology, and the goal of complete eradication of the malignancy probably applies to younger patients who will live long enough for tumors to regrow to symptomatic levels, and to patients with rapidly growing cancers, these concepts must be tempered by the surrounding clinical circumstances of each case. In summary, the intensity of treatment depends not only on the required log-cell kill to eradicate disease but also on the age of the patient and any other medical conditions that might shorten life expectancy. The purpose of cancer therapy is to achieve a normal life expectancy in the cancer patient with no more than minimum impact of the cancer on daily living.

The relatively long time required for repopulation of most tumors is also the reason why most patients must remain disease free for 5, 10, or even 15 years in order to know for certain that they have been cured. Before this time, the cancer may simply be regrowing at tumor burdens that are not clinically detectable.

8.3c   The success and failure of multimodality therapy

A high likelihood of metastatic spread by the time of diagnosis combined with a high probability that the patient will live long enough after surgery for micrometastatic disease to regrow and cause relapse has caused an evolution of treatment strategy toward multimodality therapy. This strategy combines surgery for decreasing the tumor burden, radiotherapy for control of regional micrometastatic disease and unresectable lesions, and systemic chemotherapy for micrometastatic disease in other organs. This strategy is justified by cure rates less than 100% following surgical resection alone for even the earliest stages of many cancers (Table 8–2). In other words, only the lowest stage and grade of some solid tumors can be assumed to be cured by surgery alone. For example, malignant ovarian cells from even grade I stage 1a tumors may be found outside the tumor margin in adjacent tissue, and survival rates are 94% to 98% at 6-year median follow-up rather than 100% (Colombo 1994; Young 1994). One conundrum of clinical oncology is how to determine which early-stage patients require multimodality therapy and which do not. All patients with intermediate stages or high grades must be assumed to have metastasized by time of diagnosis, and these patients are generally treated with a multimodality approach. Curing the cancer patient requires the combined use of effective local, regional, and systemic therapies, which is the rationale behind multimodality therapy.

Because of this evolution in the biologic and clinical understanding of cancer biology from the beginnings of surgical oncology (Halsted 1894–1895), surgical techniques have been modified to spare tissue and preserve function (Baum 1976; Fisher et al. 1977; Fisher, Bauer, Margolese et al. 1985; Fisher, Redmond, Fisher et al. 1985; Fisher, Osborne, Margolese et al. 1993; Charlson 1985; Haskell et al. 1990). Radical resection and exenteration (complete removal of a block or region of the body) are being phased out because they cause significant morbidity and mortality without cure. Tissue-sparing operations have developed for breast cancer (Fisher, Wolmark, Fisher, and Deutsch 1985; Fisher et al. 1993; Margolese et al. 1987), ovarian carcinoma (Colombo 1994), sarcoma of the limb (Morton, Antman, and Tepper 1993), and prostate carcinoma (Walsh, Lepor, and Eggleston 1983; Walsh et al. 1990; Jewett 1970). These conservative surgeries play an important role in increasing the quality of life of the patient by preventing the malformations and dysfunction that result from radical surgery. Given that the purpose of surgery is primarily tumor debulking (reducing tumor burden) and

Table 8–2. *Cure rates for early-stage cancers (node negative where indicated) following single-modality surgery*

If these diseases were truly localized at time of diagnosis, then cure rates should be 100%. The fact that cure rates are less than 100% indicates that many cases are systemic disease at time of treatment. Surgery is curative in true stage I disease (which is difficult to prove at diagnosis) and in those patients whose life expectancy is less than the time required for the disseminated microscopic disease to grow to clinically detectable levels.

Malignancy	Long-term survival	Reference
Breast carcinoma (node negative)	77–87%, 3-year survival	Fisher et al. 1977; Fisher, Redmond, Fisher et al. 1985; Fisher, Redmond, Poisson et al. 1989
Breast carcinoma (stage I)	38%, 10-year survival	Baum 1976
Stomach (node negative)	81–98%, 5-year survival	Haskell, Selch, and Ramming 1990b; Wu et al. 1992
Kidney	60–76%, 5-year survival	Myers, Fehrenbaker, and Kelalis 1968; deKernion and Berry 1980; Neuwirth, Figlin, and deKernion 1990
Ovary (stage I)	96%, 5-year survival	Berek and Hacker 1990
Non-small cell lung	57–86%, 5-year survival	Figlin et al. 1990
Uterus	76–80%, 5-year survival	Barber and Brunschwig 1968; Berman and Berek 1990

relief of symptoms, the effectiveness of radiotherapy and chemotherapy become critical for accomplishing long-term, disease-free survival (so-called "cures").

As an example, a patient is diagnosed with a 6 cm tumor that is resected successfully with clear margins. This tumor of approximately 110 g contains about 10^{11} cells. However, some regional lymph nodes dissected during the resection were found to contain malignant cells, so the patient is at high risk for relapsing from distant metastases later in life even though thorough staging procedures do not detect any disseminated disease at the present time. Because 10^{10} cells would not be clinically detectable (Haskell 1990; Haskell et al. 1990), it must be assumed that this many cells have migrated to distant sites prior to surgery. This case can be represented in terms of tumor burden (Figure 8– 4, top). At diagnosis, there are 1.1×10^{11} malignant cells in the body, of which 10^{11} are removed by surgery and 0.1×10^{11} are left behind in unknown locations. Assuming a 2-log cell kill per cycle of chemotherapy, then 6 cycles of therapy to reduce the tumor burden to less than 1 cell should cure the patient (Figure 8– 4, top). However, our experience teaches us that this typical cancer patient is usually not cured and if middle aged or younger has a high probability of recurrence. Why is this? Several other variables contribute to treatment failure.

To achieve the log-cell kill curve in Figure 8– 4 (top), it was assumed that tumor burden does not change between cycles of cytotoxic therapy. However, this is not the case, for tumors, like any renewing tissue, will try to replace the lost tissue between cycles of therapy. Tumor repopulation makes it more difficult to cure 10^{10} cells than simply 6 cycles of a therapy with a 2-log cell kill. It takes only 7 doublings of the tumor cell population to replace the 99% of cells lost from the 2-log cell kill

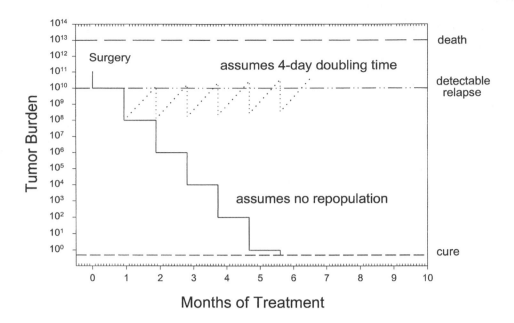

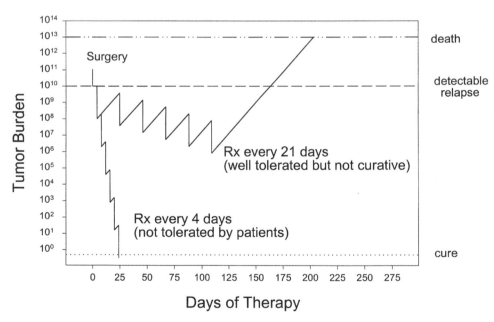

Figure 8–4. (Top) Surgical resection reduces tumor burden from 11×10^{10} to 1×10^{10}, which is the maximum tumor burden that is not clinically detectable. Six cycles of cytotoxic therapy that achieves 2-log cell kill per cycle given every 28 days will reduce this tumor burden to <1 cell, assuming the surviving tumor cells are not proliferating between cycles of therapy. However, we know this assumption is wrong, and the cytoreduction in the tumor stimulates the surviving malignant cells to proliferate. This example shows that a tumor doubling time of 4 days is sufficient to replace the 2-log cell kill of one cycle by the time of the next one, so there is actually a net increase in tumor burden over time in spite of chronic therapy every 28 days. (Bottom) Given this tumor that doubles every 4 days, cytoreductive therapy given every 4 days can achieve a cure in six cycles.

of therapy ($2^7 = 128$). If the tumor doubled rapidly, say every 4 days, the 2-log cell kill would be replaced in just 28 days (4 days × 7 cell divisions). Receiving chemotherapy every 4 weeks would only control the cancer, for the surviving cancer cells would replace the killed ones by the time of the next cycle (Figure 8–4, top). To produce any net decrease in tumor burden with each cycle, the patient must be treated every 21 days (Figure 8–4, bottom), and to be cured in 6 cycles, the patient must be treated every 4 days (Figure 8–4, bottom). The final cycles of therapy might need to be spaced even more closely together if increased dose intensity (dose per unit time) is required to eradicate the last 1 to 10 surviving cells (Wilcox 1966).

So, why not treat the patient every 4 days? Remember that cytotoxic therapy also produces dose-limiting toxicities from which the patient must recover before receiving a subsequent course of therapy. Administering additional therapy prior to recovery can exacerbate already suppressed dose limiting organ function to such an extent that it can never recover. The acute toxicity to the dose-limiting proliferative tissue is generally what limits dose intensity. Repopulation must occur to replace the log-cell kill in critical proliferating tissues like bone marrow and gastrointestinal epithelium, which manifest as the acute dose-limiting toxicities, before the next cycle can be tolerated. The clinician uses laboratory tests to check that tissue function has recovered before administering the next course. Subsequent courses can be delayed while waiting for evidence of recovery, such as a return of peripheral blood cell counts to adequate levels. It generally takes these tissues about 2 to 3 weeks to recover, so chemotherapy with most drugs is scheduled every 3 to 4 weeks. Cytotoxic dosing every 4 days for 6 cycles would likely cause life-threatening marrow suppression and/or gastrointestinal dysfunction. A few drugs can be tolerated more frequently, such as weekly taxol or gemcitabine, although these are exceptions. The delayed or protracted toxicity of some drugs like the nitrosoureas may require 6 to 8 weeks for recovery. These drugs must offer some benefit to the patient over the others to justify their use, knowing that the extent of tumor repopulation between cycles will be greater than for other drugs. Scheduling of cytotoxic therapy is a trade-off between the recovery time of the dose-limiting tissue and the repopulation rate of the malignancy. If the tumor can repopulate as fast or faster than the dose-limiting proliferative tissue, then cure with cytotoxic therapy is "clearly impossible" (Schabel 1969).

(**Figure 8–4,** *cont.*) However, this intensive regimen of cytotoxic therapy cannot be tolerated by human patients due to life-threatening, acute toxicities to gastrointestinal mucosa and bone marrow, which must recover before administering subsequent cycles. Administering cytotoxic therapy every 21 days to allow time for recovery of normal tissues between cycles could achieve a cure with a large number of treatment cycles, but patients cannot usually tolerate the required number of cycles because of cumulative dose-dependent toxicities to nonproliferative tissues (heart, brain). In the graph, it has been assumed that patients can tolerate 6 cycles of therapy. This is not sufficient therapy to eradicate all malignant cells, and the surviving tumor will regrow and kill the patient if he or she lives long enough.

Note that in our example, treatment every 21 days can achieve a net reduction in tumor burden with each treatment cycle (Figure 8–4, bottom). So, even though one cannot treat every 4 days to cure in 6 cycles, why not continue to treat every 21 days for as many cycles as are required to achieve a cure? The reason is that chronic therapy with cytotoxic modalities, either drugs or radiation, is limited by cumulative dose-dependent toxicities. These chronic toxicities may occur in different tissues than the acute toxicities, such as nerve or heart. Because the damage is to nonrenewing tissues, the toxicity of the therapy accumulates because damaged cells cannot be replaced. These toxicities are insidious and particularly troublesome clinically, for they are often delayed and are only slowly reversible, or not reversible at all. Because these chronic toxicities depend on cumulative dose, they limit the maximum number of treatment cycles that a given patient can tolerate. Although treatment every 21 days can in theory cure the patient in spite of acute, dose-limiting toxicities, in practice chronic toxicities generally limit the maximum number of tolerated cycles to a level below that required for cure. Healthy dose-limiting tissues can only tolerate a limited number of cycles of cytotoxic therapy, and eradication of every malignant cell becomes theoretically impossible once the tumor burden grows larger than can be killed by the maximum tolerated number of cycles of therapy with its characteristic log-cell kill.

This line of reasoning points to the typical requirements for cancer eradication in this example case when the cancer was detected relatively early in its course at 1.1×10^{11} cells. First, there needs to be surgical debulking to microscopic disease or less (about 10^{10} cells); otherwise, the increased tumor burden simply adds to the required number of cycles of therapy (more log-cell kill). Second, there must be sufficient log-cell kill to eradicate the tumor burden in about 6 cycles of therapy in spite of the fact that the tumor will repopulate during the typical 21-day recovery period between cycles. Given a more typical doubling time of 3 weeks for a solid tumor, you can calculate that a 2.1 log-cell kill per cycle is required.

If these requirements cannot be met, then residual cancer will remain in the body after the 6 cycles of cytotoxic therapy. Whether or not "cancer cure" still results is a matter mainly of how long the patient lives compared to how long the cancer will take to repopulate to clinically detectable levels and cause a relapse (at least 10^{10} cells). If chemotherapy reduces the tumor burden to 32 cells and this residuum doubles every 3 weeks, it will take only 21.5 months (86 weeks) for it to complete the 28.5 doublings required to reach detectable tumor burden, and just another 7.5 months for it to reach a lethal tumor burden of 10^{13} cells (Haskell 1990; Haskell et al. 1990). Even with this highly effective regimen for a relatively early detected tumor, the patient endured about 5 months of therapy to experience only 24 months of complete remission and 32 months of overall survival. If the patient died in a car crash or from a heart attack after 1 year, it would be a cancer cure, but if the patient lives 25 months, the case would be a treatment failure. The explanation for why many patients live many more than 2 years after multimodality therapy but ultimately relapse means that tumor doubling times during repopulation must be slower than 3 weeks. It cannot be due to log-cell kills more than 2.1, otherwise 6 cycles would be curative.

When is the optimal timing of radio- or chemotherapy after surgery? Understanding the answer to this question requires an understanding of tissue renewal itself. There are two cell types with different roles in tissue renewal, whether malignant or normal (Pierce and Speers 1988, Chapter 1). Progenitors (also called transient amplifying cells) are cells that divide rapidly during a relatively short period of time to generate tissue mass, but then terminally differentiate after a finite number of population doublings into the mature cell for the tissue. Stem cells divide rapidly and indefinitely, but only periodically, in order to produce more progenitor cells, or in order to maintain their own numbers; otherwise, they remain in a nonproliferative state that is relatively resistant to cytotoxic chemo- and radiotherapy. When tissue mass decreases through cell loss, the stem cell content of the growth fraction increases rapidly and transiently to produce cells that restore the tissue to its steady-state size (Al-Dewachi et al. 1980; Pierce and Speers 1988; Potten and Chadwick 1994; Potten et al. 1994). Tumors also appear to have a maximum size that is tightly regulated through homeostatic mechanisms: when tumor burden is lost, proliferation in the remaining tumor burden increases, usually via increasing growth fraction, in order to reestablish homeostatic tumor mass (Wilcox et al. 1965; Laster et al. 1969; Schabel 1969, 1975; Shackney et al. 1978; Dorr and Fritz 1980; Vaughan, Karp, and Burke 1984; Fisher, Saffer, Rudock et al. 1989a, 1989b; Cameron, Hardman, and Skehan 1990). This is the process called repopulation, and when stem cells enter the growth fraction, they become more sensitive to chemo- and radiotherapy.

In modern multimodality therapy, surgery is often used to reduce solid tumor burden to the point at which repopulation is stimulated and the remaining tumor is sensitized to cytotoxic therapy. This use of cytoreductive surgery without intent to cure is called "debulking," and the use of *properly timed* radiotherapy and/or chemotherapy after surgery to kill malignant cells during repopulation is called adjuvant therapy (Fisher, Gunduz, and Saffer 1983; Wittes 1986; Forbes 1990). This approach to chemotherapy and radiotherapy is far superior clinically to waiting for relapse to occur before administering postsurgical chemo- or radiotherapy, and it has become the dominant idea in cancer treatment. The superior efficacy of adjuvant therapy compared to therapy at relapse is evident in animal models, in which the most sensitive tumors were the ones smaller than the clinically detectable size (Wilcox et al. 1965; Laster et al. 1969; Schabel 1969; Shackney et al. 1978; Fisher et al. 1983; Fisher et al. 1989a, 1989b) and chemotherapy was only curative after surgical debulking (Shapiro and Fugmann 1957).

To make sure that the tumor is exposed during repopulation and to decrease acute toxicity, a dose of cytotoxic therapy is split into multiples and administered over time (usually on different days), a strategy called fractionation. With fractionated dosing, the malignant cells that were out of cycle during the first fraction may be at a sensitive point during the second or third. Dose fractionation originated in the finding that radiation doses which sterilize rams are equally effective but better tolerated by the overlying skin when fractionated (Regaud and Ferroux 1927). Thus, differences in growth fraction or cell cycle time between the target tissue and the dose-limiting normal tissue can be exploited by dose fractionation. Fractionation prevents toxicity caused by exceeding threshold levels but still delivers the same total dose. As an example, hyperfractionated radiotherapy (exposure 2 to 3 times per day, each below the maximum tolerated dose for the CNS) delivers 6,141 cGy instead of the 5,800 cGy dose in 200 cGy fractions (Batzdorf, Black, and Selch 1990). Unfortunately, fractionation cannot be used to protect very slowly renewing tissues such as kidney and liver from toxicity in which cellular proliferation is slower than in most cancers.

Some tissues may be treated with fractionated adjuvant therapy because of a high likelihood, based on prior cases, that they contain metastatic cells, *even though cancer is not yet detectable*. This type of adjuvant therapy, based on the predictable nature of the natural history of cancers, is called "prophylactic therapy." This is an unfortunate term for this type of adjuvant therapy because it wrongly implies that metastasis can be prevented by treating the tissue with cytotoxic therapy before any cancer cells have arrived. Instead, its mechanism must involve the log-cell kill of micrometastatic disease that is already present in the tissue but has not yet grown to sufficient size to be clinically detectable. Radiotherapy of the brain, or direct injection of drug into the brain, is often part of multimodality therapy for leukemia and small cell lung carcinoma to "prevent" brain metastases, which previous clinical experience shows are often the first evidence of relapse and sometimes the cause of treatment failure (Holmes, Livingston, and Turrisi 1993; Schiffer 1993; Trigg 1993).

Early detected tumors have probably just completed a period of maximum growth rate called log-phase growth, and the growth rate is just beginning to decelerate as tumor burden approaches homeostatic size (Dorr and Fritz 1980). As tumor burden increases toward this limit, growth rate will slow, growth fraction decrease, the proportion of stem cells in the growth fraction decrease, and the tumor will become refractory to chemo- and radiotherapy ("refractory" means reversible resistance as physiologic conditions change, distinct from "resistant," which is not reversible). It is easy to understand why chemotherapy and radiotherapy are most effective against small tumor burdens (Wilcox et al. 1965; Laster et al. 1969; Schabel 1969; Shackney et al. 1978). Not only is a lower log-cell kill required, but also a large growth fraction containing cycling stem cells maximizes sensitivity to cytotoxic therapies. In patients with advanced stages of cancer, the tumors exist in so-called plateau (stationary) phase growth, in which growth fraction is low, consisting of a few progenitor cells and rare stem cells. Consequently, these tumors are refractory to cytotoxic therapy. However, a sudden reduction in tumor burden will trigger repopulation, increasing the growth fraction in part by activating stem cells as the tumor attempts to restore homeostatic mass. The growth fraction increases, the stem cell contribution to the growth fraction increases, and the tumor becomes more sensitive to cytotoxic therapy (Wilcox et al. 1965; Laster et al. 1969; Schabel 1969; Shackney et al. 1978; Gunduz, Fisher, and Saffer 1979; Dorr and Fritz 1980; Fisher, Gunduz, and Saffer 1983; Fisher, Saffer, Rudock et al. 1989a, 1989b; Vaughan et al. 1984; Cameron, Hardman, and Skehan 1990). Repopulation can be induced in advanced tumors with some cytotoxic drugs or radiation. In animal tumor models, cyclophosphamide induces a large log-cell kill in slowly growing malignancies and sensitizes remaining tumor cells to cell cycle–specific drugs to which it was originally refractory (Schabel et al. 1965; Schabel 1969). This activity may explain why cyclophosphamide is one of the most active drugs against human and animal solid tumors (Griswold et al. 1968). This phenomenon, which occurs in leukemia as well, is important for determining the schedule for chemotherapy cycles (Vaughan et al. 1984).

The fact that gastrointestinal epithelium and blood cells reappear following radio- or chemotherapy indicate that stem cells are more resistant than progenitors to cytotoxic therapy. The relative resistance of the stem cell pool is due to several characteristics. Stem cells can express mechanisms of drug resistance not expressed in progenitors (Drenou et al. 1993), and most of them are quiescent at any one time. However, stem cells can be killed by cytotoxic therapy if treatment is timed to expose stem cells while in the cell cycle, such as during repopulation or after growth factor stimulation (Molineux et al. 1994). Malignant stem cells would behave similarly, and this may be an important factor in the success of adjuvant therapy. Because curing cancer may be closely linked to ablating normal renewing tissues, so-called high-dose therapy is being tested in which patients are exposed to cytotoxic therapy at levels above the maximum tolerated dose and then "rescued" by transplantation of replacement tissue from a donor (e.g., bone marrow transplantation after treatment with myelosuppressive drugs or total body irradiation). The idea is to administer a high enough dose to kill malignant stem cells, acknowledging that normal stem cell function will also be destroyed.

8.3d Complicating factors that decrease log-cell kill

The average-sized human produces about 40 kg of intestinal epithelium each year, but it is rare indeed for a tumor to produce 40 kg of new tumor tissue in its entire lifetime. Most patients are diagnosed with 10^{11} to 10^{12} tumor cells (10 to 100 g of tumor, about 1/20 of a pound). If the log-cell kills in the malignant and normal cell populations are equal, then the tumor should be eradicated by fewer cycles of therapy than the normal tissue because normal cells far outnumber tumor cells. Because clinical experience shows this is not the case, there must be mechanisms that skew cytoreduction toward the 40 kg of intestine and away from the 100 g of tumor. These mechanisms are the subject of this section.

To eradicate a cancer (or normal renewing tissue), the stem cells that can repopulate all of the cellular elements of a tumor must be killed. Only 1 in 500 cells are estimated to be stem cells in the normal mucosa (Winton and Ponder 1990), yet cloning experiments indicate that 25% of the tumor cells may be malignant stem cells (Pierce and Speers 1988). This exaggerated capability for tissue renewal in the tumor translates into a 125-fold difference in stem cell content and changes the stem cell differential between the normal and malignant tissues from 800 (40 vs. 0.05 kg) to about 5. Thus, eradicating a clinically advanced tumor requires an intensity of therapy that would nearly eradicate the intestine.

Additional physiologic variables further skew log-cell kill toward normal tissues and away from malignant tissue. The growth fraction of a tissue is the proportion of proliferation-competent cells that are progressing through the cell cycle at any given time (Mendelsohn 1960). Only the cells in the growth fraction are sensitive to cytotoxic therapies; quiescent cells will not be sensitive unless they enter the growth fraction during repopulation before the cellular damage has been repaired. Unfortunately, the growth fraction is usually lower in the malignant than the dose-limiting normal tissue. Slowly growing solid tumors have growth fractions that average less than 10%, although there may be large interpatient variability, but even the high end of this range is significantly below the 30% to 60% growth fraction in gut epithelium or hematopoietic tissue of the bone marrow (Skipper and Perry 1970; Charbit, Malaise, and Tubiana 1971; Malaise, Chavaudra, and Tubiana 1973; Wright et al. 1973; Al-Dewachi et al. 1974, 1975, 1979, 1980; Rijke, Plaisier, and Langendoen 1979; Smith and Jarvis 1980). Because only cycling cells are affected by cytotoxic therapy, the 6-fold or greater difference in growth fraction neutralizes the remaining 5-fold advantage that the normal tissue seemed to have over the cancer based on stem cell content (see preceding paragraph).

Although high growth fraction confers sensitivity to chemo- and radiotherapy, it does not confer curability. *In fact, it is an important clinical conclusion that sensitivity does not correlate with curability* (Bingham 1978; Haskell 1990; Fisher et al. 1993). This is why *complete remission is necessary but not sufficient for cure of the cancer patient.* Of greater importance may be the nature of the cells in the growth fraction. Cells in the growth fractions of curable tumors (testicular cancer, neuroblastoma, childhood leukemias) might include all of the stem cells, whereas the growth fraction in incurable yet highly chemosensitive cancers like small cell lung carcinoma may contain exclusively progenitors but few, if any, stem cells. Because cytotoxic therapy kills any cycling cells and progenitors contribute to tissue mass, all of these tumors shrink dramatically in response to cytotoxic therapy, but only the small cell carcinoma relapses (repopulates) because its stem cell population has not been affected.

The cancer has yet another advantage over the normal renewing tissue: poorly functional vasculature. Adequate and pervasive blood flow throughout the tumor is essential for delivery of sufficient molecular oxygen for effective radiotherapy and sufficient drug levels for effective chemotherapy. Tumor growth may exceed the rate of formation of new blood vessels, so malignant cells may lie far from a blood supply (Dorr and Fritz 1980). The blood vessels in tumors are also very dynamic, opening and closing in seemingly random fashion in response to paracrine factors produced by the tumor cells (Chaplin and Trotter 1990). Thus, entire regions of the cancer may transiently lose blood perfusion, although the duration is brief and the cells remain in the growth fraction. But a transient loss of perfusion during a cycle of therapy will diminish the log-cell kill in this region of the tumor. Because this physiologic response does not occur in the vasculature of normal tissues, it further diminishes the effectiveness of the therapy (Simpson-Herren and Noker 1990). Note that this physiologic mechanism is nonspecific, conferring resistance to all drugs, and in many cases radiation too, because the critical drugs (and oxygen) cannot get to the malignant cells. Even the most sophisticated biotechnology approaches to cancer therapy will not work if the therapy cannot be delivered to the cancer foci. The poor perfusion is also one of the main reasons that cancers show lower growth fractions than their normal counterpart tissues. Tumor cells far away from the blood supply become dormant, entering the G_0 phase of the cell cycle in which they are resistant to cytotoxic therapy but from which they can become mitotically active again (Schabel et al. 1965; Pittillo, Schabel, and Skipper 1970). This means that the most drug-refractory cells (because they are not progressing through the cell cycle) are exposed to the lowest levels of anticancer drugs and radiation damage (due to hypoxia).

A related problem of delivery of the therapeutic agents occurs when cancers metastasize into, or arise within, tissues which have blood-tissue barriers formed by endothelial cells with specialized tight junctions that exclude noxious chemicals. One such barrier, called the blood-brain barrier, not only protects brain tissue from drug exposure but also can inhibit drug penetration from the blood into a tumor in the brain. Only a few drugs can penetrate these barriers, such as nitrosoureas, and these drugs are often used to treat CNS tumors (Batzdorf et al. 1990). Some cancers are not sensitive to nitrosoureas and must be treated by direct injection of drug into the cerebrospinal fluid. But only a few drugs are suitable for intrathecal injection; many are not because of acute or delayed neurotoxicity. A similar endothelial barrier for chemotherapeutics exists in the testes (Smith and Haskell 1990). Radiotherapy is often the only treatment for cancers behind these pharmacologic barriers, especially when surgical resection is risky.

Another cellular variable that reduces the effectiveness of many cytotoxic therapies is the slower than normal cell cycle time of most malignant cells, which is consistent with the protracted natural history of most cancers (11 years to reach 10^{10} cells, a 1 g tumor [Haskell 1990]). Slowed cell cycle times decrease the therapeutic index of cell cycle phase–specific chemotherapeutics and radiotherapy by decreasing the probability that a cycling cell will be exposed during the sensitive phase of the cell cycle (Skipper, Schabel, and Wilcox 1967). For example, a 6-hour infusion of a S phase–specific drug might provide 8 hours of systemic exposure: 6 hours of infusion plus 2 hours circulating in the blood before being cleared from the body via excretion and metabolism. This exposure time represents 50% of the 16-hour cell cycle time of rapidly dividing progenitors in gastrointestinal mucosa and bone marrow, but only 10% of the 80-hour cell cycle time in an adenocarcinoma (Skipper and Perry 1970; Charbit et al. 1971; Malaise et al. 1973; Wright et al. 1973; Al-Dewachi et al. 1974, 1975, 1979, 1980; Rijke, Plaisier, and Langendoen 1979; Smith and Jarvis 1980). The more rapidly dividing cell will more likely be exposed during S phase, and the log-cell kill from the 6-hour infusion will be greater in the rapidly dividing normal tissues than in the tumor. A continuous infusion of drug for 38 hours would be needed to achieve the same log-cell kill in the adenocarcinoma that the 6-hour infusion achieves in the normal tissues, and an infusion of 78 hours would be required to expose every adenocarcinoma cell in the growth fraction during S phase. Continuous infusions of many days' duration with some drugs have been tolerated by patients without irreversible toxicity, but other drugs are extremely toxic to normal tissues when administered on these schedules.

Drug resistance can also be due to biochemical mechanisms found in specific cell subpopulations of the tumor. Genetic instability and phenotypic selection generate drug-resistant cells in tumors even before they have been exposed to drug – the Goldie-Coldman hypothesis (Goldie and Coldman 1979). These cellular phenotypes of drug resistance likely arise in untreated tumors via the same genetic mechanisms responsible for the heterogeneity of other phenotypic traits (Berek and Hacker 1990, ovarian cancer), and therefore cellular drug resistance is fundamentally an extension of the unit character concept of tumor progression (Foulds 1965). Every cycle of chemo- or radiotherapy kills nonresistant cells, including cells in small unit characters that have yet to express a resistant phenotype. But resistant cells survive the initial exposure and by proliferating come to represent an ever greater proportion of the tumor with each cycle of therapy. When the drug-resistant population represents a proportion greater than the log-cell kill of the therapy, then each successive cycle of therapy will be less effective until at some point refractory disease will develop that is no longer responsive to the treatment (Figure 8–5). This is the most likely reason why relapsing cancers fail to respond a second time to drugs that initially induced complete remission. To cure tumors under these conditions, theoretical principles indicate that subsequent courses of therapy need to be of increasing intensity (Wilcox 1966), which the patient is most likely unable to tolerate, or with drugs that do not show cross-resistance patterns.

Some cellular resistance is specific for one particular drug or drug class. For example, resistance to 5-fluorouracil or methotrexate is due to overexpression and mutation of the target enzymes of these drugs or to decreased drug transport across the plasma membrane into the cell (Allegra 1990; Grem 1990). In these cases, resistance to methotrexate will not result in resistance to unrelated drugs like Taxol. Malignant cells can also express a multidrug-resistant phenotype, meaning that development of resistance to one drug brings cross-resistance to other drugs which have not been used (Ueda et al. 1986; van der Bliek and Borst 1989; Chabner 1990c; Chin, Pastan, and Gottesman 1993). High levels of glutathione transferases rapidly conjugate electrophilic drugs (quinones, alkylating agents) for clearance out

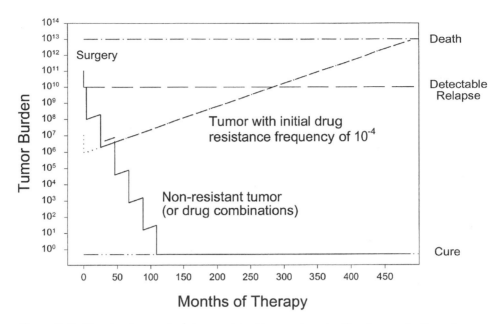

Figure 8–5. The development of drug-resistant tumors is a significant complication of cytotoxic therapy. It is caused by the overgrowth of tumor cell subpopulations that have adapted to drug exposure. Six cycles of adjuvant therapy administered to the patient as often as the doubling time of the tumor will eradicate micrometastatic disease ($<10^{10}$) after surgery (see Figure 8–4). However, the Coldman-Goldie hypothesis predicts drug-resistant cells exist with a frequency as high as 10^{-4}, so micrometastatic disease could contain 10^6 cells resistant to single agent chemotherapy even before the first cycle is administered. The resistant cells survive every cycle of therapy, grow at the doubling time until they dominate the cancer, and cause a relapse that will not respond to therapy which was effective in the patient in the past. This example assumes the same doubling time of 21 days for resistant and nonresistant cells. This phenomenon can be averted in some patients by using combination chemotherapy against micrometastatic disease because the probability that a cell is resistant to three drugs from noncross-resistant drug families is much smaller ($10^{-12} – 10^{-18}$).

of the cell and excretion from the body. Modifications to topoisomerase II cause resistance to doxorubicin, amsacrine, and etoposide. Increased capacity for DNA repair confers cross-resistance to alkylating agents like cyclophosphamide, melphalan, and nitrogen mustard. For example, a nuclear repair enzyme called alkyl transferase removes O-alkyl groups from guanine bases in DNA that result from exposure to alkylating drugs used for cancer as well as carcinogens (see Chapter 3). So-called *mer*– cells that lack this enzyme are extremely sensitive to alkylating agents; high levels of this enzyme confer resistance to many alkylating agents, not just the drugs which selected for the *mer*+ clones during chemotherapy (Ikenaga 1994).

Perhaps the most thoroughly studied cellular mechanism of multidrug resistance involves a family of proteins encoded by the *mdr* genes ("multidrug resistance" genes), the best known of which is the P-glycoprotein (and also P170) (Ueda et al. 1986; van der Bliek and Borst 1989; Chabner 1990c; Chin et al.

1993; Thorgeirsson, Gant, and Silverman 1994; Gottesman et al. 1995). These plasma membrane proteins utilize energy in ATP to pump natural products out of the cell (Cornwell et al. 1987). The *mdr* phenotype causes cross-resistance to many of the drugs commonly used in treating cancer, such as doxorubicin, Taxol, the vinca alkaloids, VP-16, and mitomycin-C. Vinca alkaloid resistance distinguishes the *mdr* phenotype from topoisomerase II–based resistance (Chabner 1990c). The drug resistance profile of this dominant phenotype could change during cytotoxic therapy because *mdr* genes are susceptible to mutations that change the substrate specificity of the pump. Cancers of the kidney, large bowel, liver, pancreas, and adrenal gland express *mdr* proteins as the normal differentiation program of these tissues to excrete toxic substances (Chin et al. 1993), and immature leukemic cells, like their normal hematopoietic stem cell counterparts, express P170 (Drenou et al. 1993). Other cancers, such as breast cancer, express ectopic P170; these tissues do not normally express the protein (Chin et al. 1993). In contrast to rodent tumors, overexpression in human tumor cells is most often due to transcriptional enhancement (Thorgeirsson et al. 1994; Glazer and Rohlff 1994; Gottesman et al. 1995).

Strategies to reverse drug resistance have been tested, but the biochemical modulation is not tumor specific and the result is generally increased drug potency; that is, less drug is required for a certain level of biologic effect, but no gain in therapeutic index or effectiveness. In addition, some modulators of drug resistance exhibit pharmacologic effects alone and are therefore not clinically useful (Sikic et al. 1994; Patel and Rothenberg 1994), but these prototype modulators of resistance can nevertheless provide structural leads for developing pure antagonists of drug resistance that are pharmacologically inert (Hait and Aftab 1992). A quite different clinical approach is to use the genes responsible for cellular drug resistance to genetically engineer increased drug tolerance in the dose-limiting tissue, usually bone marrow (Ward et al. 1994). However, the increase in dose gained with this approach is modest because higher doses are limited by more serious toxicity in some other tissues not easily modified with the resistance gene.

Drug resistance resulting from molecular characteristics of the malignant cell is very different in nature from that resulting from physiologic characteristics of the tumor such as fluctuations in blood flow or growth fraction. The first mechanism will not reverse during the course of treatment and therefore represents true drug resistance in the tumor. The second mechanism can reverse during therapy as physiologic conditions in the tumor change and therefore represents a drug-refractory tumor that is not necessarily drug resistant. Under this mechanism, cells could be killed by the therapy if only they could be exposed to it during the correct phase of the cell cycle. Cellular mechanisms of drug resistance can be studied in the laboratory (Greenberger, Cohen, and Horwitz 1994), but physiologic drug resistance is difficult to model in vitro. Perhaps the most difficult tumors to treat with cytotoxic therapy are those with rapid cell cycle time but small growth fraction. For example, higher grade ovarian cancer has a higher mitotic index, yet a worse chance of cure (Berek and Hacker 1990). Likewise, radio-resistant astrocytomas exhibit the highest mitotic index (Batzdorf et al. 1990). These findings seem like a contradiction to the theory of cytotoxic chemotherapy's preference for rapidly dividing cells. However, the high mitotic index exists in a very small growth fraction, and this fraction is so small that its eradication does not affect tumor mass and it can be quickly replaced during repopulation.

The Goldie-Coldman and Foulds hypotheses suggest that drug-resistant subpopulations can arise spontaneously via genetic instability and natural selection prior to exposure to chemotherapy, or during therapy as a result of positive selection of resistant clones (Skipper, Schabel, and Lloyd 1978; Norton 1979, 1990; Norton and Simon 1986). Combination chemotherapy uses simultaneous or closely timed exposures to several drugs that do not share mechanistic or structural similarities to minimize the impact of cellular drug resistance. By combining drugs which require different mechanisms of resistance, one is increasing the odds that every malignant cell of the tumor will be sensitive to at least one drug to which it is exposed (Schabel et al. 1980). Goldie and Coldman (1979) estimated that drug resistance occurs at a frequency of about 10^{-5}. In a tumor of 10^{12} cells, then, one would expect 10^7 resistant cells regardless of the drug selected. However, assuming independent genetic events and unrelated mechanisms of resistance, cells resistant to two unrelated drugs will be present with a frequency of 10^{-10} ($10^{-5} \times 10^{-5}$), so only 100 of these cells would be expected in this tumor. If three unrelated drugs are used, the frequency would be 10^{-15}, implying it is nearly impossible for any tumor to contain such a phenotype. Even the largest nonlethal tumor burden of 10^{12-13} cells would not usually contain any cells resistant to three drugs, and it is interesting that combination regimens developed clinically usually contain three to four drugs. The key to successful combination chemotherapy is selecting drugs for the combination that have different mechanisms of resistance in the tumor cells. A combination of ten drugs would be ineffective against a *mdr*+ tumor if all ten were substrates for the P170 pump. Combinations of drugs with different cell cycle phase specificities provide the additional advantage that tumor cells are more likely exposed to at least one cytotoxic compound during a sensitive phase of the cell cycle. If the drugs are carefully selected not only according to complementary mechanism of action but also to complementary dose-limiting toxicities, an increase in log-cell kill over monodrug therapy is usually achieved without an increase in toxicity (Chabner 1990c). In fact, the major side effects of combination chemotherapy are generally predictable from the toxicities of the individual agents used alone (Smith and Haskell 1990). However, in some cases (Haskell 1990; Haskell, Selch, and Ramming 1990a; Morton, Cochran, and Lazar 1990), single drug therapy is just as effective as combination therapy, and adding more drugs to the regimen increases toxicity but not efficacy – to the detriment of the patient. A fixed frequency of drug resistance also implies that the probability of drug resistance increases with increasing tumor burden, another factor that makes advanced disease more difficult to treat.

8.4   The evolution of treatment for intermediate-stage breast cancer

Modern multimodality treatment of breast cancer reflects the newer understanding of the biologic basis of cancer. As a summary to this chapter, we examine how three patients representative of each stage of breast cancer are treated and why. In selecting which modalities to use for a particular case, it is important to remember the strengths and weaknesses of each. Surgery, most effective for bulky and/or localized disease, is local therapy and used whenever a large proportion of the tumor burden can be resected or needs to be resected for immediate relief of life-threatening conditions. In contrast to fractional log-cell kill with radio- and chemotherapy, the 100% cell kill with surgery makes it illogical in most cases to treat *localized* disease without using surgery. Regional therapy with radiation is used to control disseminated cancer confined to a radiation field of the body and is most effective against low tumor burden found in microscopic disease. If metastases are present or micrometastases are likely based on the natural history of the disease in previous patients, then systemic therapy is required, and chemotherapy has a chance of curing disseminated disease. However, its systemic nature also exposes all of the healthy tissues of the body, and systemic therapy goes hand in hand with systemic toxicity, which is more severe than the local toxicity of radiotherapy or surgery. The benefits of these therapies must be weighed against the toxicities and risks, and the decision to use them is on a case-by-case basis. Clinical experience with a wide variety of cancers indicates that single modality therapy is rarely curative, and multimodality therapy, which combines local, regional, and systemic therapy by using two or even all three of the modalities in the patient, either sequentially or simultaneously, is emerging as the best treatment strategy. Multimodality therapy protocols include treatment with surgery and/or radiotherapy initially to dramatically decrease the local or regional tumor burden and provide quick relief of symptoms, and carefully timed chemotherapy is then continued for an extended period of time to cure the micrometastatic disease.

A patient presenting with stage I breast cancer has the best prognosis (Figure 8–6a). The small tumor can be completely resected with clear margins and with no sign of cancer in local lymph nodes (Figure 8–6d). Single modality radiotherapy and chemotherapy are not generally used because of the chance they will leave a residuum (Figure 8–6g and j). Greater than 90% of true stage I patients will be cured surgically. The low probability of local or distant micrometastatic disease means that the risks and toxicities of radio- and chemotherapy outweigh the benefits, and routine follow-ups are generally indicated with no postsurgical treatment. The fact that patients with node-negative early-stage breast cancer treated only with surgery have less than 100% chance of disease-free survival provides strong evidence for the idea that cancers can disseminate prior to reaching clinically detectable size. Although adjuvant chemotherapy would be effective treatment for

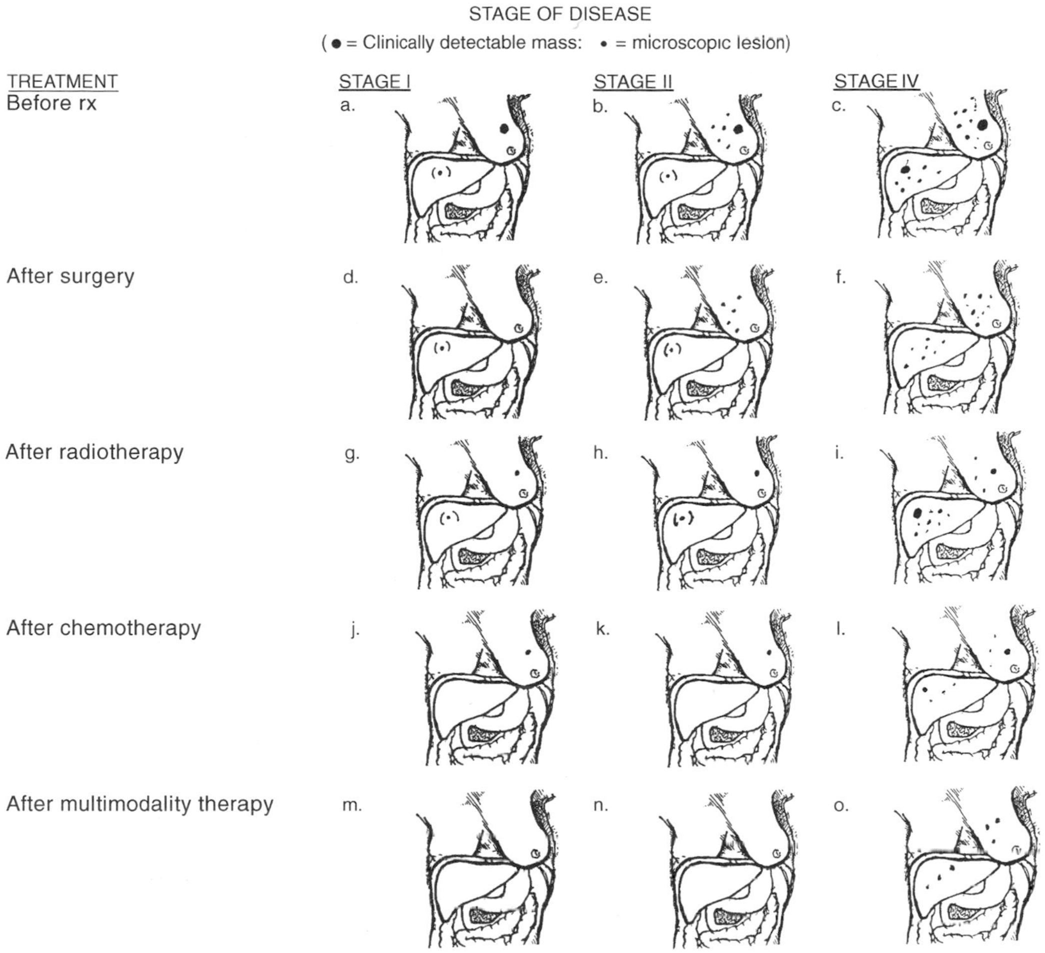

Figure 8–6. The strengths and weaknesses of each modality are illustrated using breast cancer as an example. (a,b,c) Stage I, II, and IV indicate three stages of disease that are possible at diagnosis; only stage IV has clinically detectable metastatic lesions, but micrometastatic disease might be present in some patients staged to I and II disease. (d,e,f) Surgery removes the obvious tumor lesions but cannot be used to eradicate locally invasive or systemic disease. Radiotherapy (g,h,i) and chemotherapy (j,k,l) are best at eradicating microscopic disease that is localized to a region or is systemic or in unknown locations, respectively. The cytotoxic modalities work poorly against bulky disease (note partial responses of primary tumors and gross metastatic lesions in g–l). When combined in an integrated approach to treatment, localized bulky disease is treated surgically, regional disease after tissue-sparing surgery with radiotherapy, and micrometastatic disease with properly timed adjuvant chemotherapy (m,n,o). A minority of patients with stage II disease may not have disseminated cancer and therefore don't benefit from chemotherapy, but it is difficult to identify these patients with current prognostic markers, so all patients are treated. Conversely, the majority of patients with stage I disease are cured surgically, so adjuvant chemotherapy is not usually given to these patients, even though it is recognized that a few do have undetectable systemic disease at time of diagnosis.

the micrometastatic disease responsible for distant relapse (Figure 8–6m), exposing all of these patients to toxic modalities when the overwhelming majority will not benefit is considered unethical. Tests are urgently needed that can detect the high-risk patients in this good prognosis group. Oncogenes linked to unusually aggressive behavior of breast cancer cells (Chapter 5) may provide a prognostic marker for stratification of stage I patients. In node-negative patients, radical resection with regional lymph node removal (radical mastectomy) does not provide better long-term survival or lower distant relapse rates than conservative resections that remove only enough nodes for pathologic sampling and spare as much of the healthy breast tissue as possible (lumpectomy) (Haskell et al. 1990). Reconstructive surgery, using prostheses or implants, helps compensate for deformities resulting from cancer surgery. Proper surgical planning and technique includes a conscious effort to minimize morbidity and disfigurement and retain tissues that will be necessary for reconstructive surgery of good cosmetic quality. There is a trend toward combining the resection of the tumor with reconstruction during a single operation, which has been shown to reduce psychological morbidity in breast cancer patients (Dean, Chetty, and Forrest 1983).

A patient presenting with stage II breast cancer has a much greater chance of local and distant micrometastatic disease (Figure 8–6b). Although these cancers appear to be local-regional disease that extend no farther than regional lymph nodes, again radical mastectomy does not provide any better long-term survival than lumpectomy. In a direct comparison, lumpectomy or total mastectomy produced the same rates of survival, disease-free survival (lack of relapse), and distant disease–free survival (lack of distant metastases), regardless of age, tumor size, and number of positive nodes (Fisher et al. 1989, 1993). The only clinical difference between the two techniques is a higher incidence of local relapse following lumpectomy that can be controlled with radiotherapy and does not affect survival rates (Fisher et al. 1993). Optimal treatment is currently a multimodality approach and includes lumpectomy as the first choice to remove the primary tumor (Figure 8–6e), regional radiotherapy to control local relapse (Figure 8–6h), and adjuvant chemotherapy to eradicate the micrometastatic disease responsible for relapse and shortened survival (Figure 8–6k). The multimodality approach has a much higher chance of curing the patient than each modality used alone (Figure 8–6n) because single modality therapy will likely leave residual disease no matter which is selected (Figures 8–6e, h, and k). Note that each individual patient is treated based on the clinical evidence in the average patient, and tests for high-risk cancers might also benefit this patient group by identifying the minority of patients who will never relapse with distant disease and do not require systemic chemotherapy.

Patients with stage IV disease present with clinically detectable metastatic disease, which is usually an early manifestation of extensive, noncurable disease (Figure 8–6c). Surgery is used sometimes to debulk the tumor burden (Figure 8–6f), unless there are so many detectable lesions that removing a few will not significantly decrease the tumor burden. Radiotherapy and chemotherapy alone will achieve only partial remission because of the large tumor burden and the unfavorable cytokinetics (Figures 8–6i and l). In this case, the purpose of multimodality therapy becomes prolongation of life and ultimately palliation – alleviation of pain and suffering, and stabilizing a patient's condition while the disease completes its natural course, for example, removing intestinal obstruction and isolated brain lesions (Batzdorf et al. 1990). Even with highly effective multimodality therapy, usually the best that can be achieved is cancer control because a relatively large tumor burden remains after maximum tolerated doses of cytotoxic therapy (Figure 8–6o) and this residuum will regrow relatively quickly.

8.5 Summary

In conclusion, three modalities are commonly used today in combination to treat cancer: surgery, radiotherapy, and chemotherapy. With the exception of skin cancer, surgical cure rates for early detected cancers (Table 8–2) indicate that cancer is generally a disseminated disease at the time of diagnosis, and in general single modality surgery is curative only in patients with life expectancies shorter than the time required for residual micrometastatic disease to grow to detectable levels. These data have changed the purpose of cancer surgery from "complete resection" of all involved tissue to control of local disease, reduction of bulky tumor burden, and stimulation of repopulation for optimal adjuvant therapy, all while sparing normal tissue. Radio- and chemotherapy are indicated if there is a likelihood of local-regional dissemination and systemic metastases, respectively, *even if there is no evidence of disease from standard staging evaluations with x-rays or CT scans.* The dose of adjuvant therapy is fractionated over time (Dutreix et al. 1971; Wittes 1986; Forbes 1990) to expose the remaining malignant cells several times during repopulation, their most susceptible time, and to attempt to expose all of the tumor's growth fraction during a sensitive phase of the cell cycle. If the adjuvant therapy is delayed until the disseminated disease becomes symptomatic or clinically detectable, then one has lost the window of opportunity to treat the cancer with radio- and chemotherapy under optimal conditions: lower tumor burden, better tumor perfusion, larger growth fraction with a stem cell component, lower probability of drug and radio resistance, and fewer dormant or quiescent cells. Combination chemotherapy has a better chance than single drug therapy because of the likelihood of preexisting drug-resistant cells. Cures with adjuvant therapy are possible as long as the healthy dose-limiting tissues can tolerate the required number of cycles, although the slow doubling times of most tumors require many years of disease-free survival, for example 10 to 15 years for breast cancer patients, to be certain of cure (Mueller and Jeffries 1975; Haskell et al. 1990).

The minimal size required for clinical detection of a solid tumor is 10 billion cells, approximately a 1 cm³ tumor (Haskell et al. 1990a; Haskell 1990). Early detection and diagnosis increase the probability of cure with multimodality therapy by decreasing the probability of dissemination, minimizing required log-cell kill, decreasing the probability of resistance to cytotoxic therapy, and increasing the probability of optimal cell cycle kinetics for log-cell kill. The emphasis on early detection of cancer from warning signs like unusual bleeding, a change in bowel habits, a persistent cough, or an unusual or changing mole can make an important contribution to cancer cure. It cannot be overemphasized that the chances of curing cancer are much higher when two events coincide: the patient goes to a physician early during the course of the disease when symptoms are just appearing and the physician quickly recognizes the symptoms, establishes the diagnosis, and begins treatment. Delays by either person make cures ever less likely. Many late-stage patients cannot tolerate surgical debulking because of poor overall health caused by the cancer and must receive chemotherapy or radiotherapy under highly unfavorable biologic conditions that limit cytoreductive therapy to palliative purposes (Charlson 1985).

Cytotoxic therapy encounters problems when cancers repopulate faster than dose-limiting normal tissues, when cancers have smaller growth fractions, and when the log-cell kill is significantly greater in normal tissue than the cancer. Unfortunately, one or more of these appears to be the case in most malignancies. This is not to say some cancers cannot be cured; in fact, some of the less common ones are cured in a very high number of cases. Methotrexate or etoposide monotherapy cures 95% of cases of stage 1 choriocarcinoma (Berkowitz and Goldstein 1990). Adjuvant chemotherapy after surgical debulking is given with curative intent in patients with advanced seminoma of the testis (Smith and Haskell 1990). Nonseminomatous tumors of the testis are also chemosensitive, allowing the "luxury" of delaying chemotherapy after surgery until relapse is detected, thereby sparing surgically cured patients the drug toxicity (Smith and Haskell 1990). Neuroblastoma, Ewing's sarcoma, retinoblastoma, and acute lymphocytic leukemia in children and Hodgkin's lymphoma in adults are additional examples of malignancies treated with the intention of cure. The biologic basis for the curability of these cancers has received little attention. The vast majority of mechanistic research on cancer pharmacology is dedicated to answering why malignant cells are resistant to cytotoxic agents. Perhaps a more important question is why cytotoxic therapy works so well against a few types of cancers. It is hoped that you will one day study these curable malignancies and discover what makes the difference between curable and the more common, incurable cancers. For example, methotrexate-sensitive leukemic cells accumulate intracellular drug faster than their resistant counterparts (Whitehead et al. 1987), and polyglutamation of methotrexate that traps the drug inside the cell is less efficient in hematopoietic progenitors then leukemic cells (Fabre, Fabre, and Goldman 1984; Koizumi et al. 1985). This differential handling of methotrexate between relatively resistant and sensitive cells explains why *leucovorin rescue,* which is the administration of this folate analogue after high-dose methotrexate, decreases bone marrow but not tumor log-cell kill: there is time for leucovorin to compete with methotrexate in the resistant cells but not in the sensitive leukemic cells that accumulated high levels of polyglutamated methotrexate before leucovorin rescue (Allegra 1990).

Curing cancer is obviously a tremendously complicated goal. The development of curative strategies for several cancers, many of them childhood malignancies, is a tribute to slow but steady progress. However, the research and regulatory communities need to provide better chemotherapies and radio-modulating agents with significantly improved therapeutic indices. Because it is the ratio of only two variables, therapeutic index can only be increased by either (1) increasing the MTD (decreasing cellular damage in healthy tissue or its consequences) without losing efficacy (the log-tumor cell kill), or (2) increasing efficacy without lowering the MTD. Another approach used for treating some cancers, called biotherapy, is the subject of Chapter 9.

9

Biotherapy

Ralph E. Parchment

9.1 Introduction

Chapter 8 describes the conventional modalities of surgery, chemotherapy, and radiotherapy, which are cytoreductive strategies aimed at eradicating malignant tissue. In contrast, this chapter describes therapeutic approaches that aim to alter the biology of the malignancy so its growth can be controlled, a high quality of life can be maintained, and the patient can live a normal life span without risk of death from cancer or its complications. In other words, biotherapies try to harness what is understood about normal and malignant tissue renewal to control the latter without affecting the former (Figure 9–1). Malignant tissue renewal is assumed different from the normal, either qualitatively or quantitatively. The theory that malignancy caricatures the process of tissue renewal (Chapter 1) suggests that differences are primarily quantitative, that is, exaggerations of certain aspects of normal renewal, so the most successful biotherapies will be those that restore some balance between malignant stem cell renewal and differentiation. Malignant tissue will not be eradicated completely, but its impact on the patient will be negligible. Biotherapy is therefore particularly useful for malignancies in older patients that are relatively refractory to cytoreductive therapies, such as carcinoma of the prostate and kidney.

Some biotherapies target ectopic or overexpressed antigens on malignant cells with cytolytic immune modulators (both cellular and humoral) or ectopic or overexpressed receptors with ligand-conjugated toxins. These approaches target progenitors and differentiated cells in the cancer because they express the targets. However, stem cells do not express differentiation markers and therefore are not direct targets, so cure in the sense of eradicating the malignant stem cells is likely impos-

250

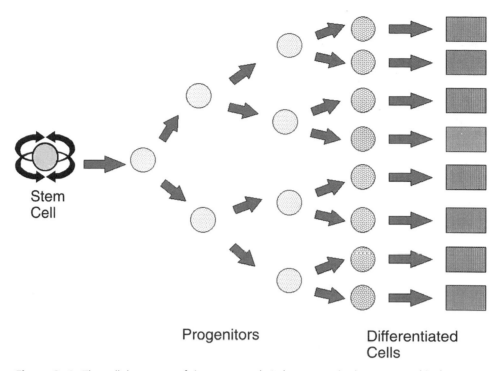

Stem
Cell

Progenitors   Differentiated
                               Cells

Figure 9–1 The cellular nature of tissue renewal. At homeostasis, tissue renewal is the result of balance between production of new differentiated cells and the loss of older ones through sloughing or apoptosis (cell death). New specialized cells (squares) are produced via the differentiation of immature precursor cells (dense stipple pattern), which in turn have been produced via the finite proliferation and maturation of progenitor cells (light stipple pattern). Although usually quiescent, stem cells can enter the cell cycle periodically to produce committed progenitor cells and replenish the stem cell population as needed. Malignancy is a caricature (exaggeration) of normal tissue renewal in that cell production and differentiation become disconnected and cell production and loss are not in balance (see Chapter 1). Biotherapy aims to control malignant tissue renewal by targeting its differences from normal tissue renewal, which are quantitative rather than qualitative differences.

sible. Nevertheless, if stem cell renewal is slow enough, biotherapy can shrink the tumor to a nonsymptomatic size and maintain this reduced tumor burden for decades or longer, which can equal the life span of older patients. Although these biotherapies are discussed in this chapter, these strategies are clearly cytoreductive in nature because they attempt to destroy malignant cells.

The following sections emphasize biotherapies that are FDA approved for treating human cancer or for treating side effects of cytotoxic chemotherapy (Table 9–1). Although not yet approved, cellular therapy is discussed in this chapter because it might be relevant to understanding the antitumor activity of interleukin-2. Many other biologic strategies are under development at this time, but it is too early to know if they are in fact generally clinically effective. Although they are scientifically appealing and may be showing clinical promise, it

Table 9-1. *Biotherapies currently approved by the FDA for treating cancer or its complications*

The small "r" in front of the common name indicates a recombinant protein produced via genetic engineering of a competent cell, either bacteria (*E. coli*), yeast (*S. cerevasiae*), or mammalian cells with the expected differences in glycosylation.

Common name	Commercial product	Indicated use
rInterferon- α_{2a}	Roferon-A	hairy cell leukemia
rInterferon- α_{2b}	Intron-A	AIDS-related Kaposi's sarcoma
rInterleukin-2	Proleukin (aldesleukin)	metastatic renal cell carcinoma
BCG extract	TICE	bladder carcinoma in situ
rG-CSF	Neupogen	chemotherapy-induced neutropenia
		supports hematopoietic transplants
rGM-CSF	Leukine	supports bone marrow transplants
rErythropoietin	Epogen	anemia
	Procrit	

Sources: Adapted from Vadhan-Raj, S., 1996. Appropriate use of hematopoietic growth factors; and Kim, B., 1996. Biological therapy: Interferons, interleukins, and monoclonal antibodies. In: *Cancer Management: A Multidisciplinary Approach*, R. Pazdur, L. R. Coia, W. J. Hoskins, and L.D. Wagman, eds., pp. 581–592. Huntington, NY: PRR, Inc.

is premature to include them in a textbook about the biologic basis of cancer therapy. See recent reviews of investigational biotherapy for more detailed descriptions of the emerging therapeutic strategies that are not covered in this chapter (*Cancer Medicine* [Holland, Frei, Bast, Kufe, Morton, and Weichselbaum 1993], Friedman, Grimley, and Baron 1996; Thrush et al. 1996).

9.2 Targeting differentiation markers for delivery of cytotoxic agents

Although not usually categorized as "differentiation therapy" or biotherapy, it is possible to take advantage of the biologic function of the differentiated cells in neuroendocrine tumors by administering a radioactive precursor compound that is concentrated to high levels in the malignant tissue and the corresponding normal tissue (Figure 9–2). The high levels of radioactivity that accumulate in the differentiated tumor cells not only expose these cells which contain it but also neighboring cells, even though the neighbors might be undifferentiated and do not accumulate the drug. For example, I^{131} is used to treat tumor residuum and metastases from thyroid carcinoma (Dworkin, Meier, and Kaplan 1995; Sweeney and Johnston 1995). The normal function of the thyroid epithelium is to incorporate iodine into precursors of thyroid hormone, so I^{131} can be delivered with a high degree of tissue selectivity to cancerous thyroid tissue as long as there is some degree of thyroid differentiation in the tumor. Likewise, I^{131}-iodobenzylguanidine is useful for treating metastatic neuroendocrine tumors (Gelfand 1993; Sloan et al. 1996). A similar strategy for targeting differentiation markers is also being employed with toxin-conjugated antibodies and ligands for cell surface receptors (Friedman, Grimley, and Baron 1996; Thrush et al. 1996).

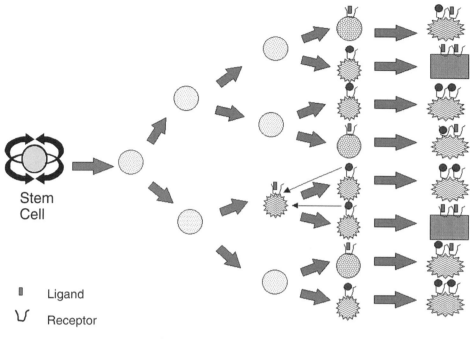

Figure 9–2. Targeting differentiation markers for delivery of cytotoxic agents that reduce tumor burden. Malignant tissue contains a proportion of terminally differentiated cells, albeit a much lower proportion than is found in the normal counterpart tissue. The differentiation program leads to expression of receptors for specific ligands such as growth factors (indicated by the square ligand and its receptor). Cytotoxic chemicals and high energy radioactive nuclides can be attached to the ligand (indicated by circles) and delivered to tumor cells via receptor binding, thereby killing them (indicated by the abnormally shaped cells). Radiation from high energy nuclides may also damage neighboring cells, even though they do not interact with the toxin (indicated by the two arrows). Furthermore, a phenomenon called the bystander effect may lead to the death of neighboring, undifferentiated cells (target negative) when more differentiated cells die. Cell surface targets do not have to be receptors. Toxins and radionuclides can also be delivered to nonreceptor targets using monoclonal antibodies, and radiation can also be delivered in the form of radioactive biochemical precursors of substances produced by the differentiated cells (e.g., I^{131} and thyroid cancer). Of course, any normal tissue that expresses the same target will also be damaged by the therapy, and this strategy may work poorly against anaplastic tumors which express little if any differentiated features, including the target. Also note that malignant stem cells will only be affected if they happen to be in proximity to another cell that expresses the target.

9.3   Trophic factors in tissue renewal and therapy

As described in detail in Chapter 1, tissue renewal is maintained by the balance between several subpopulations of cells: stem cells, progenitors committed to expansion and differentiation, and terminally differentiated cells that perform the specialized function of the tissue (Figure 9–1). Tissue mass can be increased or

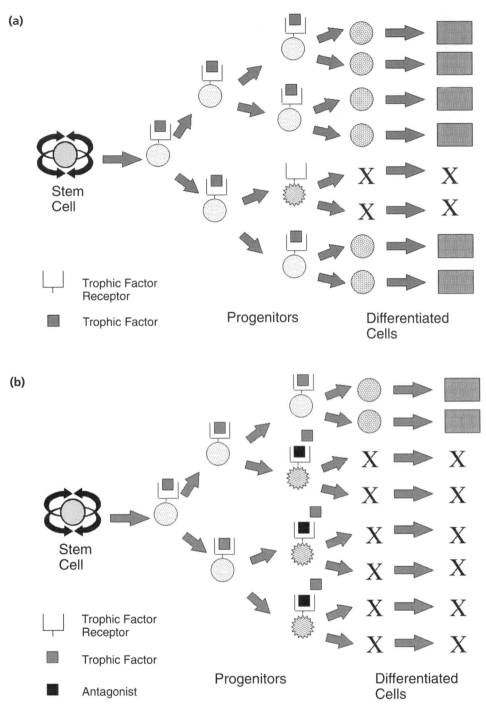

Figure 9–3. Trophic factors as targets for cancer therapy. (a) Trophic factors (squares) that serve as targets for cancer therapy are endocrine or paracrine mediators that block apoptosis in the progenitor population of a renewing tissue. They may interact with cell surface or intracellular receptors to promote progenitor survival so that differentiated cells can be generated. Progenitors that fail to receive adequate exposure to the specific trophic factor die via apoptosis (indicated by the abnormally shaped cells) and therefore do not produce any differentiated cells. Homeostasis usually occurs at a level of trophic factor that cannot block apoptosis of all of the progenitors. This creates a reserve for rapidly increasing cell production **(Figure 9–3,**

decreased by altering the number of differentiated cells produced by each progenitor and/or by altering the number of progenitors that produce differentiated progeny. Although there are exaggerated stem cell numbers and activity in malignancies (Pierce and Speers 1988), both normal and malignant tissue have mechanisms that maintain homeostasis and stimulate repopulation after loss of differentiated tissue mass (Al-Dewachi et al. 1980; Schabel 1975; Mayo et al. 1972; Cameron, Hardmand, and Skehan 1990; Fisher et al. 1989a, 1989b; Potten and Chadwick 1994; Potten et al. 1994).

Trophic factors prevent apoptotic cell death in progenitors and often in terminally differentiated cells as well, which results in an increase in tissue mass (Figure 9–3a). When the progenitors cannot survive, new cell production is lost, senescent and sloughed terminally differentiated cells cannot be replaced, and the tissue atrophies. Inhibition of apoptosis is a universal function of trophic factors from all renewing tissues: hemato- and lymphopoietic (Abbrecht and Littell 1972; Borthwick et al. 1996; Hassan and Zander 1996; Jelkmann and Metzen 1996); endometrial (Hopwood and Levison 1976; Sandow et al. 1979); breast (Ferguson and Anderson 1981a, 1981b; Anderson, Ferguson, and Raab 1982); and prostate (Kerr and Searle 1973; Sandford, Searle, and Kerr 1984; van Steenbrugge et al. 1984).

Trophic factors can be endocrine in nature (humoral) or paracrine (locally produced and secreted). For example, soluble factors are present in blood plasma after cytoreduction of normal and tumor tissue that initiate and then halt cell division in the affected tissue (Fisher et al. 1989a, 1989b; Potten and Chadwick 1994; Potten et al. 1994). Some of these factors have been isolated and identified, but it seems likely that many more remain to be characterized. The trophic factors include sex steroids, produced by the gonads and adrenal gland under pituitary control, and several cytokines produced by specialized cells in the kidney and lymphoid system (Table 9–2).

Antagonists of some of these trophic factors are used in oncology to control tumor growth; a few of these trophic factors are used to minimize therapy-induced toxicity and avoid delays in scheduled cytotoxic treatment (Table 9–1). There are additional trophic factors that cause enlargement of an organ by increasing cell size (hypertrophy) rather than cell number, and these factors probably do not play a role in balancing cell production and cell loss in renewing tissues.

---

cont.) if required and a mechanism for physiologic response to the environment. Biotherapies approved for use in treating hematologic side effects of cancer therapy are trophic factors for blood cell progenitors in the bone marrow. (b) Drugs that block trophic factor binding to its receptor (squares), or that interfere with the action of the trophic factor, will increase the number of progenitor cells that die via apoptosis (indicated by the abnormally shaped cells), substantially diminish new tissue production, and result in reduced tissue mass over time. Note that stem cells do not die as a result of antagonizing trophic factor activity, although some evidence indicates that stem cell renewal and their production of progenitors may be slowed by trophic factor withdrawal.

Table 9–2. *Tissue-specific trophic factors, their target tissue, and their tissue of origin*

As a general rule, these factors cause a trophic effect on the target tissue by inhibiting apoptosis of progenitors and their progeny. The *mpl* ligand likely inhibits apoptosis, but this has not been proven. Generally these trophic factors are tissue specific.

Trophic factor	Target tissue	Tissue of origin
testosterone	prostatic epithelium	testis
estrogen	mammary epithelium	ovary
	uterine epithelium	
G-CSF	neutrophil progenitors	???
Interleukin-3	myeloid progenitors	???
mpl ligand	platelet progenitors	???
erythropoietin	erythroid precursors	kidney
interleukin-2 and -4	several lymphocyte subtypes	???

9.3a Antagonism of trophic factors to decrease tumor mass

Cancers from tissues that depend on trophic factors often retain the factor dependence of their tissue of origin. Therapies that block (antagonize) the biologic activity of trophic factors cause atrophy of both the malignancy and its normal counterpart (Figure 9–3b). These treatments can induce tumor shrinkage and provide years of control of malignant disease. Because of the tissue specificity of the trophic factor, the toxicity of antihormone therapy is mild compared to conventional cytotoxic chemotherapy. This approach is known by many names: endocrine therapy (more appropriately antiendocrine therapy), hormonal therapy, hormone ablation therapy, or antihormonal therapy.

There are three requirements for effective control of cancer with antiendocrine therapy. First, the cancer must originate from a tissue that is not critical for life. For example, inhibiting the trophic factor for prostate epithelium and causing atrophy (involution) of both prostate cancer and the normal prostate does not cause life-threatening toxicity, whereas inhibiting the trophic factor for white blood cell production would quickly lead to a life-threatening neutropenia. Second, the malignancy must retain responsiveness to the trophic factor, including receptors, a signal transduction mechanism, and the hormone-response genes. Tumors from trophic factor-dependent tissues can evolve into factor-independent phenotypes that are refractory to antihormonal therapy. This evolution is probably the result of genetic instability of the Goldie-Coldman type (1979) and positive selection of subpopulations that gain a growth advantage (Isaacs and Coffey 1981; Isaacs et al. 1978, 1982). For some reason, some malignant stem cells eventually adapt to trophic factor deprivation and may at any time begin to produce progenitors that no longer require trophic factor for expansion (Isaacs and Coffey 1981). The biologic explanation for the appearance of "hormone-resistant" ("hormone independent") malignancy remains an enigma because hormone independence never occurs in normal tissue deprived of trophic factor. Third, antihormonal therapy is most effective when the trophic factor is tissue specific. For example, insulin and growth hormone, which have systemic effects on many target tissues, would not be suitable targets for antiendocrine chemotherapy because hormone withdrawal would cause serious complications in many organ systems. The less the tissue specificity of the targeted trophic factor, the more complicated the side effects, and the less effective the biotherapy.

In contrast to progenitor cells, repopulation of atrophied tissue in response to a new source of trophic factor proves that stem cells survive for long time periods without trophic factors. For example, androgen withdrawal causes involution of the prostate, but the secretory epithelium rapidly repopulates with androgen replacement therapy even after a dormancy of many months (Huggins and Hodges 1941; Kerr and Searle 1973; Leav et al. 1978; Isaacs and Coffey 1981; Isaacs et al. 1982). This cycle of regrowth and atrophy can be repeated many times (Sandford et al. 1984). The human hematopoietic stem cell can survive in the absence of human- and myeloid-specific trophic factors in an immune-deficient mouse, without producing any surviving progenitors, yet it will produce human leukocytes when human trophic factors like GM-CSF and IL-3 are administered to the animal (Lapidot et al. 1992; Murray et al. 1994, 1995). However, an ability to survive in the absence of trophic factor does not mean the stem cell won't respond to the factor. In fact, trophic factors may accelerate repopulation of the tissue by stimulating stem cell proliferation or increasing growth fraction. Therefore, therapies that block trophic factor activity should only be expected to achieve tumor control initially via large reductions in tumor burden following apoptosis and then subsequently via slowed stem cell activity.

Even though they do not eliminate the malignant stem cell, antiendocrine therapies are quite useful clinically and serve a very important role in oncology, especially against slower-growing tumors that because of their slow growth rate are usually unresponsive (refractory) to conventional cytotoxic chemotherapy. Although malignancy is presumed to remain after treatment, antihormonal therapies can control disease for years, and can "cure" patients that have a shorter life span than the time required for evolution of the hormone-independent state. A reduction in tumor burden with mild, non-life-threatening toxicity may be sufficient in many of these patients to reach their life expectancy with minimal interruption in quality life. The antiendocrine therapy may also slow repopulation by the stem cell, so continuing to take the antiendocrine chemotherapeutic even after the patient enters remission (maintenance therapy) offers a theoretical advantage. However, chronic trophic factor deprivation might also accelerate the development of trophic factor independence. The issue of intermittent versus chronic antiendocrine therapy should be resolved soon by ongoing clinical trials.

Trophic factor withdrawal can be accomplished surgically by removing the organ that produces the trophic factor (Table 9–2), as long as the organ is not critical for survival. More often a combined chemotherapeutic and surgical approach is used: surgical resection of hormone-producing tissues, such as testes or ovaries, plus chemotherapeutics to block any possible effects of the lower yet constant levels of hormone secreted by other tissues that cannot be easily resected.

9.3b Antagonism of steroid trophic factors

In the reproductive system, steroid hormones serve as trophic factors. In the male, the androgenic steroids maintain the cellularity and specialized differentiation of prostatic epithelium. "Androgen ablation therapy," which means eradicating the supply of testosterone and/or dihydrotestosterone, or antagonizing their activity, results in rapid apoptosis of prostatic epithelium and atrophy of the prostate gland (Huggins and Hodges 1941; Kerr and Searle 1973; Isaacs and Coffey 1981; Isaacs et al. 1982; Sandford et al. 1984; van Steenbrugge et al. 1984). However, the stem cells of the epithelium survive androgen ablation and repopulate the epithelium in response to androgen replacement (Leav et al. 1978; Sandford et al. 1984). In the female, progesterone and estrogens cause analogous increases in cell number and maintain differentiated epithelium in breast and uterus (Horwitz, Koseki, and McGuire 1978; Sandow et al. 1979; Anderson, Ferguson, and Raab 1982; Anderson et al. 1989; Helgason et al. 1982; Going et al. 1988; Potten et al. 1988). This knowledge has been used to develop antiendocrine therapies for cancers originating in these tissues.

9.3c Carcinoma of the prostate

Charles Huggins was the first to show that removal of the testes (orchiectomy) in prostatic carcinoma patients induced clinical regressions with high frequency (Huggins and Hodges 1941), a discovery for which he was awarded the Nobel Prize in medicine in 1966 (Wasson 1987). Today, antiendocrine therapy plays a major role in the treatment of prostatic carcinoma (Neuwirth and deKernion 1990). Most prostate cancers at early stages of their natural history are composed primarily of androgen-dependent progenitors and differentiated epithelium, and androgen-independent stem cells are at most only a minor component (Smolev, Coffey, and Scott 1977; Smolev et al. 1977; Isaacs et al. 1978, 1982). Thus, androgen ablation therapy causes clinical regression, and the atrophy of the tumor tissue occurs quickly enough to provide prompt relief from bone pain in metastatic disease (Neuwirth and deKernion 1990). Androgen ablation therapy relieves symptoms without life-threatening toxicities, and remission may be of sufficient duration to achieve normal life expectancy in older patients. However, just as normal prostatic epithelium stem cells survive androgen ablation (Leav et al. 1978; Sandford et al. 1984), malignant stem cells also survive (Isaacs and Coffey 1981; Isaacs et al. 1982). But, in contrast to normal stem cells, these malignant stem cells will begin to repopulate the tumor despite the absence of testosterone, and at relapse more intensive antiandrogen therapy is usually ineffective (Neuwirth and deKernion 1990). Therefore, "curative" tumor control can be achieved in older patients with antiandrogen therapy but not in younger patients.

Repopulation of the prostatic carcinoma and the presence of differentiated cells under conditions in which the normal prostate remains atrophied (Kraljic, Kovacic, and Tarle 1994) lead one to try to explain the difference between stem cells and progenitor cells from normal versus malignant prostatic epithelium. Androgen hypersensitivity might develop via a Goldie-Coldman-type mechanism or via environmental pressure, mutation, and natural selection. This hypothesis proposes that the relapsing cells are surviving and/or proliferating in response to very low but detectable blood levels of androgen that persist after orchiectomy, perhaps via a mutated androgen receptor that binds testosterone with greater affinity. However, there is legitimate disagreement about the importance of these low levels of circulating androgens in these patients.

Building on the pioneering work of Huggins, modern androgen ablation is achieved pharmacologically as well as surgically in the patient with prostatic carcinoma. Most of the testosterone is made in the testes by Leydig cells, although a small amount is also produced in the adrenal glands. When androgen levels are low, the hypothalamus secretes pulses of luteinizing hormone-releasing hormone (LHRH), and this in turn stimulates release of luteinizing hormone (LH) and follicle-stimulating hormone (FSH) from the pituitary gland that costimulate androgen synthesis by the Leydig cells. Therefore, antiandrogen therapy can be accomplished in several ways (Neuwirth and deKernion 1990): surgical removal of the androgen source (testes) with orchiectomy; inhibiting the response of the pituitary to low androgen levels using chronic exposure to estrogen, diethylstilbestrol (DES), or LHRH agonists like Leuprolide (Santen, Manni, and Harvey 1986); inhibiting the action of testosterone on the target organs using cyproterone acetate (Barradell and Faulds 1994) or the androgen receptor antagonist flutamide (Klein 1996); or in emergency situations, using ketaconazole, a relatively nonspecific inhibitor of the cytochrome P-450 that produces testosterone (English et al. 1986; Clarke et al. 1990). In most cases, androgen ablation is achieved clinically by combination orchiectomy and receptor antagonists. In rare cases, the source tissue of low androgen levels, the adrenal glands, might be surgically removed. Clinical trials are addressing which treatment or treatment combination is the best. In relapsed cancer, LHRH agonists show no activity, and hypophysectomy is not beneficial (Neuwirth and deKernion 1990).

9.3d Carcinoma of the breast

Estrogen is a steroid that acts as a trophic factor for the epithelium in the glands of the breast (Ferguson and Anderson 1981a, 1981b; Anderson et al. 1982, 1989; Going et al. 1988; Potten et al, 1988), so antiendocrine therapy plays an important role in breast cancer therapy (Goldhirsch and Gelber 1996). Antiestrogen therapy in premenopausal women is more complex than androgen ablation therapy in men because estrogen and progesterone levels fluctuate during the menstrual cycle (Ferguson and Anderson 1981b; Going et al. 1988) and estrogen levels drop dramatically at menopause (Anderson et al. 1982; Potten et al. 1988). Similar to antiandrogen therapy for prostatic carcinoma, antiestrogen therapy for breast carcinoma does not eliminate the malignant stem cell but rather slows its proliferation and promotes apoptosis of progenitors and their progeny. As expected with this mechanism, relapse during estrogen ablation therapy can be due to estrogen-independent cells (Reddel et al. 1988; Fisher et al.1993), although relapsing tumors may still respond to antiestrogen therapy administered at higher doses (Manni and Arafah 1981; Muss, Smith, and Cooper 1987).

Estrogen ablation in premenopausal women is achieved by removing the primary tissue source of estrogen (ovaries) in a procedure called an oophorectomy. In postmenopausal women, estrogen production by the ovaries is very low, and there is no need to remove them. The possible trophic activity of the low estrogen levels in the absence of ovarian production (from diet or the adrenal gland) is blocked pharmacologically with chronic tamoxifen therapy (Vogel 1996), an antagonist of estrogen, binding to its intracellular receptor in breast tissue (Parczyk and Schneider 1996). Infrequently, adrenal production of estrogen may be high enough to warrant treatment with aminoglutethimide to inhibit the enzymes that produce estrogen, but the greater toxicity of aminoglutethimide therapy relegates it to the role of salvage therapy after tamoxifen failure (Fisher et al. 1993).

Tamoxifen is recommended by some as the treatment of choice when the tumor is expected to be estrogen responsive, when the toxicity of cytotoxic chemotherapy is not tolerable, or when the age or health of the patient indicates only a short period of tumor control will be required (Jordan 1988; Fisher et al. 1993). Clinical cure is more likely in patients with slower growing tumors or in older patients with shorter life expectancies. As predicted from the mechanism, maintenance therapy (constant exposure to tamoxifen) is more effective at controlling tumor regrowth than brief exposure (Fisher et al. 1993). Adjuvant therapy with tamoxifen after surgery is effective in premenopausal women and in postmenopausal patients with estrogen receptor–positive tumors (Haskell et al. 1990). Five years of tamoxifen are superior to 1 to 3 years of treatment (Fisher et al. 1993), and current clinical trials are addressing whether indefinite tamoxifen is still better. It is unclear whether concurrent tamoxifen plus chemotherapy is better than either alone (Fisher et al. 1993). A 1985 Consensus Conference on Breast Cancer at the National Institutes of Health concluded that the optimal adjuvant therapy (either chemotherapy or tamoxifen) was not yet identified. A similar conference in 1990 failed to resolve this confusion. However, chemotherapy reduces the annual odds of death from relapse over the initial 5 to 10 years by 25% in premenopausal patients, whereas tamoxifen reduces these odds only in postmenopausal patients.

Although tamoxifen is employed as an estrogen antagonist, under some conditions tamoxifen demonstrates estrogenic activity (Horwitz, Koseki, and McGuire

1978; Boccardo et al. 1981; Noguchi et al. 1988; Fornander, Rutqvist, and Wilking 1991). This ambivalent activity may contribute to the slow repopulation in breast carcinoma that leads to relapse and to an increased risk of uterine cancer (Fornander et al. 1991; Jordan, Gottardis, and Satyaswaroop 1991; Spinelli et al. 1991). Tamoxifen is extensively metabolized by the liver to other chemical species, which might contribute to its confusing array of biologic activities (Katzenellenbogen et al. 1984; Campen, Jordan, and Gorski 1985; Loser, Seibel, and Eppenberger 1985; Loser et al. 1985; Lyman and Jordan 1985; Murphy et al. 1990). It is known that there are substantial species differences in metabolism and the pharmacologic activity of the metabolites (Jordan 1982; Jordan and Robinson 1987; Phillips et al. 1996).

In theory, all breast cancers containing the estrogen receptor ("ER+" tumors) should respond to antiestrogen therapy, but only 50% to 60% of ER+ tumors respond regardless of whether antiestrogen therapy is surgical or pharmacologic (Osborne et al. 1980). It is poorly understood why 40% to 50% of ER+ tumors fail to respond, but this could be due to signal transduction defects in premenopausal women (Fisher et al. 1993). Because estrogen binding to the estrogen receptor regulates expression of progesterone receptor (Horwitz et al. 1978), dysfunctions in estrogen signal transduction can be detected in women with adequate estrogen levels by an absence of progesterone receptors ("PR–") in ER+ tumor cells. This hypothesis is consistent with, but does not completely explain, response rates of only 74% in ER+/PR+ tumors (Osborne et al. 1980; Fisher et al. 1993). Another confusing result is why 10% of the ER– PR– tumors respond to tamoxifen therapy (Fisher et al. 1993). It is possible that assay cutoffs used by clinical laboratories to define "receptor negative" may be misleading (Encarnacion et al. 1993; Fisher et al. 1993).

Because these confusing clinical results might be caused by weak estrogenic activity of tamoxifen or its metabolites, second-generation estrogen receptor antagonists without any estrogenic activity are on the horizon (Osborne et al. 1995). These "pure" estrogen receptor antagonists show superior antitumor effects to tamoxifen against human tumors engrafted into animals. Recently, media attention has focused on a large clinical trial to test whether prophylactic tamoxifen (administered before any evidence of disease in order to prevent disease) reduces the incidence of breast carcinoma in high-risk patients (Jordan 1988; Boone, Kelloff, and Malone 1990; Bush and Helzlsouer 1993). Certainly, a pure antiestrogen superior to tamoxifen would be desirable for prophylactic therapy should this trial find therapeutic benefit for women at high risk for breast cancer.

Tamoxifen causes a brief tumor flare (a transient "worsening" of disease) 1 to 2 weeks after therapy is initiated, so tamoxifen can only be recommended for patients who can wait the 6 to 8 weeks required to ascertain the tamoxifen responsiveness of the tumor (Fisher et al. 1993). In addition to the tumor flare, tamoxifen benefits must be weighed against other toxicities, including a menopauselike condition, increased risk of uterine cancer, and increased risks of blood clots

(Fisher et al. 1993). Tamoxifen is definitely contraindicated in patients with rapidly relapsing breast cancers (aggressive disease) or liver metastases, and in those failed by antiendocrine therapy, because these patients need a rapid regression of the lesion (Haskell et al. 1990) and because estrogen receptors are less likely in metastases than in primary lesions (Fisher et al. 1993).

9.3e   Carcinoma of the uterus

Estrogen and progesterone also affect the proliferative status and cellularity of the uterine epithelium during the menstrual cycle (Hopwood and Levison 1976; Sandow et al. 1979), so antiendocrine therapy has been tested against carcinoma of the endometrium of the uterus (Gronroos et al. 1987; Grenman et al. 1988). Stage III/IV endometrioid cancer that is ER+ PR+ responds to the progesterone antagonist medroxyprogesterone (Rendina, Donadio, and Giovannini 1982). Other antiprogestins like megestrol acetate and 17-alpha-hydroxyprogesterone are indicated in recurrent or metastatic endometrial carcinoma and achieve a higher response rate than cytotoxic chemotherapy (Berman and Berek 1990).

9.3f   Interferon-alpha: A protein antagonist of trophic factors in the immune system

Interferon-alpha is FDA approved for the treatment of hairy cell and chronic myelogenous leukemia as well as melanoma and Kaposi's sarcoma (Kim 1996). Although its clinical activity is well established (Vedantham, Gamliel, and Golomb 1992; Bouroncle 1994; Gutterman 1994), the mechanism responsible for this activity has been actively debated. On the one hand, interferon-alpha may be clinically effective because it functions as a trophic factor for the immune system, reversing immunosuppression caused by the leukemia and enhancing the antitumor response of the immune system. On the other hand, it may be active because it exerts a direct antiproliferative effect on the leukemic cells. However, it now seems clear that the activity of interferon-alpha against hairy cell leukemia results from a direct effect on the malignant cells (Vedantham et al. 1992). This conclusion is consistent with the finding that coadministration of prednisone, which suppresses the immune system, reduces the severity of side effects of interferon-alpha therapy, but not its efficacy (Fosså, Gunderson, and Moe 1990). This direct effect on tumor does not appear to be cytotoxicity because the tumor burden declines more slowly than with cytoreductive chemotherapy. Although unrelated to tumor regression, the immune system stimulation is an added benefit for the patient because opportunistic infections are the leading mortality-associated complication of hairy cell leukemia (Vedantham et al. 1992).

B-cell growth factor and tumor necrosis factor (TNF) are trophic factors for hairy cell leukemia, as is interleukin-2 (IL-2) after the cells respond to B-cell growth factor. Interferon-alpha binds to a cell surface receptor, which leads to a diminished proliferative response to these trophic factors (Vedantham et al. 1992). Interferon-alpha inhibits IL-2 responsiveness by decreasing the number of cell surface IL-2 receptors; however, inhibition of TNF responsiveness is independent of the number of cell surface TNF receptors (Vedantham et al. 1992). Interferon also stimulates the hairy cells to differentiate to a mature phenotype, which is less responsive to growth factors. Clinical benefit is derived from the slowed natural course of disease progression as a result of reduced proliferative responses to these trophic factors (slower rate of increase in tumor burden). A soluble form of the interferon-alpha receptor is found at elevated levels in body fluids of patients with hairy cell leukemia, and pharmacologic doses of interferon-alpha may be required to overcome the neutralizing effects of this factor (Novick, Cohen, and Rubenstein 1992). Toxicity is mild (mostly flulike symptoms).

Resistance to interferon-alpha may occur in some cases of hairy cell leukemia. In general, resistance is associated with hairy cells at differentiation stages slightly before or after the stage of B-cell differentiation that responds to BCGF, TNF, and IL-2, and in fact, resistant cells generally do not express cell surface IL-2 receptors (Vedantham et al. 1992). These resistant cells might be at a stage of differentiation that utilizes other trophic factors which have receptors that are not regulated by interferon-alpha. Moreover, the presence of a cell surface receptor for a negative regulator does not necessarily imply responsiveness to that factor. For example, chronic lymphocytic leukemia cells also express surface receptors for interferon-alpha, but they lack the intracellular pathway that inhibits response to trophic factors after receptor binding of interferon (Vedantham et al. 1992). Molecular defects explain other cases of interferon resistance: for example, blocked mRNA translation of cell surface receptors for interferon-alpha (Platanias et al. 1992) or in an experimental system blocked signal transduction from normal cell surface receptors (Ozes et al. 1993).

9.3g Interleukin-2 and tumor-reactive lymphocytes

Clinically, IL-2 is effective therapy for metastatic renal cell carcinoma, a malignancy refractory to radiotherapy and cytotoxic chemotherapy (von Rohr and Thatcher 1992; Tartour, Mathiot, and Fridman 1992). Although "all-natural," pharmacologic doses of IL-2 can cause life-threatening side effects, such as capillary leak syndrome (Kim 1996).

The mechanism of action of IL-2 in renal cell carcinoma remains controversial (Borthwick et al. 1996). Some investigators propose that its clinical activity is due to its trophic effects on the tumor-reactive T-lymphocyte pool. Lymphocytes exposed to antigen plus trophic factor will proliferate and terminally differentiate (Vazquez et al. 1991; Yoshino et al. 1994), whereas exposure to either antigen or trophic factor alone induces apoptosis (Mercep et al. 1989; Takahashi, Maecker, and Levy 1989; Parry, Holman, Hasbold, and Klaus 1994; Parry, Hasbold, Holman, and Klaus 1994). This control of the immune response permits proliferation of lymphocytes only during an immune reaction or thymic development (Mercep et al. 1989). IL-2 therapy increases the number of circulating cytotoxic T-cells and NK cells (natural killer cells), the latter having been shown to recognize renal carcinoma cells in vitro (Hermann et al. 1991; Hayakawa et al. 1992).

IL-2 is also used to stimulate ex vivo proliferation of specialized T-lymphocytes such as lymphokine-activated killer cells (LAK cells) and tumor-infiltrating lymphocytes (TIL cells). LAK cell therapy involves isolation of lymphocytes from the cancer patient, specific expansion in vitro of clones that respond to tumor antigen, and then reinfusion of these expanded clonal populations (Rosenberg 1988). The patient is then treated with additional IL-2 to maintain the activation of the killer phenotype while in the patient. TIL cells are isolated from lymphatic drainage of the tumor, or from tumor specimens directly, and then expanded in vitro with IL-2. Cytolytic T-lymphocytes can be found in TIL cell preparations. TIL cells exhibit the useful behavior of homing to the site of tumor after reinfusion into the patient (Whiteside, Jost, and Herberman 1992; Alexander et al. 1993). After genetic modification to express an antitumor factor or recognize a tumor antigen, reinfused cells can be used to "deliver" high levels of the antitumor effect specifically to tumor-containing sites (Culver et al. 1991; Wang and Rosenberg 1996).

Regardless of these clinical proofs of immune system modulation and the resulting benefit to the patient in boosted immunity, other results suggest that these immunologic effects are not related to the antitumor activity of IL-2. For example, LAK cell activity does not relate to clinical response (Hermann et al. 1991), and TIL cells show only weak responses to tumor antigens in the presence of IL-2 (Alexander et al. 1993). Furthermore, IL-2-induced leukocytosis occurs earlier in nonresponders than in responders (Bergmann et al. 1993), and increases in T-cell subsets of the leukocyte fraction from peripheral blood do not correlate with tumor responses (Hermann et al. 1991). Direct effects of IL-2 on tumor are possible because renal carcinoma cells express cell surface receptors for IL-2 that mediate IL-2-dependent inhibition of proliferation in vitro and G_o/G_1 arrest (Yasumura et al. 1994).

9.3h Corticosteroids

Not only are there trophic factors that offset apoptosis, but there are also factors which can induce apoptosis in progenitors and differentiated cells even though they are not receptor antagonists of the trophic factors (they do not inhibit trophic factor interactions with specific receptors). Pharmacologic levels of corticosteroids produce lymphocytopenia and thymic atrophy, and analogues that do not exhibit mineralocorticoid activity (prednisone, prednisolone, dexamethasone) have been developed to treat lymphocytic leukemia and lymphoma. These compounds induce apoptosis of lymphoma and lymphocytic leukemia with a T-cell phenotype, as well as normal T-cells, at particular stages of differentiation by binding to an intracellular glucocorticoid receptor (Robertson et al. 1978; Wyllie and Morris

1982). Even as a single agent, prednisone induces complete remissions in 70% of children with acute lymphoblastic leukemia (Berenson and Gale 1990). These drugs are also useful in oncology for controlling inflammatory reactions caused by tumor or treatment, such as intracranial pressure from brain tumors (Fosså et al. 1990).

9.3i Treating hematologic side effects of cytotoxic therapy with hematopoietic trophic factors

A frequent dose-limiting toxicity of cytotoxic chemotherapy is suppression of new blood cell production by the bone marrow, leading to dose reductions or treatment delays that compromise the efficacy of the treatment (see Chapter 8, section 8.3c). Trophic factors for the erythrocyte and neutrophil lineages are used clinically to accelerate recovery from anemia or neutropenia (Vadhan-Raj 1996). The factor for the red cell lineage is called erythropoietin (Epo), which is synthesized and secreted into the blood by interstitial cells in the kidney cortex (Koury, Bondurant, and Koury 1988; Lacombe et al. 1988). Epo production and secretion into the blood is regulated by blood oxygen tension (Abbrecht and Littell 1972; Schuster et al. 1987). The recombinant form of Epo is called Epogen, which is approved for the treatment of anemia associated with renal failure (Epo deficiency) (Jacobs et al. 1985; Recny, Scoble, and Kim 1987) (Table 9–1).

Two trophic factors are used clinically to treat neutropenia (Table 9–1). G-CSF (granulocyte-colony stimulating factor) is available commercially in recombinant form as Neupogen, which is used to stimulate neutrophil recovery after cytotoxic chemotherapy. The recombinant form of GM-CSF (granulocyte/macrophage-colony stimulating factor) known as Leukine is used to accelerate repopulation of myelopoietic tissue in the bone marrow after transplantation. Thrombocytopenia often accompanies neutropenia and is just as serious a toxicity. A recombinant factor specific for the megakaryocytic lineage called *mpl*-ligand may prove useful for treatment of this condition (Methia et al. 1993; Bartley et al. 1994; de Sauvage et al. 1994; Gurney et al. 1994, Kaushansky et al. 1994; Lok et al. 1994; Wendling et al. 1994).

9.4 Differentiation therapy

In this book, the term "differentiation therapy" has been reserved for treatment that seeks to alter the balance between proliferation and differentiation in a tumor, and perhaps even to reset homeostasis in the tumor so stem cell expansion is inhibited by many fewer terminally differentiated cells and tissue renewal is no longer exaggerated. In fact, the most effective cancer therapy would likely be one that can restore the normal balance between proliferation, differentiation, and apoptosis to tissue renewal in the tumor – a strategy that targets the fundamental

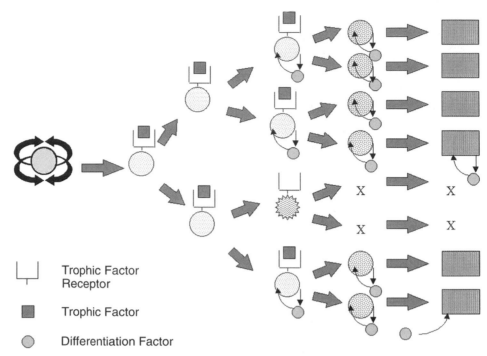

Figure 9–4. Coupling of trophic factor stimulation and terminal differentiation in tissue renewal. Trophic factors (squares) not only inhibit apoptosis (indicated by the abnormally shaped cells), but also trigger the production of factors that stimulate terminal differentiation of the proliferating progenitors and sometimes enhance the differentiated function of the mature cells. These factors could be autocrine or paracrine in nature. When production of the differentiating factors is not properly triggered, or when cells are desensitized to their effects, progenitors continue to proliferate well beyond the normal number of cell divisions (perhaps "indefinitely"). However, the progenitors in these malignancies may still respond to exogenously supplied differentiating factor, especially at the higher pharmacologic levels used for cancer treatment. Also, biotherapy may down-regulate trophic factor receptors either directly, or indirectly as a result of promoting differentiation to a stage that does not express the receptor. All of these mechanisms could result in some degree of tumor control.

nature of malignancy. Because normal tissue already expresses this mechanism, toxicity of specific therapy would be expected to be mild at most.

Proper response to trophic factors probably includes secretion of and response to other soluble factors that induce terminal differentiation in the dividing progenitor (Falk and Sachs 1980; Liebermann, Hoffman-Liebermann, and Sachs 1982). By coupling trophism and differentiation, progenitors will divide a finite number of times before differentiation, and tissue expansion is automatically limited whenever repopulation is stimulated (Figure 9–4). Malignancies characteristically have a defect in this coupling of progenitor expansion and differentiation (Lotem and Sachs 1982; Symonds and Sachs 1982), which probably results in too many rounds of progenitor cell division (infinite number?) before differentiation begins. Malignancies that result from this loss of coupling may respond to exogenously supplied differentiating factor, or perhaps pharmacologic levels of the factor if uncoupling is due to decreased sensitivity. Accordingly, replacement therapy with the differentiating factor or biotherapy that bypasses the defect can be used to achieve long-term control of malignancy, and this approach to cancer therapy is called differentiation therapy (Reiss, Gamba-Vitalo, and Sartorelli 1986; Pierce and Speers 1988).

Differentiation therapy probably maintains smaller tumor size by promoting differentiation of progenitors after fewer rounds of cell division. The increased abundance of differentiated tissue in the tumor might trick the cells into signaling the stem cell compartment that homeostasis has been reached, resulting in some tumor control via the slowing of stem cell repopulation. This control of tumor size might last for years with constant therapy and allow a normal life expectancy with only minimal trouble from tumor. However, these therapies should not be expected to eradicate the malignant stem cell and therefore cannot be used for making a patient "tumor free."

9.4a   Retinoic acid for acute promyelocytic leukemia

The retinoic acid receptor-alpha gene was implicated in acute promyelocytic leukemia (APL), and all-trans retinoic acid (ATRA) has become important for treating this subset of acute myeloid leukemias (Flynn et al. 1983; Daenen et al. 1986; Huang et al. 1988; Fenaux 1993; Wu et al. 1993; Eardley, Heller, and Warrell 1994). The activity of retinoic acid in APL has been impressive, especially when used for relapsed or chemoresistant APL (Liang et al. 1993; Tallman 1994). As with all biotherapy, resistance can develop, and the duration of the remission is relatively short (Gallagher et al. 1989; Ikeda et al. 1994). Genetic alteration of the retinoid receptor (Trayner and Farzaneh 1993), or altered intracellular drug metabolism or pharmacokinetics, may lead to resistance during drug exposure. Greater clinical activity might be obtained in the future by combining retinoic acid with other differentiating agents (Chen, Licht, Wu et al. 1994) or using more active analogues of retinoic acid (Tsurumi et al. 1993).

9.5   Biotherapy with uncertain mechanism: BCG therapy for in situ bladder carcinoma

There is one more approved biotherapy: a standardized extract of a particular attenuated strain of bovine Mycobacterium tuberculosis called bacillus Calmette-Guérin (BCG) (Dorr and Fritz 1980). This extract induces remission of localized bladder carcinoma when administered directly into the bladder ("intravesical therapy"), and response rates with BCG are higher than response rates with instilled cytotoxic drugs (Redman, Kawachi, and Schwartz 1996). Because side effects are mild and the probability of responses from localized disease is quite high, this therapy is often tried first before partial or complete bladder resection as long as the patient is at low risk for metastatic spread. Of course, local BCG therapy is useless against metastatic bladder cancer.

The mechanism of action of BCG is controversial. It is often described as an immune modulator or stimulator, and the extract does induce a granulomatous reaction in the bladder wall, which is an inflammatory response to the presence of the mycobacterial chemicals. It is possible that the inflammatory reaction causes tumor regression via the "bystander effect." This term means that cytotoxicity in a target cell can lead to cytotoxicity in neighboring cancer cells even though they were not originally affected by the therapy (Takamiya et al. 1992). In the case of BCG therapy, tumor cells could be killed by bactericidal cytotoxic mediators released from inflammatory cells. Inflammation might also cause tumor regression via the direct action of negative regulators like interferons and interleukins secreted by the inflammatory cells, in which case BCG therapy may be just a way to obtain a localized exposure to the same biotherapeutics that are active against metastatic renal cell carcinoma.

9.6   Future directions

Biotherapy is a relatively new strategy for treating malignancy, and the field is extending into the arena of molecular modification as a form of therapy. In addition to TIL cell delivery of cytotoxic biologicals to the tumor (section 9.3g), several approaches are being actively pursued, and some of these are described briefly. See recent review articles for more detailed discussions (Rosenberg 1992; Chen, Shine, Goodman, et al. 1994; Kato et al. 1994; Ettinghausen and Rosenberg 1995).

Antisense oligonucleotides are small DNA derivatives that can enter an intact cell and hybridize via Watson-Crick base-pairing to mRNA and its gene to inhibit expression of the encoded protein. This strategy is powerful for molecular targets found only in malignant cells but not the normal counterpart: a mutated gene sequence (e.g., *ras*) or a gene chimera produced by chromosomal translocation (e.g., *bcr-abl*). The important and crucial question is whether the target is critical for survival of the malignant cells. If it is not critical, or if there are other pathways that can compensate for loss of the protein, then inhibiting its expression will make little difference in disease progression. There is also a rapidly increasing effort in genetic modification of the malignant cell either to make the tumor hypersensitive to a cytotoxic therapy or to induce an immune response. An example of the former is treatment of glioma by intrathecal injection of a replication-defective adeno- or retrovirus engineered to contain the herpesvirus thymidine kinase gene. The virus infects many cells in the brain, and expresses the herpes thymidine kinase–derived enzyme. Then exposure to antiviral thymidine analogues results in high-level exposure of infected cells to these inhibitors of DNA synthesis. Because the phosphorylated forms of the drugs are only cytotoxic to proliferating cells, it is hoped that only glioma cells will die while normal, terminally differentiated neurons will be spared. An example of genetic modification of tumor cells to enhance immune response is the genetically engineered expression of foreign proteins on the surface of melanoma cells that are sterilized and then reinjected back into the same patient. The idea is to elicit an immune response to the modified melanoma cells and in the process begin to elicit an immune response to other epitopes on the endogenous melanoma cells.

More research on the stem cell and its biologic characteristics is obviously needed before treatments for the underlying cause of the disease can be developed: the abnormal tissue renewal controlled by the malignant stem cell. At present, too little is understood about the factors that control normal tissue renewal, or exaggerated tissue renewal in tumors, to proceed toward therapeutic development.

9.7 Summary

Several strategies currently used for treating cancer exploit knowledge about trophic factors and inhibitory factors that antagonize trophic factors. Detailed studies of steroid-responsive tissues teach us that trophic factors are required for progenitor survival and optimal growth of stem cells, whether malignant or normal, and trophic factor ablation induces apoptosis of tumor cells derived from these tissues, achieves durable remissions, and alleviates pain from metastases with relatively mild toxicity compared to cytotoxic chemotherapy. These therapies may be effective against tumors that do not respond to conventional cytotoxic modalities, and they may even be curative in older patients with slowly growing tumors that are refractory to cytotoxic chemotherapy whose life expectancy is shorter than the time required for tumor repopulation. However, antitrophic factor therapy does not eradicate the cancer completely because the malignant stem cell, like its normal counterpart, survives in the low hormone environment. If the patient lives long enough, these stem cells will repopulate the tumor with cells adapted for growth in the low hormone environment. Because both malignant and the counterpart normal tissue will atrophy, hormone ablation therapy is most effective against cancers originating from tissues that are not critical for life.

Cancer immunotherapy is based on difficult-to-prove mechanisms of action because clinically effective substances such as interferon-alpha and IL-2 exert direct effects on both the immune system and malignant tissue. The interpretation of inflammatory cells in and around the tumor is important (Russell et al. 1976;

O'Sullivan and Lewis 1994): do these cells indicate necrosis-induced inflamma-tion and, with carcinomas, low-grade infections from disrupted epithelial func-tion, or instead, do they indicate an immune response against the tumor? Even after accepting the assertion that there is an immune response against tumors, the fact that a tumor grows to clinically detectable size in an immunocompetent host implies that the immune system cannot destroy it. Any tumor cured by immune surveillance will never be detectable clinically. Thus, considerable research efforts have been devoted to "engineer" an immune response that can reject established tumors and achieve effective therapy.

Description of selected tumors

G. Barry Pierce

A.1 Adenocarcinoma of the breast

Adenocarcinoma of the breast usually presents as a small painless mass (Figures A–1a,b). The disease rarely occurs in the second and third decades but then increases in incidence through and past the menopause. Because of the seriousness and great range in age incidence of the disease, any mass in the breast must be viewed as cancer until proved otherwise (see letter in Introduction).

Patients are fearful of cancer and cancer surgery, but their only hope is early diagnosis and adequate initial therapy prior to the occurrence of metastasis. Chemotherapy and endocrine therapy for the patient with metastasis is palliative.

The stem cell population of the breast that will give rise to the milk-producing acini are located in the small ductules and persist in the fibrotic stroma of the senile breast. These stem cells are the target in carcinogenesis (Figures A–2a–c). The majority of adenocarcinomas of the breast arise in these same small ductules. They may present in several ways: the vast majority develop as small solitary painless nodules, but some within the ductules may present with a bloody discharge from the nipple as the nodule develops prior to the development of the mass. As the adenocarcinomas grow, they infiltrate surrounding tissues and eventually become fixed to the deep tissues and to the skin. This may cause dimpling of the skin or, if it invades the nipple, retraction of the nipple (Figure A–1a). With further growth and development it may spread first to the lymph nodes in the axilla, and shortly thereafter to the lungs, pleura, and to the bones of the spine and pelvis.

The patient often undergoes extirpational biopsy (lumpectomy) to establish the

(a)

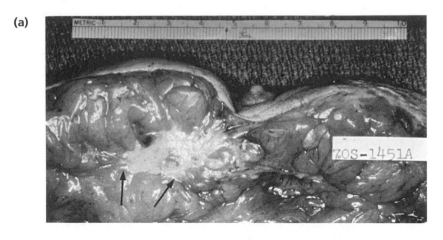

(b)

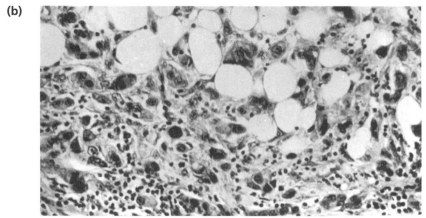

Figure A–1. (a) This breast, removed surgically, was cut through the margins of the nipple and through the adenocarcinoma. The tumor is several centimeters in diameter with gray streaks of invading cancer radiating into the fat (arrows). Invasion to underlying muscle and nipple caused retraction of the nipple. This is a bad prognostic sign. (b) Anaplastic adenocarcinoma of the breast. Huge irregularly shaped nuclei of cells infiltrating fat and cords of anaplastic cells are the clue to the diagnosis.

diagnosis, which is made by examination of a frozen section of the lesion. If the lesion is cancer, the breast and the lymph nodes of the axilla may be removed. Thorough examination of these specimens establishes the stage of the disease and the type of adenocarcinoma involved. Staging is most important, because if spread has not occurred, the patient is usually cured. If more than two lymph nodes are involved, the prognosis for the patient becomes very poor. If fewer than two axillary lymph nodes are involved the prognosis is good, with a 50% chance of cure.

In the recent past the treatment of choice for adenocarcinoma of the breast has been radical mastectomy. The breast, the muscles underlying the breast, and the contents of the armpit were removed in an attempt to extirpate all tumor cells. It is now known that lumpectomy (removal of the carcinomatous lump) followed by

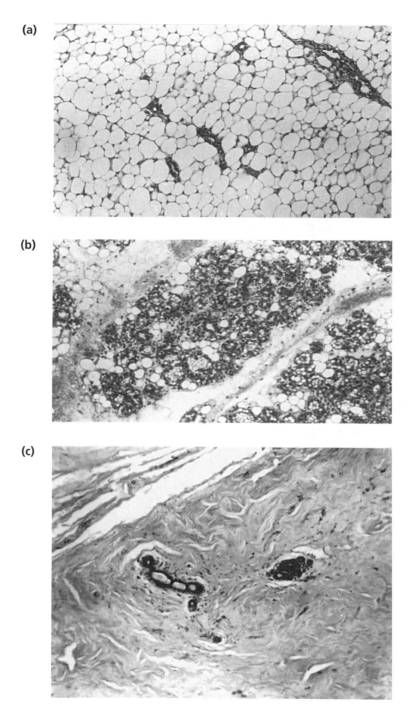

Figure A–2. (a) Photomicrograph of breast of virgin mouse. Note all of the epithelial elements of the breast are included in the primitive tubules embedded in the fat that characterizes the breast at this stage. (b) Photomicrograph of mouse breast in mid-pregnancy. The stem cells in the primitive tubules have multiplied in response to the hormones of pregnancy and become hyperplastic. The epithelial cells have displaced fat cells. Glandular acini are forming. (c) Photomicrograph of breast of old mouse. As in the virgin state there are only a few ducts present, but unlike the virgin state these are embedded in dense scar tissue. The epithelial cells (stem cells) in these ducts must be the target in carcinogenesis because there are no other cells present.

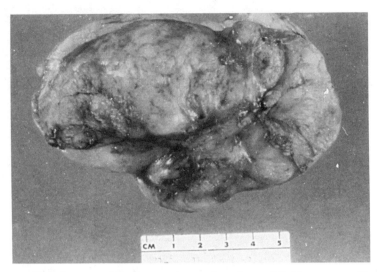

Figure A–3. Fibroadenoma of the human breast, female age 30. The mass, unlike adeno-carcinoma, which is stone hard on palpation, is rubbery in consistency. This is such a sub-jective distinction that an extirpational biopsy is peformed. The mass is encapsulated and was shelled out of the organ with blunt dissection. The smooth capsule suggests invasion has not occurred; however, final diagnosis is made microscopically. The patient is cured by extirpation of the mass that was easily shelled out of the breast. It is benign and the patient has been cured by simple resection.

irradiation and chemotherapy is as effective as radical mastectomy in terms of cure. Importantly, the patient is spared the psychological and physical trauma of a mutilating surgical procedure.

If distant metastases are present, then palliative procedures are invoked. These are designed to increase the quality and quantity of life and can include radiother-apy to painful bony metastasis and hormonal therapy. Chemotherapy is also used for the patient with metastasis, but the efforts are palliative only.

Contrast this gloomy picture with that of the individual with a fibroadenoma of the breast (Figure A–3). The lesion is encapsulated, has no evidence of anapla-sia, and simple removal results in cure.

A.2 Adenocarcinoma of the prostate

Adenocarcinoma of the prostate is the most common malignancy in men over 65 years of age. The disease is slowly progressive from its site of origin usually beneath the capsule of the posterior part of the gland. It grows and compresses and obstructs the urethra and causes symptoms of bladder obstruction, hematuria (blood in the urine), and/or infection. It may invade outward to involve perineural lymphatics and presacral lymph nodes (Figure 1–12). In its late stages the tumor metastasizes via the bloodstream to liver, lungs, and bone. High intra-abdominal pressure, which occurs

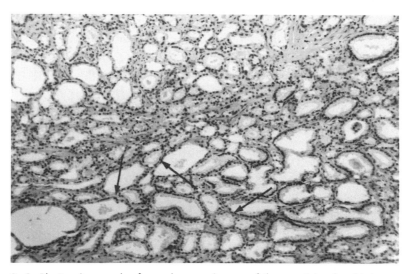

Figure A–4. Photomicrograph of an adenocarcinoma of the prostate gland taken at very low power. Evidence of lobular pattern is still present, but the epithelial stromal relationships are abnormal. No stroma separates acini (arrows). Adenocarcinoma of the prostate is usually not very anaplastic, and the diagnosis is made at low magnification based on the abnormal epithelial stromal relationships.

in lifting, for example, may reverse the direction of blood flow in the vertebral venus plexus, and tumor emboli flow from the prostate to the vertebral bodies and then from one vertebra to another. Adhesion factors facilitate specificity of metastasis for particular tissues. This adenocarcinoma has a propensity to metastasize to bone.

The tumors are usually composed of small innocuous-appearing cells arranged in acini that lie back to back with no stroma separating them, as is usual in normal glandular acini (Figure A–4). Similar cells are found in perineural lymphatics and in the metastases where they often elicit an osteoblastic response (formation of new bone). Invasion and metastasis of carcinoma of the prostate is associated with synthesis of acid phosphatase and prostate specific antigen (PSA), which can be measured in the serum and used diagnostically.

Islands of small anaplastic carcinoma cells can also be found in the prostates of elderly men, and it is said that most men over the age of 75 have such latent foci. Little is known about them because they are asymptomatic and discovered incidentally. If the hosts live long enough, invasive adenocarcinoma would likely develop from them. Unfortunately, literally nothing is known about them.

A.3   Adenocarcinoma of the colon

Adenocarcinoma of the colon is a common disease of individuals 55 to 70 years of age. In men the lesion usually occurs in the sigmoid colon or rectum and at least 65% of them are within reach of the examining finger. A routine physical in an individual over the age of 50 is incomplete if rectal examination has not been done. The lesion invades circumferentially and results in an ulcerated napkin ring–like constriction of the colon that causes bowel obstruction and bleeding (Figures 1–7 and 1–8). In women the adenocarcinomas are more common in the cecum and ascending colon where large fungating lesions occur that ulcerate and bleed, but do not cause bowel obstruction. This is because there is less constriction of the colon and the fecal stream in the right colon is liquid, whereas that in the left is solid. Bowel obstruction is a surgical emergency, and a person with a complete obstruction will die in about a week if untreated. In addition to obstruction or bleeding, a multiplicity of symptoms, many nonspecific, can herald the disease. For example, anemia and colicky (spastic) abdominal pain can be symptoms.

Etiologic agents include ingestion of carcinogens and production of carcinogens endogenously. Multiple polyposis (Figure 1–11) and ulcerative colitis predispose to adenocarcinoma and should be considered precancerous lesions.

Adenocarcinomas of the colon are slow growing and take long to develop. Those that arise in polyps do so in the superficial mucosa, invade the pedicle of the polyp, and then the bowel wall. Ulceration is common; the central part of the polyp sloughs, leaving an ulcer with a raised margin and a base of malignant tissue that, if left untreated, invades the bowel wall.

The association of polyps and carcinoma of the bowel stems from the observation that multiple congenital polyposis has a 100% incidence of adenocarcinoma of the colon before the age of 40. It is not uncommon to find polyps adjacent to or near existing adenocarcinomas in the colon. Single polyps are often found in areas of the colon predisposed to adenocarcinoma. Malignant change in polyps is frequently seen.

Grades of malignancy of colonic adenocarcinoma span a wide range: from some with well-formed glands to others in which there are plugs of tumor cells (lacking glandular lumina) (Figure 1–10). Columns of single cells form in some, and still others are composed of signet ring (mucin-containing) cells unattached to each other that penetrate the bowel wall and cause a diffuse thickening of the colon. If the lesion is confined to the mucosa, the cure rate is about 90%, and if there has been involvement of the whole thickness of the bowel wall, the cure rate is about 70%. Penetration into lymphatics occurs early, and when lymph node metastases are present the cure rate is only about 30%. Thus, staging is most valuable in assessing prognosis.

Carcinoembryonic antigen and glycoprotein antigen is synthesized by the tumor cells and can be assayed in the serum. It is particularly useful in detecting recurrences as metastases grow and produce the antigen.

A.4   Squamous cell carcinoma

Squamous cell carcinomas develop wherever squamous epithelium occurs normally or from squamous metaplasia that may occur in the respiratory tract, renal pelvis, or bladder. The tumors develop most commonly in skin, bronchus, and cervix. Each behaves differently in these locales. Those in the skin are usually removed before they metastasize, but given time they can spread to regional lymph nodes and later to distant sites. In the bronchus they metastasize early to regional lymph nodes and distant sites. Those in the cervix invade locally in an extremely aggressive fashion, and usually destroy the adjacent ureters, rendering the host incapable of excreting liquid wastes, and cause death from uremia. For reasons unknown, they rarely metastasize to distant organs.

There are no explanations why tumors that are histologically similar should vary so in their biologic behavior. Cure rates are about 10% for squamous cell carcinoma of bronchus, 50% for cervix, and 100% for skin. Squamous cell carcinomas of the mucous membranes such as the mouth, vulva, glans penis, bladder, and bronchus have a much worse prognosis than those of the skin, presumably because they are much more advanced when diagnosed. Squamous cell carcinoma of the lung is discussed in section A.7.

Squamous cell carcinomas of the skin ulcerate early (Figure A–5) and because ultraviolet light is a carcinogen for squamous cells, the tumors are most numerous on sun-exposed parts of the body. The histologic patterns vary from well-differentiated, slow-growing tumors with numerous squamous pearls (areas of fully differentiated squamous epithelium) (Figure A–6) to anaplastic, fast-growing masses with little or no evidence of squamous differentiation. The tumors penetrate the basement membrane and invade and destroy local structures. Because of early diagnosis, they are usually identified before metastases have occurred. If the tumors are excised with wide margins, the individuals are cured. If not, they recur and then are more prone to metastasize, usually to regional lymph nodes and ultimately to distant organs.

Studies of squamous cell carcinoma of the cervix are illustrative of the pathogenesis of squamous cell carcinoma in general, so cervical carcinoma is discussed here. Squamous cell carcinoma of the cervix is most common in women 45 years of age or older and develops from carcinoma in situ at the junction of squamous and glandular epithelium of the cervix (Figures A–7 and A-8). The lesion is first evident as a white patch, and with progression it becomes invasive and ulcerates, but seldom metastasizes to distant organs. Invasive squamous cell carcinoma occurs about 10 years later than carcinoma in situ.

Invasive carcinoma, carcinoma in situ, or inflammation may all present as leukoplakia, which means "white patch." Therefore, presence of a white patch on examination does not necessarily mean a diagnosis of cancer. The diagnosis of cancer is made using the Papanicolaou technique (Pap smear). In this procedure, the cervix is scraped and the surface cells are smeared on a slide and examined microscopically. The surface cells of a carcinoma in situ are anaplastic and appear so on the slide (Figure A–8). The presence of such anaplastic cells does not distinguish in situ and invasive lesions. The pathologist alerts the gynecologist to take an extirpative biopsy to make this important distinction. The tumor is considered in situ if the basement membrane, which normally delineates the junction between epithelium and stroma, is intact.

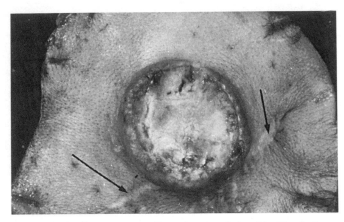

Figure A–5. Photograph of squamous cell carcinoma of the skin. The lesion measures about 1.5 to 2 cm in diameter and has a raised, rolled edge (typical of malignant ulcers). This edge is a reflection of invasion and proliferation of the tumor cells beneath the skin. The base of the ulcer is crusted and composed of necrotic cancer and stromal cells. Note the surgical scar at the arrows. This lesion is a recurrence of an incompletely removed carcinoma. It is such inadequately treated lesions that are prone to metastasize because the surgery opens lymphatics and small venules, chief avenues for metastasis.

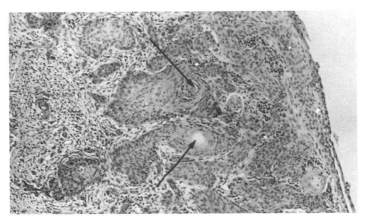

Figure A–6. Photomicrograph of squamous cell carcinoma of skin illustrating plugs and columns of anaplastic cancer cells invading stroma. Note the squamous pearls (arrows), good examples of squamous differentiation.

Histopathologically, carcinoma in situ is anaplastic, about grade III on the Broder scale of I to IV. Carcinoma in situ develops in the basal layer and the transformed cells, instead of differentiating in orderly successive steps until senescence occurs and the dead cells are sloughed from the surface of the mucous membrane, form layer after layer of cells closely resembling the dark-staining basal cells. Thus, the full thickness of the epithelium shows no evidence of maturation (Figure A–8). Why the lesions do not invade is not known, but through the use of the Papanicolaou technique, carcinomas in situ are easily identified and resected. As a

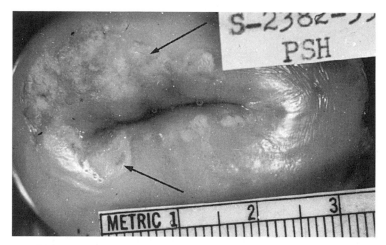

Figure A–7. Photograph of cervix with leukoplakia (white patch, arrows). This white patch is in reality a carcinoma in situ of the cervix, but leukoplakia could be dysplasia, inflammation, or an invasive carcinoma. The term is merely descriptive.

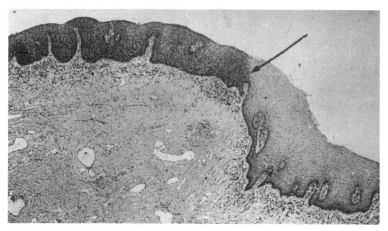

Figure A–8. Photomicrograph of carcinoma in situ. Note the sharp line of demarcation (arrow) between the carcinoma in situ (to the left) and normal squamous epithelium (on the right). Note the progressive differentiation of squamous cells in the normal epithelium, from the dark-stained basal (stem cell layer) to the terminally differentiated cells at the surface. This differentiation is lacking in the carcinoma in situ, and the surface cells are as undifferentiated as the basal layer of cells. Notice there is no evidence of invasion.

consequence, fewer cases of invasive carcinoma develop now than prior to the use of this simple technique.

Thus, carcinoma in situ is believed to be a stage in the development of squamous cell carcinoma. Less clear is the role of dysplasia in the cervix in the generation of carcinoma in situ and squamous cell carcinoma. Dysplasia is characterized by hyperplasia of the basal cells with inappropriately differentiated hyperchromatic squamous cells occupying the more superficial layers of the epithelium. Thus, the orderly pattern of differentiation is not present. It is generally believed that these dysplastic lesions are precancerous (Figure 1–3). From an experimental stand-point, dysplastic lesions occur first when carcinogens are painted on the cervix of animals, followed by lesions that resemble carcinoma in situ, and finally by inva-sive carcinoma.

The cell biology of squamous cell carcinomas was clarified in a study using rats carrying transplantable squamous cell carcinomas of the skin (Figure A–9). The relationship of the anaplastic tumor cells and the apparently differentiated ones in the squamous pearls was studied. Rats were injected with H-thymidine to identify tumor cells synthesizing DNA. Autoradiography was performed on the tissues at 1 hour and at 96 hours after administration of the thymidine. The autoradiogra-phy showed that labeled nuclei were found only in the anaplastic cancer cells at the first time point (Figure A–10), indicating that the pearls were composed of postmitotic differentiated cells and the proliferative pool of the tumor lay in the anaplastic cells. Samples taken at 96 hours showed the presence of labeled cells not only in the anaplastic tumor but in the pearls, indicating that the anaplastic cells labeled in the first hour migrated to or invaded the pearls as well. Autoradi-ography with electron microscopes showed that the anaplastic cells lacked the fea-tures of squamous epithelium and that the labeled cells of the pearls were well-differentiated squamous cells with typical membrane and cytoplasmic characteris-tics. It was concluded that the anaplastic cells had migrated into the pearls and had differentiated. When the squamous pearls were dissected from the tumor and transplanted to histocompatible hosts, they failed to produce tumors, whereas equivalent amounts of anaplastic cells formed tumor on subcutaneous transplan-tation. It was thus concluded that in the process of spontaneous differentiation, anaplastic cancer cells gave rise to benign squamous cells no longer dangerous to the host (Pierce and Wallace 1971).

A.5 Teratocarcinomas

An emphasis disproportionate to the incidence or clinical importance is given these tumors because of the impact their study has made on oncology. Teratomas, literally benign tumors resembling monsters or malformed babies (terata), commonly occur in the ovaries. They may contain almost any tissue found in the body, and recogniz-able organs, such as fingers, eyes, or jaws with teeth, may be present in them. Terato-carcinomas, like teratomas, contain tissues representing each of the three embryonic germ layers (endoderm, mesoderm, and ectoderm). As the term implies, teratocarci-nomas are malignant. For reasons unknown they are more commonly found in the testes than in the ovary (Figure A–11). They contain embryonal carcinoma (Figure A–12a), an exceedingly anaplastic carcinoma that is interspersed with the other tis-sue of the tumors (Figure A–12b) (Pierce et al. 1978). Embryonal carcinoma cells resemble the inner cell mass cells of the blastocyst (Chapter 1).

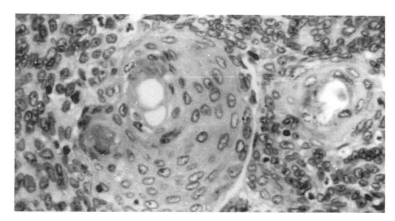

Figure A–9. Photomicrograph of squamous cell carcinoma of the skin of a rat to illustrate the masses of dark-stained anaplastic tumor cells surrounding islands of well-differentiated squamous cells with keratotic centers. These are referred to as pearls.

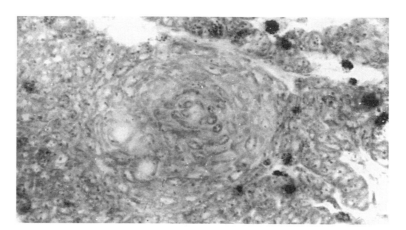

Figure A–10. Autoradiogram of squamous cell carcinoma of a rat injected 1 hour previously with tritiated thymidine. The tumor was fixed, sectioned, and the sections were covered with photographic emulsion and stored in the dark. The photographic emulsion was exposed by decay of the tritiated thymidine at sites of its incorporation into DNA (see black dots over nuclei at arrows). No labeled cells were seen at 24 hours, indicating pearl cells did not synthesize DNA because they were differentiated and postmitotic.

Teratocarcinomas usually present as a painless enlargement of the testis in the third decade of life. They are so lethal that survival is measured in two years; the patient is either dead or cured two years after diagnosis and treatment. Most of the tumors have metastasized at the time of diagnosis to the periaortic lymph nodes and to the liver, lungs, and other organs. Teratocarcinomas and pure embryonal carcinomas are extremely sensitive to chemotherapy; cure can be achieved in over 80% of cases.

With the exception of two strains of inbred mice, teratocarcinomas are ultrarare

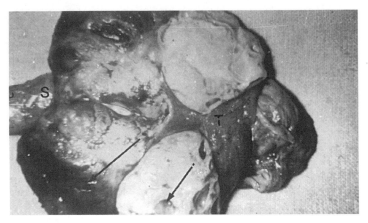

Figure A–11. Photograph of a human testis cut in half and opened like the wings of a butterfly. The spermatic cord is at S, normal testis T, and numerous nodules of teratocarcinoma; note epithelium-lined cysts (short arrow) and choriocarcinoma (long arrow).

in lower animals. Strains 129 and LT have high incidences of testicular teratocarcinoma and ovarian teratocarcinomas, respectively. They have proved to be useful models of the human tumors (Pierce, Shikes, and Fink 1978). As in humans, the differentiated tissues are well formed in the murine tumors but haphazardly arranged, so it is not uncommon to find islands of brain, gut, muscle, bone, and gland all mixed up together (Figure A–12b–d) and with embryonal carcinoma (Figure A–12a). Organization of tissues is seen in the tumors more often than can be accounted for by chance (Figure A–12c). In other words, gastrointestinal glands may be surrounded by a double layer of smooth muscle, or masses of cartilage may be ossifying and marrow may form in the developing bone. A tooth may lie in juxtaposition to membrane bone (Figure A–12d). Structures named embryoid bodies, because they resemble early preimplantation mouse embryos, are frequently found in them, particularly in the ascites form (Figure A–12e). These structures have proved useful in studies of the tissue interactions in teratocarcinomas.

Dixon and Moore (1952) proposed that embryonal carcinoma cells were multipotent cells that differentiated into the somatic tissues of the teratocarcinomas. This was proved correct when embryonal carcinoma cells were shown to differentiate into the three germ layers, which in turn differentiated into the somatic tissues characteristic of the tumors (Pierce and Dixon 1959). These tissues were benign and of no danger to the host (Pierce, Dixon, and Verney 1960). These observations were confirmed by cloning embryonal carcinoma in vivo. Single embryonal carcinomas transplanted into the intraperitoneum of mice gave rise to teratocarcinomas containing the 12 or more tissues characteristic of the tumors plus embryonal carcinoma (Kleinsmith and Pierce 1964). Because the histiotypic potential of embryonal carcinoma so resembled that of the inner cell mass of the blastocyst, which is the only other tissue capable of giving rise to the three germ layers, it was postulated that embryonal carcinoma was the neoplastic equivalent of inner cell mass cells.

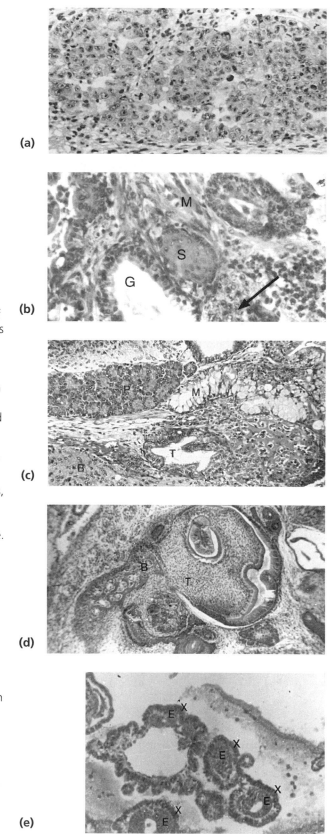

Figure A–12. (a) Photomicrograph of focus of embryonal carcinoma of testis with no evidence of differentiation. (b) Photomicrograph of teratocarcinoma of a mouse illustrating embryonal carcinoma at long arrow, smooth muscle (M) representing mesoderm, gland (G) representing endoderm, and squamous epithelium (S) representing ectoderm. (c) Photomicrograph illustrating tissues of a teratocarcinoma of a mouse. Note trachea with ciliated epithelium, T; brain tissue, B; pancreas, P; and mucinous gland, M. (d) Photomicrograph illustrating a tooth (T) found in a teratocarcinoma of a mouse. It lies next to membrane bone (B), a normal relationship. (e) The embryoid bodies from the ascites of a testicular teratocarcinoma of the mouse. The embryoid bodies contain embryonal carcinoma representing embryonic epithelium (E) of the embryo. Entoderm overlies the epithelium whether benign or malignant at X. Compare this structure to the embryo in Figure 1–16. Embryonic epithelium is overlain by endoderm in the embryo. An embryoid body such as this placed in the subcutaneum of an appropriate mouse gives rise to a teratocarcinoma composed of a minimum of 12 differentiated tissues representing all of the germ layers. Thus, embryonal carcinoma is the multipotential stem cell precursor of teratocarcinoma.

There are three principal sources of embryonal carcinoma cells: primordial germ cells, inner cell mass cells, and transplantation of preimplantation embryos. Stevens (1967) proved that the spontaneous teratocarcinomas of strain 129 mice originated from primordial germ cells. On the 12th day of gestational age, he transplanted genital ridges into the testes of adult animals. If the genital ridges contained germ cells, they produced islands of embryonal carcinoma cells visible 7 days after transplantation. Tumors several centimeters in diameter were formed in the ensuing two weeks that contained the multiplicity of differentiated tissues characteristic of teratocarcinoma. If the genital ridge did not contain germ cells (Stevens transplanted genital ridges from animals with a congenital absence of germ cells when in the homozygous state), teratocarcinomas did not occur. The primordial germ cells may be construed as the stem cells of the species, and when the ultrastructure of the primordial germ cells was compared to that of the embryonal carcinoma cells to which they gave rise, they proved to be equally undifferentiated (Pierce, Stevens, and Nakane 1967). This eliminated the need for the concept of dedifferentiation as a mechanism in the genesis of these tumors and led to the idea that tumors in general probably originate from the stem cells in any differentiated tissue. Preimplantation mouse embryos of appropriate strain transplanted into the testes of adult mice also give rise to teratocarcinoma. The most widely used teratocarcinoma cell line, OTT6050, originated in such a manner (Stevens 1970).

Strains of low-grade embryonal carcinoma cells, which are called ES cells (embryonic stem cells), have been developed by culturing inner cell mass cells in vitro where they become spontaneously transformed (Evans and Kaufman 1981; Martin 1981). This confirmed the idea that embryonal carcinoma cells are the neoplastic equivalent of the inner cell mass. ES cells form beautiful teratocarcinomas in vivo, differentiate well in vitro, and are less progressed than any of the transplantable spontaneous teratocarcinomas.

Another important observation was made by Stevens (1967) in the conduct of the transplant experiments that give rise to teratocarcinomas. As many as 11 or more islands of embryonal carcinoma were found in individual testicular tubules by the 7th day after transplantation of genital ridges. It is widely accepted that most tumors are monoclonal in origin. To rationalize these observations and ideas, it is clear that the monoclonality of tumors must result from selection of the fittest clone. Thus, in the embryonal carcinomas, one of the eleven clones would dominate, be selected, and the resultant tumor would be monoclonal. The significance of the observation is that selection of cells is an important process in the latent period during carcinogenesis and later during progression of the established tumors.

Brinster (1974) was the first of many who have shown that embryonal carcinoma cells injected into the blastocyst of mice are regulated by the blastocyst to behave as normal inner cell mass cells, take part in embryonic development, eventuating chimeric animals. The animals are chimeric because their tissues are composed of embryo-derived cells and embryonal carcinoma-derived cells. A majority of chimeras, in which aneuploid embryonal carcinoma cells were used, were chimeric in only a few tissues. If an animal was chimeric at all, it was in coat color and then possibly in some of the internal organs. These animals were almost always sterile (Mintz and Illmensee 1975). In contrast, chimeras produced by injection of ES cells into blastocysts give a much greater percentage of chimeras, many of which are fertile. These are used in the construction of transgenic mice (Brinster et al., 1984).

The technique of mutating ES cells or transfecting them with genes of various kinds provides a powerful tool for studying molecular events in genetics. For example, if gene X is transfected into ES cells and a fertile male chimera is produced with them, in one backcross a pure inbred line carrying gene X will be produced. This is of great value in modeling a variety of genetic diseases.

Finally, how does one rationalize teratocarcinoma with other carcinomas and sarcomas? It can be stated unequivocally that all of the lessons learned from the study of teratocarcinoma extrapolates to all other kinds of cancer. The phenotypes of tumors like those of their normal counterparts differ only in the potency of their stem cells. Embryonal carcinoma cells have the multipotency of inner cell mass cells. Multipotent tumors differ from unipotent ones, such as squamous cell carcinomas, in that the potential of the stem cell originating the squamous cell carcinoma has been reduced during development to a single tissue. Carcinogenesis does not alter the histiotypic potential of the responding stem cell – it adds the malignant phenotype to it.

Why some embryonal carcinomas differentiate into somatic tissues and others do not is not known. Retinoic acid, applied to cells that normally do not differentiate in vitro or in vivo, causes them to differentiate (Strickland, Smith, and Marotti 1980). Therefore, they have the potential to differentiate, but for reasons unknown do not express it. Apparently, environment controls the expression of the determined state. These observations form the basis for the chemical induction of differentiation as a form of therapy for cancer.

A.6   Liver cell carcinoma

Liver cell carcinoma is rare in North America, but it is common in the Third World. In North America, individuals with this disease often have cirrhosis of the liver and the clue to hepatoma is an unexplained deterioration in the individual's condition as a complication of cirrhosis. Weight loss, a right upper quadrant mass, or abdominal pain may be present. The role of cirrhosis as an etiologic agent is unexplained, although the multiple nodules of regenerating liver in the cirrhotic condition represent an ever-present hyperplasia, which in turn could present more cellular targets for the carcinogenic event. Macronodular cirrhosis is a hepatitis-related condition and indicates a role for hepatitis B virus in the etiology of hepatoma. Hepatitis B virus antigens are present in over 80% of the patients with hepatocellular carcinoma, and the tumor is more often present in hepatitis B–asso-

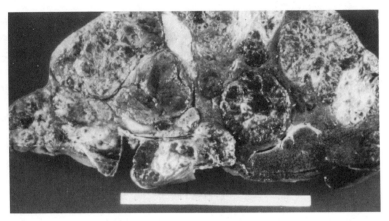

Figure A–13. Photograph of metastatic embryonal carcinoma to the liver. Metastatic carcinoma is the most common tumor of the liver in North America. It is easily distinguished from the normal liver parenchyma, and even in a case such as this with massive replacement of liver by embryonal carcinoma, death does not result from liver failure, but rather from cachexia and infection.

ciated cirrhosis than in other conditions. Hepatitis B virus is accepted as an important etiologic agent in the production of the disease (Sell and Pierce 1994). Hepatocellular carcinoma synthesizes alpha fetoprotein, which is a useful marker for the tumor.

Aflatoxin, which is produced by the mold *Aspergillus flavus,* is a potent liver carcinogen in the Third World, where food is often contaminated by the carcinogen. Butter yellow (P-dimethyl aminoazobenzene) causes liver cell carcinoma without cirrhosis, indicating that the cirrhotic process itself is not essential for the production of the disease.

The tumors may be single or multifocal in origin, usually well demarcated from the surrounding normal liver tissue, and have either of two histologic patterns. One is composed of well-differentiated-appearing hepatocytes (parenchymal liver cells) arranged in trabeculae (fibrous cords of connective tissue) separated by sinusoids (minute blood vessels). Many of the cells are functional because some of the nodules are bile stained (the bile canaliculae of the tumor do not connect with the normal bile ducts) and contain liver enzymes that can be demonstrated histochemically. The second pattern is anaplastic adenocarcinoma, which arises in small bile ducts. These bile duct carcinomas are rarely associated with cirrhosis and very rarely associated with parasitic infections, such as liver flukes. The tumors metastasize widely, especially to the lung.

Because the vascular drainage of the entire gastrointestinal tract passes through the liver, it is not surprising that the liver is a common site for metastatic tumors, which are much more numerous than primary ones (Figure A–13). The liver is the primary organ of detoxification of xenobiotics. In this process, poisons are oxidized to inactive forms and solubilized so they can be excreted in either the bile or the urine. Procarcinogens can be metabolized to highly active carcinogens by these processes, and not only liver cells, but those of the bladder, are exposed to the carcinogens if they are excreted in the urine. We supposedly live in a sea of synthetic chemical carcinogens that require metabolic activation, and it is interesting that the liver does not have a much higher incidence of primary carcinomas than other organs.

A.7   Lung cancer

Lung cancer is epidemic in North America. More than 30% of men dying of cancer will die of lung cancer. The incidence is increasing in women, and in either sex the death rate equals the incidence. This is a slight overstatement, but it is accurate enough to grimly portray our inability to cure people with the disease and to stress the need for prevention. When one thinks of epidemic, one thinks of infectious diseases, but in the case of lung cancer, the epidemic is caused by inhaled pollutants. The most important of these is cigarette smoke. Numerous studies have implicated cigarette smoking in the dramatic increase in incidence of lung cancer from 1930 to 1970. Compounding the effect is carcinogenic synergism that occurs when multiple carcinogens are applied simultaneously. Inhalation of asbestos by a smoker is a case in point. As a rule, if a smoker has a 10-fold increased risk of lung cancer and an asbestos worker has a 10-fold increased risk, then the cigarette smoker exposed to asbestos has a 100-fold increased risk (Selikoff, Seidman, and Hammond 1980). These numbers are approximations, but sufficiently accurate to make the point that if no one smoked, the risk for lung cancer would be less than 5% of the current incidence. For the first time in the past 10 years, the combination of low tar cigarettes and education has resulted in a decline in the epidemic in the younger age groups.

Other carcinogenic agents include exposure to irradiation (the uranium miners of Colorado), chromates and nickel as industrial hazards, and finally automobile exhaust, which contains carcinogenic products of combustion.

The airway has been designed to clean, moisten, and warm inhaled air. In addition, it has a mechanism to lubricate and sweep particulate matter and fluid from the smaller to larger airways (the cough sensitive portion of the lung) (Figure 1–1) where the contaminants can be expelled by coughing. Anything that interferes with ciliary movement or secretion of mucus perturbs the housecleaning mechanism of the lung and results in chronic inflammation. Cigarette smoking bombards the bifurcations of the bronchi with hot dry gases, particulates, and carcinogens. This leads to squamous metaplasia, which in turn is believed to be the site of origin of squamous cell carcinoma of the bronchus. To give a measure of the hazards, a smoker is at high risk with a 20-pack-year smoking history ("pack years" equal the packs smoked per day times the number of years of smoking). The cancer risk varies directly with the number of cigarettes smoked, but with at least the square of the duration of smoking. Thus, longtime smokers do not protect themselves by merely cutting back on the number of cigarettes smoked per day. To accomplish this goal, smoking must be stopped.

Squamous cell carcinoma of the lung usually arises in major airways in smoking men 45 years of age or older. The lesions are painless but associated with a cough. They grow around the airway, destroying the ciliated epithelium and eventually forming an ulcer, which may bleed (hemoptysis). Secretions are dammed up behind the cancer and become infected, leading to chronic bronchitis, or to bronchopneumonia. The infections respond to therapy, but immediately recur. The recurrence of symptoms in a middle-aged man with a long history of smoking leads to the suspicion of squamous cell carcinoma of the lung. The tumors are generally not far advanced at the time of first diagnosis, but complete removal by widespread resection is impossible because of the vital structures that lie in close proximity to the bronchi. Later, the tumors are extremely invasive and eventually involve the lymph nodes at the bifurcation of the trachea. Blood vessels are invaded and tumor cells are disseminated to the liver, brain, adrenals, and bone. Death is usually due to lung abscess and pneumonia, often caused by yeast and fungi.

Squamous cell carcinoma of the lung and small cell (or oat cell) carcinoma are often associated with paraneoplastic syndromes, the result of secretion of biologically active molecules that cause disease. For example, ACTH, which stimulates the adrenal cortex to produce toxic amounts of cortisone and Cushing's syndrome, may be produced by the tumor cells. Similarly, parathyroidlike hormones among other trophic hormones may be produced.

The cells of oat cell or small cell carcinoma of the lung resemble lymphocytes at the light microscopic level, but the cells have secretion granules typical of neuroendocrine cells. This type of lung carcinoma is extremely prone to metastasis. It is said that when the first oat cell divides, one of the daughters metastasizes. The tumors originate in the neuroendocrine cells of the airway, and as a result often synthesize clinically significant amounts of hormone such as serotonin. They are usually inoperable at the time of discovery.

Adenocarcinoma of the lung is much less common than squamous cell carcinoma or small cell carcinoma. Lung adenocarcinomas develop silently in the periphery of the lung; although slow growing, they, too, have a poor prognosis because of the stage of the disease at the time of diagnosis.

A.8   Malignant melanoma

Melanomas develop most commonly in the skin of the lower extremities or other sunbathed areas, but they can occur wherever pigmented cells are found in the body (mucous membranes, eye, central nervous system). About one-third of the tumors develop from preexisting moles or "beauty spots" (pigmented nevi).

The clinical course varies from extremely malignant lesions that invade and

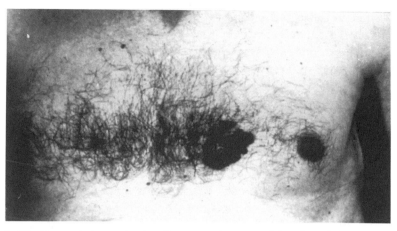

Figure A–14. Gross photograph of malignant melanoma of the skin of the chest of a man, illustrating various degrees of pigmentation and an irregular margin.

metastasize and cause death in a matter of months to indolent ones that pursue a protracted course over years. The best prognostic indicator is the stage of the disease at the time of diagnosis. This is determined by microscopic examination of the lesion and actual measurement of the depth of penetration of malignant cells into the dermis. Melanomas that have not invaded beyond the superficial dermis are cured at the 100% level by widespread excision, but when the lesion has penetrated into the subcutaneous tissue, the cure rate is less than 15%. This dramatic change in prognosis occurs when the tumor has penetrated to a depth of only 1.5 mm.

Malignant lesions are usually discovered accidentally. Change in color, shape, size, or consistency of a preexisting nevus are considered danger signs. Invasive melanomas are almost always inflamed. Therefore, itching or mild inflammatory change in a mole must also be construed as a danger sign (Figure A–14).

Three kinds of melanoma are described, and although none is as important as the stage of the disease at the time of diagnosis, they are of some interest. The first is melanoma in situ that develops in a Hutchinson's freckle. Hutchinson's freckles are brownish, slightly raised spots about 3 cm in diameter on exposed surfaces of elderly patients. Over prolonged periods of time these freckles develop melanoma in situ and eventually invasive melanoma (Figure A–14). Like other carcinomas in situ, melanoma in situ is cured by local removal. Superficial spreading melanomas are the most common and are slightly raised, variably colored lesions that may show evidence of growth in one area and regression in another. The actively growing areas are light brown, whereas the regressing areas may be blue or inflamed. Melanoma cells are found at the epidermal-dermal junction and spread throughout the epidermis. Nodular melanoma carries the worst prognosis and has a tendency to infiltrate into the subcutaneous tissues, resulting in a more advanced stage at the time of diagnosis.

Metastasis of malignant melanoma may occur early and extensively via the lymph nodes and blood vessels, with widespread involvement of many organs.

Often the metastases are amelanotic and rapidly growing. Under very rare circumstances, metastases can be dormant and the patients survive asymptomatically for many years after the removal of the primary tumor, only to have metastases eventually grow in the liver, for example. The reasons for this dormancy are not known.

A.9   Retinoblastoma and neuroblastoma

Retinoblastoma and neuroblastoma, malignant tumors of children derived from related neuroblasts, are considered together because of their common histogenesis and morphologic similarity. Retinoblastoma occurs in the retina of one eye in nongenetic cases and bilaterally in genetic ones. About 5% of the cases are familial and transmission is by an autosomal gene. The tumors occur in the posterior chamber of the eye and push the retina forward into the vitreous (Figure A–15). They are composed of masses of small darkly staining, rapidly proliferating retinoblastoma cells that differentiate into rosettes typical of developing normal neuroblasts. The tumors metastasize, but cure can be achieved with appropriate therapy in a large number of cases.

Neuroblastoma, as the name implies, is derived from neuroblasts in the autonomic nervous system, most commonly in the adrenal medulla (Figure A–16). The tumors are composed of masses of rapidly growing, highly invasive, dark-staining cells that may form rosettes resembling primitive neuroblasts of the embryo (Figure A–17a). They invade and destroy the adrenal gland. The liver and other contiguous organs may be invaded directly. Distant metastases occur early and involve the lymph nodes, lung, liver, and bones, especially the skull.

In the embryo, neuroblasts differentiate into adult nerve cells termed ganglion cells. In some tumors the neuroblastoma cells also differentiate into the same terminally differentiated ganglion cells in which case the tumors are called ganglioneuroblastoma (Figure A–17b). It is apparent from studies of neuroblastoma in vitro and in vivo that this tumor is a caricature of the process of development of nerve cells.

Neuroblastoma may undergo spontaneous regression, in which case all of the neuroblastoma cells either differentiate into ganglion cells or disappear. Spontaneous regression is rare to the point that it is a medical curiosity. If infants are to be saved from this disease, early and effective therapy must be undertaken. Interestingly, very young children with the disease have a better prognosis than older children.

Much experimentation has been done on spontaneous and induced differentiation of neuroblastoma cells.

A.10   Wilms' tumor (nephroblastoma)

Wilms' tumor is a rare, rapidly growing tumor of infants or children under the age of 5 years. Hematuria is rare because the renal pelvis (the expanded proximal end

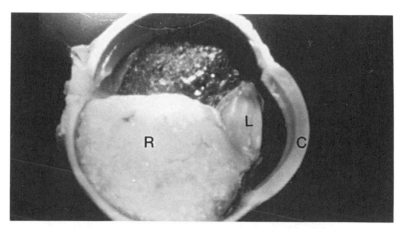

Figure A–15. Photograph of bisected eye with retinoblastoma (R) pushing the retina forward into the posterior chamber of the eye. The lens is at (L), the cornea at (C).

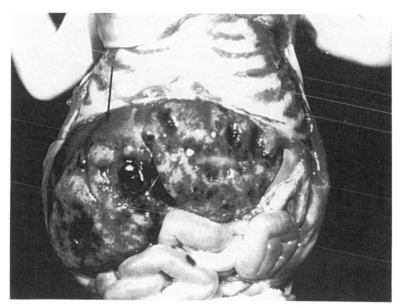

Figure A–16. Photograph of abdomen of child who died with neuroblastoma that developed in the right adrenal gland. Note massive tumor that has infiltrated the liver (arrow). The absence of fat indicates cachexia.

of the ureter that receives urine) is not involved. It usually presents as a large abdominal mass that invades locally and metastasizes widely. It is briefly discussed here not because of its clinical importance, but because of its potential for new types of therapy.

The tumor is composed of masses of undifferentiated spindle cells and epithelial elements arranged in tubules and even in structures resembling primitive glomeruli (small structures in the malpighian body of the kidney made up of

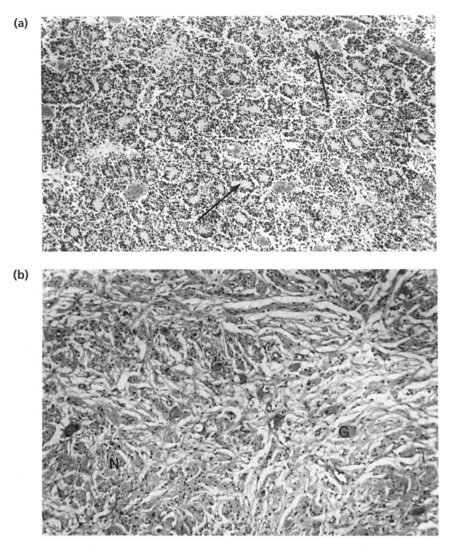

Figure A–17. (a) Photomicrograph of a neuroblastoma of the adrenal gland illustrating the abortive differentiation and organization of primitive neuroblasts into structures resembling embryonic spinal cords (arrows). It is of interest that these cells pallisaded around central accumulations of nerve fibers, migrated to the center of the embryonic organ, divided, and then migrated back out, and if the environment was appropriate would differentiate into mature ganglion cells. (b) Part of ganglioneuroma to show tangles of nerve filaments (N) and ganglion cells (G) derived from tissue illustrated in Figure A–17a.

capillaries). Occasionally muscle or cartilage is present as an integral part of the tumor.

The kidney is derived from the mesoderm, which under the influence of the ureteric bud differentiates into the tubules and glomeruli of the kidney. Because the stem cells of the tumor are multipotential, the normal counterpart must be the multipotent mesenchymal cells capable of forming the adult kidney. Nephrogenesis is still in progress at birth. Therapy includes surgical removal of the tumor with irradiation and chemotherapy. The tumors are responsive and significant cures can now be achieved.

This is an ideal tumor on which to test the idea that the appropriate embryonic field should regulate its closely related cancer.

A.11   Sarcomas

Sarcomas are malignant tumors that originate in connective tissues such as muscle, bone, fibrous connective tissue, cartilage, or fat. Although they can and do develop in the connective tissue stroma of epithelial organs, they are a much less common cause of malignancy than the carcinomas of those organs. To emphasize the point, a liposarcoma can occur in the breast but it is much less common than an adenocarcinoma of the breast. Sarcomas developing in the stroma of deep-seated organs such as stomach or colon are difficult to distinguish preoperatively from their respective carcinomas and are usually discovered at surgery.

The mesenchymal tissues originate from multipotent mesodermal precursors of the embryo. This potential is parceled out to produce fibroblasts, lipoblasts, chondroblasts, osteoblasts, and hematoblasts, which in turn develop into the mature tissues characteristic of their lineages. Whether or not the potentials of cells of these lineages are irreversibly reduced, as appears to be the case with most epithelial cells on differentiation, is not known. The fully differentiated fibrocyte on injury synthesizes DNA, divides, and aids in repair. Thus, terminal differentiation as it occurs in many epithelial tissues may not occur in the mesenchymal ones.

There are connective tissue metaplasias in which cartilage will form in otherwise normal-appearing scars that are usually composed of fibrous tissue. Bone tissue may form in muscles after injury. Similarly, if bone residues after grinding and extensive washing are applied to chondrocytes in vitro or in vivo, these cells develop into bone. Because bone can be derived normally from cartilage it may mean that cartilage cells per se still have the potential to produce bone in the appropriate environmental circumstances. The mechanisms of these reactions are not known, but may relate to the interesting observations of Taylor and Jones (1980) in which 5-azacytidine applied to 3T3 cells in vitro causes them to differentiate into a wide variety of mesenchymal tissues. These effects have been related to hypomethylation of the DNA of the responding cells.

Sarcomas in general are soft fleshy tumors with a pseudocapsule often penetrated by invading sarcoma cells. Thus, wide excision of the lesions is required at the time of first therapy to avoid local recurrence. The sarcoma cells are able to excite a stroma that often has thin-walled vascular spaces, which are easily penetrated by invading cells, resulting in widespread metastases. Hemorrhage can result from these sinusoidal capillaries.

A.11a   Fibrosarcoma

Fibrosarcoma cells are derived from fibroblasts and occur commonly in the skin or wherever there is a preponderance of connective tissue. The majority are low grade, have few mitoses, and little evidences of anaplasia. These tumors synthesize collagen and are usually cured by local excision. Some recur after excision, but metastases are rare. High-grade fibrosarcomas occur very rarely. They are soft and composed of spindle-shaped cells, with many mitoses, which synthesize little collagen, invade and metastasize, and cause death.

A.11b   Liposarcoma

Liposarcomas are uncommon tumors that usually occur in the thighs, buttocks, or retroperitoneum. The most highly differentiated look like lipomas or encapsulated masses of yellow fat. The most poorly differentiated are soft and myxoid (gelatinous). They may recur upon removal, and after relatively prolonged courses, can metastasize and cause death.

A.11c   Chondrosarcoma

Chondrosarcomas, cartilaginous sarcomas, are most common in individuals over the age of 40. Most occur in the pelvic girdle, ribs, or the proximal femur. A mass develops centrally in the bone, or sometimes on its surface. Often the tumors are large enough to make this distinction impossible. The bone cortex is thinned, penetrated, and then the adjacent soft tissues are invaded.

The histopathologic diagnosis of chondrosarcoma can be difficult because the tumor cells have minimal anaplastic change. Thus, a few pleomorphic (many-shaped) cells, multinucleated cells, and a rare mitosis may be enough to make the diagnosis. It is usually best to sample the tumors at the growing edge where their activity is highest. Chondrosarcomas usually have a prolonged course with local recurrence. Metastases occur late.

A.11d   Leiomyosarcoma

Leiomyosarcomas are combined leiomyomas and sarcomas. They are derived from smooth muscle and occur wherever smooth muscle is found. Thus, they are most common in the uterus and in the wall of the gastrointestinal tract. The majority are low grade and have few mitoses and little evidence of anaplasia. High-grade tumors are rare and composed of pleomorphic cells, with many mitoses. They are highly invasive and metastasize to the lungs.

A.11e   Osteosarcoma

Osteosarcoma occurs most frequently in the second and third decades of life. It is a malignant tumor of bone-forming (osteocytes) and bone-destroying (osteoclasts)

(a)

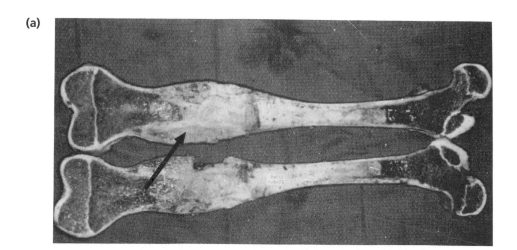

(b)

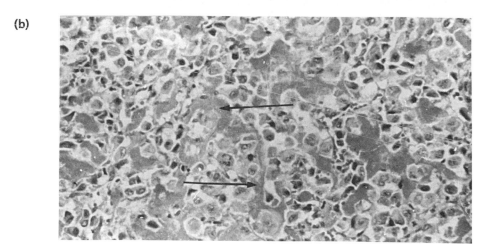

Figure A–18. (a) Osteosarcoma in the shaft of a femur of a young person. The marrow has been invaded, the cortex thinned, and the tumor is growing in the soft tissues outside of the periosteum (arrow). (b) Photomicrograph of an osteogenic sarcoma that could be subclassified as an osteoblastic tumor because masses of osteoid (bone matrix, arrows) are interspersed among myriads of anaplastic sarcoma cells.

cells. These cells probably have a common cell of origin. Some of the tumors are osteoblastic and form dense, neoplastic bone; others are osteolytic and destroy normal bone resulting in pathologic fracture. The shafts of long bones, particularly those around the knee or upper femur or upper humerus, are most commonly involved (Figures A–18a,b). The process starts with pain, and the bone may be warm to the touch, giving the impression of an inflammation. Osteosarcomas metastasize early to the lungs. At this time serum alkaline phosphatase is usually increased and serves as a marker of the tumor. The five-year survival from these tumors has been 5% to 20%, but modern chemotherapy cures more than 50% of patients.

A.12 Leukemia

Malignancies of the hematopoietic cells (blood-forming) and reticuloendothelial cells (cells throughout the body that have the power to ingest particulate matter, e.g., macrophages) are the commonest of the sarcomas, but they are seldom, if ever, classified with the sarcomas. Primitive mesodermal cells of the embryo differentiate into a hematopoietic stem cell that is capable of dividing and differentiating (parceling out its potentials) into stem cells of the erythrocyte, granulocyte, monocyte, lymphocyte, and megakaryocyte lines of cells. It would appear that the parceling out of these potentials is irreversible and each cell type is individually regulated to meet the needs of the host. The sarcomatous process leading to the development of leukemia (literally, "white blood") can affect any of these stem lines or can affect the primary multipotent hematopoietic stem cell. Thus, the disease originates in the bone marrow and secondarily involves the spleen, lymph nodes, and liver. As the malignant cells proliferate, the normal hematopoietic elements are displaced from the marrow by leukemia cells. In an effort to maintain status quo, the normal cells colonize the spleen and liver (extramedullary hematopoiesis). Later, these and other organs are also overgrown by the leukemic cells. As a result, the final reserves for producing functional hematopoietic elements are lost, and the patient presents with the picture of bone marrow failure. These signs and symptoms will include the inability to mount effective defense mechanisms against bacterial, fungal, or parasitic infections, and inability to stop capillary bleeding as a result of displacement of megakaryocytes (petechial hemorrhages – small purplish hemorrhagic spots – can occur in the skin and mucous membranes, with bleeding around the teeth, for example). As the erythroid elements are displaced, anemia results. Finally, leukemia cells will grow in the stroma of any organ, and although it is difficult to pinpoint symptoms directed at the particular organs, the organs are all enlarged, can give a sense of heaviness, and the tremendous number of proliferating leukemic cells can result in cachexia.

As mentioned, a leukemia can originate in the stem cells of any of the hematopoietic lines. Lymphatic leukemias are different clinically from myelocytic leukemias, which in turn are different from monocytic leukemias. Even greater differences can be seen between the acute and chronic forms of any of the leukemias. Thus, there are acute lymphatic leukemias and chronic lymphatic leukemias, acute myelocytic leukemias and chronic myelocytic leukemias. They have differences in age of incidence and response to therapy. For example, acute lymphatic leukemia is most common in children and acute myelogenous leukemia is more common in adults as the terminating crisis of chronic myelogenous leukemia. Acute lymphatic leukemia presents with fever, weakness, pallor, bleeding from the gums, petechial hemorrhages in the skin – in other words, hematopoietic failure. Death can occur in a few weeks. The bone marrow and other organs mentioned are overrun with malignant lymphoblasts.

Chronic lymphatic leukemia is a disease of old people characterized by over-growth of neoplastic lymphoblastic cells and their differentiated precursors. These precursors may be present in large numbers in the circulation and lymph nodes. Lymphoid deposits in the intestine as well as the bone marrow are overrun by pro-liferating lymphoblasts and their descendants. Whereas acute lymphatic leukemia is curable with cytotoxic chemotherapy (Chapter 8), chronic lymphatic leukemia may respond to chemotherapy initially, but it is not cured by it.

Chronic myelocytic leukemia is usually a disease of the elderly, although it can occur at any age. There are excessive accumulations of neoplastic granulocytes and their precursors, principally in the bone marrow, spleen, liver, and blood. Medical historians have noted that the blood of chronic lymphatic leukemias contained so many white cells that it appeared creamy, hence the name leukemia. The disease is insidious, and people usually are fatigued with weakness and possibly weight loss and some are even diagnosed incidentally through routine blood studies. The prognosis is invariably fatal with average survival time of 3 to 4 years. It is not uncommon for the chronic leukemic to undergo "blast crisis." This is the super-imposition of an acute myelogenous leukemic syndrome on the chronic state. The demise is usually from intercurrent infection or blast transformation.

About 90% of the cases of chronic myelogenous leukemia have an abnormal chromosome 22, which has been called the Philadelphia chromosome. The Philadelphia chromosome results from a reciprocal translocation between chromo-somes 9 and 22 (see Chapter 4). This is passed on to its descendants, and the abnormal chromosome is found not only in the granulocytic series of cells but also in lymphocytes, megakaryocytes, and monocytes.

Chronic myelogenous leukemia in the untreated stage displays a bone marrow overgrown by granulocytic elements, with an enormously enlarged spleen, but with less involvement of lymph nodes.

A.13 Hodgkin's disease

Hodgkin's disease is a chronic malignant disease of the reticuloendothelial system. The disease usually presents in localized forms. Because of the cell types involved it seems to be a caricature of a graft versus host reaction. Viruses or other infec-tious agents have not been identified as causative elements. The stem cell of origin has not been identified.

There is a bimodal incidence: young adults in their late teens to mid-30s and individuals over the age of 54 are most commonly afflicted. The diagnosis can only be made by tissue biopsy, usually of a lymph node. Reed-Sternberg cells must be present to make the diagnosis of Hodgkin's disease. Bilobed nuclei impart an owl-eyed appearance to the cells. The cytoplasm of these giant cells is eosinophilic and sometimes vacuolated. Mononuclear cells similar to Reed-Sternberg cells may be present and are also malignant. In addition, there are neutrophils, plasma cells, eosinophils, and fibroblasts, which are considered by many to be reacting to some unknown stimulus, hence the idea of a caricature of a graft versus host reaction.

Four separate subtypes of the disease have been identified: 15% are lymphocyte predominant, 40% are nodular sclerosing, 30% are of mixed cellularity, and 15% are lymphocyte depleted. The first two subtypes carry better prognoses than the last two, but the stage of the disease is also very important in generating prognosis and indicating therapy. Stage I disease involves a single lymph node chain such as those in the neck. Stage II is characterized by disease either above or below the diaphragm, stage III involves lymph nodes above and below the diaphragm with or without lesions in other tissues, and in stage IV, there is wide dissemination to organs outside the reticuloendothelial system.

The patient often presents with enlarged lymph nodes, and/or with itching, intermittent fever, night sweats, and weight loss. The enlarged lymph nodes may be in the neck, abdomen, or mediastinum. They become matted together by a fibroblastic reaction and can cause symptoms as result of compression: jaundice as a result of compression of bile ducts, wheezing or coughing as a result of compression of bronchi, and edema of an extremity as a result of obstruction of lymph. The stage and grade of the disease are made by biopsy.

Treatment involves irradiation of involved areas and chemotherapy. Cures in a high percentage of cases can be obtained, especially with the nodular sclerosing type of Hodgkin's disease.

Glossary

This is not a dictionary. Many of the terms have been defined in the context of how they are used in this text. You may wish to obtain a medical dictionary, which, of course, will be far more complete than this short list of terms.

adduct   Covalent product formed from the reaction of DNA with an activated carcinogen.

adenocarcinoma   Malignancy of glandular epithelial cells.

adhesion plaque   Fibroblasts adhere to a substratum at centers known as adhesion plaques.

adipose   Refers to fat or to fatty cells or tissues.

adjuvant therapy   Use of another treatment in addition to primary surgical treatment of cancer.

Adriamycin   Trademark name for doxorubicin (q.v.).

agonist   Pharmaceutical agent that stimulates physiologic activity.

aminoglutethamide   Inhibitor of adrenalcortical synthesis of hormones such as estrogen.

amitrole   Toxic (carcinogenic) herbicide not used on edible plants.

anaplastic   Adjective describing tumor cells that do not have normal morphology.

androgen   Substance with the potential for inducing masculine characteristics.

anemia   Reduction in the number of red blood cells.

aneuploid   Abnormal chromosome number.

angiogenic   Causing the formation of blood vessels.

antigenic   Adjective describing a substance having the competence to induce the formation of a specific antibody.

apoptosis   Programmed cell death.

ara-C   Chemotherapeutic agent, also known as cytarabine, which inhibits DNA synthesis by the inhibition of DNA polymerase.

astrocytoma   The most common brain tumor, composed of glial cells known as astrocytes.

ataxia telangiectasia   Hereditary disorder characterized by lack of muscular coordination, immunodeficiency, inadequate DNA repair, and an increased prevalence of malignancies.

atrophy   Loss of size in a tissue or organ.

autonomous   State of a tumor when it no longer responds to its usual control signals.

bacillus Calmette-Guérin (BCG)   Vaccine produced from a particular strain of *Mycobacterium bovis*. BCG is also known as *Bacille Calmette-Guérin*.

B-cell growth factor   Substance derived from T-cells that stimulates growth of B-cells in vitro.

benign   Not malignant. Benign tumors do not invade and have no metastatic colonies.

bioassay   Biologic assay for determination of drug or chemical effect on living organisms or living cells or tissues.

blastoma   Tumor comprised of cells, many of which appear less than fully differentiated, and thus resemble cells of an embryo.

bleomycin   Chemotherapeutic agent that functions by binding to and cutting DNA with resultant inhibition of DNA synthesis. Bleomycin is obtained from the funguslike bacteria *Streptomyces verticillus.*

Bloom's syndrome   Genetic disorder characterized by dwarfism, redness of the face, chromosomal abnormalities, and an elevated prevalence of leukemia.

bolus   A drug ready to swallow.

Burkitt's lymphoma   Most common childhood cancer of central Africa (and elsewhere) with a characteristic chromosome abnormality (a translocation of chromosome 8 with another chromosome). The malignancy may be caused by a herpesvirus.

cachexia   Weak and wasted appearance of a terminally ill cancer patient.

camptothecins   Chemotherapeutic alkaloids from the Chinese woody plant *Camptotheca acuminata* (family Nyssaceae); believed to inhibit topoisomerase I.

carboplatin   Chemotherapeutic agent similar in its action to cisplatin (q.v.).

carcinogen   Chemical substance, virus, or radiation that produces or incites cancer.

carcinogenesis   Production of cancer.

carcinoma   Malignancy of epithelial cells.

carcinoma in situ   Tumor that appears to be malignant but has not demonstrated its competence to transgress the basement membrane and disseminate. The tumor is confined to the epithelial cells of its origin.

carotenoid   Fat-soluble yellow or red pigments; examples include beta-carotene, which can be converted to vitamin A in the body, and the red pigment lycopene, found in tomatoes.

castration   Removal of the testes or ovaries.

cathepsin B   Hydrolytic enzyme that digests proteins and peptides.

C. elegans   Scientific shorthand for *Caenorhabditis elegans,* a small free-living roundworm that is of particular interest to developmental biologists who wish to trace the history of every single cell.

cellular heterogeneity   A malignant tumor is generally thought to be composed of a diversity of cells with differing growth rates, gene products, and vulnerability to chemotherapeutic agents – this diversity is known as cellular heterogeneity.

chemotherapy   Treatment of cancer with chemical agents (drugs).

chlorambucil   Alkylating chemotherapeutic agent of a class of drugs known as mustards.

chlorodeoxyadenosine (2-chlorodeoxyadenosine)   A chemotherapeutic agent that inhibits DNA synthesis and DNA repair.

choriocarcinoma   Epithelial malignancy of placental trophoblast cells or a similar malignancy derived from ovarian or testicular tissue.

chromosome banding   Differential staining along the length of a chromosome that permits positive identification of each chromosome as well as the identification of chromosomal aberrations such as deletions or translocations of a size detectable with the light microscope.

chronic lymphatic leukemia Chronic leukemia of the lymphocytic type, usually with abnormal proliferation of small lymphocytes. This leukemia is rare in young people.

chronic myelogenous leukemia Chronic leukemia, characterized by differentiated cells of the myeloid granulocytic type, almost always associated with the Philadelphia chromosome (q.v.). Also known as chronic myeloid leukemia.

cisplatin Chemotherapeutic agent containing platinum used in the treatment of solid cancers (e.g., testicular and ovarian cancers).

clonal expansion Production of a population of cells descended from a single cell.

collagenase Generic term for one of several enzymes with the competence to degrade collagen. Collagens are specific fibrous proteins associated with connective tissue, skin, tendon, bone, cornea, cartilage, and basement membranes.

colostomy Opening of the colon (large intestine) to the exterior of the body formed as the result of a surgical procedure.

corticosteroids Hormones elaborated by the adrenal cortex; the two groups are glucocorticoids (q.v.), which influence the metabolism of carbohydrates, fats, and proteins, and mineralocorticoids (q.v.), which affect electrolytes and water balance.

crocidolite Blue asbestos associated with lung cancer and mesotheliomas.

cruciferous vegetables Plants of a family that include broccoli, Brussel sprouts, cabbage, cauliflower, kale, mustards, radishes, and turnips.

Cushing's syndrome Overproduction of glucocorticoids by the adrenal cortex.

cyclophosphamide Alkylating chemotherapeutic agent included within the class of drugs known as mustards.

cyproterone acetate Synthetic hormone with antiandrogen (antimale hormone) properties.

cystoscope Instrument for the visual examination of the interior of the bladder.

cytochrome P-450 Intracellular protein, similar to hemoglobin, that occurs in most cells and is involved in oxygenation of many reactions; P refers to "pigment" and "450" refers to the 450 nm absorption maximum in a spectrophotometer.

cytoreductive surgery Surgical procedure that results in diminished tumor mass; also known as debulking.

cytoreductive therapy Therapy that results in diminished tumor mass.

cytotoxic Poisonous to cells; used in chemotherapy to describe a drug or agent poisonous to cancer cells.

cytotoxic T-cells Lymphocytes that function in graft rejection and the killing of virus-infected cells and tumor cells.

debulking Elimination of much of a tumor to enhance subsequent therapy because of reduced tumor load (see cytoreductive therapy).

dedifferentiation Loss of characteristic structure of cells that express their mature phenotype; also referred to as anaplasia (q.v.).

deletion Loss of genetic material from a chromosome.

determination Commitment of cells to a particular differentiation pathway. Determined cells are not visibly differentiated.

dexamethasone   Synthetic glucocorticoid (q.v.) drug.

dicotyledon   Plant having two seed leaves (cotyledons); the group includes most trees that lose their leaves and most herbs and shrubs.

diethylstilbestrol (DES)   Synthetic estrogen that may be used for treating the symptoms of prostate cancer.

differentiation   Acquisition of the mature phenotype of cells.

differentiation therapy   Cancer therapy that seeks cure by treatment of malignant cells with agents which induce expression of mature (and benign) characteristics.

diverticular disease   Sacs or pouches of the large intestine as a result of weaknesses in the muscular wall of that organ.

dormancy   State of a malignancy when its component cells are quiescent.

doxorubicin   Chemotherapeutic agent of a class of compounds known as anthracyclines derived from *Streptomyces peucetius* that interfere with DNA synthesis.

duplication   Repeated genetic material in a chromosome.

dysplasia   Abnormalities in cancer cell morphology.

E-cadherin   A specific calcium-dependent cell adhesion molecule.

ectopic   Expression (of an antigen) in an abnormal site or tissue.

endogenous   Arising from within the body.

epidemiology   Study of factors that influence the prevalence and distribution of disease.

epigenetic   Differentiation and morphology due to selective gene activation and expression or repression. Not due to changes in the genome.

epithelial cells   Cells that cover the body and line the internal surfaces. Glands may be derived from epithelial cells.

epitopes   Antigenic determinants.

Epogen   Trademark name of epoetin alpha. Epoetin alpha is human erythropoietin produced by recombinant DNA technology and is used to stimulate red blood cell production.

Epstein-Barr herpesvirus   Specific herpesvirus that has been linked with Burkitt's lymphoma (q.v.). The virus is also associated with infectious mononucleosis and nasopharyngeal carcinoma.

estrogens   Female sex hormones produced by the ovary responsible for secondary sexual characteristics and monthly changes in the lining of the vagina and uterus.

etiology   The cause of a disease.

etoposide   Chemotherapeutic agent that is a semisynthetic derivative of podophyllotoxin (obtained from the mayapple plant, *Podophyllum peltatum*). The drug functions by affecting the topologic state of DNA.

euploid   Normal chromosome number for a species or an exact multiple of the haploid number. The haploid number for humans is 23; thus, any multiple of 23 is euploid.

extravasation   Exit from a capillary or lymph vessel. Extravasation is a prerequisite for metastasis.

familial adenomatous polyposis   Genetic condition that leads to the production of thousands of small benign tumors (polyps) in the colon which have a high potential to become malignant.

Fanconi's syndrome   Rare genetic disease with a deficiency in blood cells, inadequate bone marrow development, brown facial spots, multiple muscular and skeletal abnormalities, and genital and urinary malformations.

fecal occult blood test   Many colon cancers release blood to the lumen of the colon. These tests for colon cancer are designed to reveal blood (hidden) in the feces.

fibroblast   Connective tissue cell.

floxuridine   Chemotherapeutic agent that is metabolically altered to an intermediate with the same properties as 5-fluorouracil (q.v.); a drug that inhibits DNA synthesis.

fluorouracil (5-fluorouracil, 5-FU)   Chemotherapeutic agent that inhibits DNA synthesis.

flutamide   Synthetic antiandrogen (antimale hormone) used in prostate cancer treatment.

free radical   Highly reactive constituent (component) of a molecule with an unpaired electron.

germ line   Refers to sperm or eggs, not somatic (body) cells.

Giemsa banding   Chromosome banding (q.v.), with a specific mixture of stains.

glioma   Tumor of the brain or spinal cord composed of non-neural supporting cells.

glucocorticoid   General term for adrenal cortical hormones; see corticosteroids.

GM-CSF   Granulocyte macrophage colony stimulating factor.

granulomatous reaction   Response to chronic inflammation that is characterized by the presence of granulomas. Granulomas are usually composed of aggregations of mononuclear inflammatory cells and other white blood cell types.

hairy cell leukemia   A chronic leukemia. The blood of these patients has many large mononuclear cells with cytoplasmic projections that give the "hairy" cell appearance.

hemangioblastoma   Nonmalignant tumor of proliferating blood vessel cells and angioblasts found in the brain, spinal cord, or retina.

hemostasis   Stoppage or blockage of blood flow.

hepatoma   Malignancy of the liver, also known as hepatocellular carcinoma.

homeothermic   Refers to an animal that maintains constant body temperature despite changes in the environment. The animals are referred to as warm blooded in contrast to cold-blooded animals.

homogenate   Tissue minced into extremely fine subcellular particles.

homologue   The other chromosome of a pair; only the y chromosome lacks a homologue in a metaphase spread.

17-alpha hydroxyprogesterone   Intermediate compound in the formation of male and female hormones.

hyperplasia   Abnormal increase in the number of normal cells in a tissue.

hypophysectomy   Surgical removal of the hypophysis (pituitary gland).

hypoxia   Diminished supply of oxygen to cells and tissues.

I^{131}   Radioactive iodine.

initiation   Initial preneoplastic change in the genetic material of cells caused by a chemical carcinogen. Overt cancer occurs with subsequent exposure of the initiated cells to the same or another carcinogen.

in situ   Localized, in position. A carcinoma that has not breached the basement membrane.

interferon   Member of a group of glycoproteins with antiviral activity. Interferon alpha, an interferon produced by leukocyte cultures, used in the treatment of hairy cell leukemia.

interleukin   Interleukins are nonantibody proteins; interleukin-2 (IL-2) is produced by T-cells and stimulates specific T-cell replication; it is used in the treatment of solid tumors. Interleukin-3 (IL-3) is produced by T-cells and stimulates production of bone marrow blood cells and lymphoid stem cells.

intrathecal injection   Placement (injection) of a drug under the sheath that covers the spinal cord.

intravasation   Entrance of cancer cells into capillary or lymph vessels.

invasion   Movement of cancer cells into contiguous tissue.

in vitro   Term used to indicate that a cell manipulation is outside the body. Originally, the manipulations were performed in glass dishes (hence, in vitro) but now most studies are carried out in plastic laboratory dishes.

in vivo   In the living body or organism. An experimental procedure using an intact live animal.

inversion   Chromosomal abnormality formed by a double break with subsequent reuniting of the broken parts with the sequence of genes reversed in their linear order in relation to the remainder of the chromosome. Inversions can be detected by chromosomal banding (q.v.).

Kaposi's sarcoma   Malignant tumor of presumed endothelial origin that causes pink or red spots on the skin. The disease occurs in Africa, in central and eastern Europe, and in some immunologically deficient patients.

karyotype   Photograph of metaphase chromosomes arranged by pairs in descending order of size.

ketaconazole   Antifungal drug that in high doses reduces testosterone production.

latency   Period of time between the application of a carcinogen and the appearance of a tumor.

leiomyosarcoma   Malignancy of smooth muscle. Smooth muscle is nonvoluntary muscle (e.g., the muscle of the uterus).

leukemia   Malignancy of the blood-forming tissue characterized by overproduction of white blood cells.

Leukine   Trademark name of sargramostim, which is a granulocyte/macrophage colony stimulating factor (GM-CSF) produced with recombinant DNA technology. GM-CSF enhances blood cell production in the bone marrow.

leukocytosis   Increase in the number of leukocytes (white blood cells) in the blood.

leuprolide   Synthetic gonadotropin-releasing hormone used to relieve symptoms of prostate cancer.

ligand   One molecule that binds to another.

lignin   Polysaccharide component of plant cell wall.

Lucké renal adenocarcinoma   Kidney cancer that is the most common malignancy of the North American leopard frog, *Rana pipiens*.

lumpectomy   Surgery that removes primarily the tumor mass and little of adjacent normal tissue, used in reference to breast surgery. Contrast with radical mastectomy (q.v.).

luteinizing hormone   Anterior pituitary hormone important to corpus luteum formation in the female and testicular interstitial cell development in the male.

luxury molecules   Gene products that are specific to specialized cells.

lymphocyte   Nongranular mononuclear white blood cells. Two groups include B- and T-lymphocytes. T-lymphocytes are involved in cellular immunity.

lymphocytopenia   Decrease in the number of lymphocytes in the blood.

lymphokine-activated killer cells (LAK cells)   Cytotoxic T-cells and natural killer cells activated by culture in interleukin-2 (q.v.).

lymphoma   Malignancy of lymphoid tissue. Lymphoid tissue includes the thymus and lymphocyte-producing bone marrow as well as lymph nodes, tonsils, and Peyer's patches.

lysosome   Cellular structure that contains hydrolytic enzymes.

malignant   Adjective modifying "tumor" or "neoplasm" meaning cancer. A malignant tumor is invasive and has the competence to form metastatic colonies.

mammogram   X-ray of the breast.

mastectomy   Surgical removal of the breast.

medroxyprogesterone   Progestational agent, that is, a substance that stimulates the endometrium as does progesterone.

megestrol acetate   Synthetic drug with an effect similar to progesterone.

melanocyte   Cell with the competence to produce the pigment melanin.

melanoma   The most lethal of skin cancers arising from melanocytes (pigment cells).

melphalan   Alkylating chemotherapeutic agent, also known as L-PAM (L-phenylalanine mustard), and a member of the class of drugs known as mustards.

menopause   Event that marks the permanent termination of menstruation; premenopausal refers to the time before menopause and postmenopausal refers to the time after menopause.

mercaptopurine (6-mercaptopurine)   Chemotherapeutic agent that inhibits purine synthesis, which in turn blocks DNA synthesis.

mesenchymal   Adjective describing embryonic connective tissue cells, of mesodermal origin.

mesothelioma   Tumor that arises from the lining of the body cavities (peritoneum, pericardium, and pleura).

metaplasia   Change in cell type of a mature cell.

metastasis   Anatomically distant cancer growth.

methotrexate   Chemotherapeutic agent, a folic acid antagonist, that inhibits synthesis of DNA by interfering with the enzyme dihydrofolate.

microsome   Subcellular fragment of the endoplasmic reticulum formed after disruption of the cell.

mineralocorticoid   Hormones of the adrenal cortex important in retention of water and sodium with reduction of potassium; see corticosteroids.

minute   Adjective to describe an abnormally small chromosome.

mitogen   Something that will provoke cell proliferation.

monoclonal   Population derived from a single cell.

morbidity   Disease.

mortality   Death.

mucosa   Mucous membrane that lines cavities of the body composed of a surface epithelium, basement membrane, and underlying connective tissue.

mucositis   Infection or inflammation of a mucosa (q.v.).

multimodality therapy   Therapy involving more than one therapeutic agent.

mummichog   Common name of the fish *Fundulus heteroclitus.*

myeloid   Bone marrow.

myeloma   Tumor of bone marrow.

myelosuppressive   Inhibition of blood- and platelet-forming cells in bone marrow by a drug or other toxic agent.

natural killer cells   Large lymphocytes produced in the bone marrow, distinct from T-cells, which kill virus-infected and tumor cells.

necrosis   Cell death.

neoplasm   Abnormal growth serving no useful function to the host.

neoplastic   Adjective describing abnormal growth of tissue, which, if invasive with the potential of metastasis, is referred to as cancerous.

Neupogen   Trade name for filgrastim, a granulocyte colony stimulating factor (G-CSF), produced by recombinant DNA technology, used to enhance neutrophil production.

nevi   Plural of nevus, congenital mark or blemish, a mole.

nitrosoureas   Class of alkylating chemotherapeutic agents that inhibits DNA synthesis.

nucleosome   Chromatin is composed of DNA and protein; nucleosomes are beadlike packages of chromatin arranged on a "string."

neutropenia   Reduction in the number of neutrophilic leukocytes.

nulliparity   Condition of having borne no children.

oligonucleotide   Polymer of 20 or fewer nucleotides.

oncogene   Gene with the competence, under certain circumstances, of inducing normal cells to form cancer cells.

oncogenic   Having the capacity or competence to cause cancer.

orchiectomy   Male castration; removal of the testes.

osteosarcoma   Bone cancer; it may be classified as osteoblastic, chondroblastic, or fibroblastic; also known as osteogenic sarcoma.

P-450 cytochrome   Oxygenating enzyme found in most tissues important in the detoxification of many drugs.

palliation   Provision of relief but not cure.

papilloma   Benign epithelial tumor with the appearance of fingerlike protrusions from the surface of the epithelium.

Pap test/smear   Method of staining shed (exfoliated) cells that permits diagnosis of cancer; named for Dr. George Nicolas Papanicolaou.

paraneoplastic syndrome   Symptoms produced in a patient that cannot be attributed directly to the cancer or its metastases but rather by a substance produced by the cancer; the liberation of a hormone by a nonendocrine cancer is an example.

parenchymal   Refers to the functional cell types of an organ as distinguished from connective tissue stroma.

pellagra   Disease characterized by dermatitis, diarrhea, and dementia caused by a dietary deficiency of niacin.

pharmacokinetics   Concerns the fate of drugs in the body over a period of time including absorption, organ and tissue distribution, metabolism and excretion.

phenobarbitol   Long-acting sedative that may be used as a promoter to enhance carcinogenicity.

pheochromocytoma   Usually benign neoplasm of the adrenal medulla that may produce symptoms caused by oversecretion of epinephrine.

Philadelphia chromosome   Minute (q.v.) chromosome associated with chronic myelogenous leukemia; the chromosome was jointly described by a professor from the University of Pennsylvania, Philadelphia, and an investigator at the Fox Chase Cancer Center, Philadelphia.

phosphorylation   Process of incorporation of a phosphate group into an organic molecule.

pleomorphic   Characterized by the presence of several forms.

polarity   Property of having an axis with different properties at the extremes; epithelial cells are said to possess polarity when their basal portions clearly differ from their apical portion; loss of polarity is a characteristic of anaplastic (q.v.) cells.

polysaccharides   Carbohydrates that yield more than two molecules of simple sugar upon hydrolysis; includes but not limited to starch, glycogen, and cellulose.

polysome   Synonym for polyribosome; a cell structure involved in peptide synthesis.

prednisone   Synthetic drug derived from cortisone used to reduce inflammation and to suppress the immune system.

prodrug   Drug inactive in the administered form but, after metabolic conversion within the body, becomes a pharmacologic active drug.

progesterone   Ovarian and placental hormone responsible for preparation of uterine lining for implantation of the early embryo.

prognosis   Prediction of the likely, or expected, outcome of a disease.

progression   Progressive acquisition of greater malignancy with increased growth rate, invasiveness, and metastases. Said to be preceded by initiation and promotion (q.v.).

promoting agent   Co-carcinogen that activates carcinogenesis after initiation usually by a different chemical carcinogen.

promotion   Expression of the malignant potential of initiated cells after exposure to the same or to a different chemical carcinogen.

prophylactic   Treatment taken to prevent disease.

proto-oncogene   Gene which, when activated by mutation or other change (i.e., becomes an oncogene), has the potential for causing a normal cell to become malignant. Normal proto-oncogenes are thought to function in normal growth and differentiation.

pyelonephritis   Bacterial infection of the kidney.

pyknosis   Condensation (shrinking) of the cell nucleus resulting in chromatin becoming a dense mass. The condition is associated with cell death.

radiation   Energy transmitted by means of electromagnetic waves or by a stream of particles such as electrons, protons, neutrons, or alpha particles.

radical mastectomy   Removal of the breast including underlying pectoral muscles, lymph nodes of the axilla, and associated skin and subcutaneous tissue. Contrast with lumpectomy (q.v.).

radiosensitizer   Agent that augments efficacy of radiotherapy.

radiotherapy　Treatment of disease with radiation.

refractory　Resistance to therapy.

resection　Surgical removal of a tissue mass or organ.

resectoscope　Optical instrument provided with a wire loop for the surgical removal of tissue from the bladder, urethra, or prostate with access to those tissues via the urethra.

residuum　Remainder of cells or tissues after initial treatment.

reticuloendothelial cells　Any of a group of cells of the reticuloendothelial system with the competence to ingest foreign particulate matter (phagocytosis) located at many sites in the body.

retinoblastoma　Pediatric eye malignancy that may be either sporadic or hereditary.

retinoic acid　Form of vitamin A. The all-trans retinoic acid (ATRA) isomer is also known as tretinoin.

retinoid　Vitamin A or similar type compound.

retrovirus　Small RNA virus whose genome serves as a template for production of cell DNA via reverse transcriptase. The new DNA is integrated into the host cell. Many retroviruses are believed to be oncogenic.

rhabdomyosarcoma　Malignant tumor of striated (voluntary) muscle cell progenitors.

Rous sarcoma virus　Chicken retrovirus (q.v.) that was the first virus shown to cause a malignancy; named for Nobel laureate Francis Peyton Rous.

sarcoma　Malignant tumor arising from embryonic mesoderm or connective tissue.

secondary tumor　Metastases.

seminoma　Malignant tumor of the testis.

septicemia　Infection of the blood with disease-causing organisms or their toxic substances. The condition is also known as blood poisoning.

sigmoidoscopy　Visual examination of the lower colon with an illuminated optical device.

somatic　Body; used to describe nongametic cells; in contrast to sexual cells.

spontaneous regression　Return to a nonmalignant condition either by death of tumor cells or by differentiation to a benign or nontumorous state without intentional external influence.

squamous cell carcinoma　Second most common skin cancer that occurs at about half the frequency as basal cell carcinoma. It may occur as a result of solar damage. This malignancy also occurs in the lung.

stem cell　Mother cell with the capacity to give rise to more stem cells as well as to cells that differentiate along specific pathways.

stroma　Connective tissue elements of an organ in contrast to the functional (parenchymal) cells of that organ.

suppressor　Gene whose loss of activity leads to cancer.

synovioma　Tumor of the membrane that lines joint cavities, bursae, and tendon sheaths.

tamoxifen　Drug with antiestrogen activity.

Taxol　Trademark name for paclitaxel, a chemotherapeutic drug that affects polymerization and stability of intracellular microtubules. The agent was originally derived from the Pacific yew tree (*Taxus brevifolia*).

teratocarcinoma   Malignant tumor, arising most often in the testis, comprised of pleuripotential cells that give rise to tissues of many types as well as malignant epithelial cells.

ternary complex   Comprised of three chemical components.

testosterone   Male hormone produced by the testes.

thrombocytopenia   Diminished platelets in the blood.

tissue renewal   Continuous replacement of skin, the lining of the gut, and blood cells that would otherwise be lost due to wear and senescence.

tocopherol   Any of several substances having vitamin E activity.

transfected cell   Transfer and incorporation of DNA from one cell type to another.

transformation   Change that a normal cell undergoes as it becomes malignant.

translocation   Movement of genetic material from one site in a chromosome to another site in the same or another chromosome.

tremolite   White or gray mineral of the amphibole group.

trophic   Refers to nutrition.

trophoblast   Extraembryonic cells of the mammalian embryo that attach the blastocyst (the young embryo) to the uterine wall. Trophoblast cells are essential for the nutrition of the embryo.

tubulin   The constituent molecule of microtubules; microtubules form the mitotic spindle as well as a significant portion of the cytoplasmic cytoskeleton.

tumor   Abnormal new growth of cells that may be either benign (q.v.) or malignant (q.v.).

tumor grade   Degree of differentiation as well as growth rate, as indicated by mitotic frequency.

tumorigenesis   Carcinogenesis (q.v.).

tumor-infiltrating lymphocytes (TIL cells)   Lymphoid cells that invade solid malignant tumors. TIL cells can be grown in interleukin-2, and some TIL cells have competence to lyse specific tumor cells.

tumor markers   Molecular products of cancer that may be useful in diagnosis, prognosis, and assessment of therapeutic effect and possible recurrence.

tumor necrosis factor (TNF)   Substance released by macrophages that causes, among many effects, the stimulation of T-cell production and cell death in some tumors.

typhus   Group of severe bacterial diseases caused by rickettsiae.

tyrosine kinase   Enzyme(s) that catalyze phosphorylation of the amino acid tyrosine in target proteins.

urogenital ridge   Mesodermal area in early embryo that will give rise to both the kidney and the gonads.

vinblastine   One of a class of chemotherapeutic agents known as the vinca alkaloids (q.v.).

vinca alkaloids   Cytotoxic chemotherapeutic agents, derived from the Madagascar periwinkle, *Catharanthus roseus,* that target cellular microtubules.

vincristine   One of a class of chemotherapeutic agents known as the vinca alkaloids (q.v.).

von Recklinghausen's disease   Neurofibromatosis, a hereditary condition with superficial stalked and soft tumors over the entire surface of the body with many internal abnormalities.

xenobiotics   Substances foreign to the organism.

xeroderma pigmentosum   Hereditary disease, caused by inadequate DNA repair, characterized by extreme sensitivity to the sun with multiple skin problems including malignancies.

Wilms' tumor   Hereditary pediatric kidney cancer.

zinc finger   Many DNA-binding proteins, including transcription factor IIIA, are associated with zinc. The protein with the zinc form a loop structure known as a zinc finger.

References

Abbrecht, P.H., and Littell, J.K. 1972. Plasma erythropoietin in men and mice during acclimatization to different altitudes. *J Appl Physiol* 32:54–58.

Adams, J.M., Harris, A.W., Pinkert, C.A., Corcoran, L.M., Alexander, W.S., Cory, S., Palmiter, R.D., and Brinster, R.L. 1985. The c-*myc* oncogene driven by immunoglobulin enhancers induces lymphoid malignancy in transgenic mice. *Nature* 318:533–538.

Adler, I. 1912. *Primary Malignant Growths of the Lungs and Bronchi.* New York: Longmans, Green.

Advisory Committee to the Surgeon General. 1964. *Smoking and Health.* U.S. Public Health Service Publication 1103, Washington, DC: U.S. Dept. Health, Education and Welfare.

Albini, A., Melchiori, A., Santi, L., Liotta, L.A., Brown, P.D., and Stetler-Stevenson, W.G. 1991. Tumor cell invasion inhibited by TIMP-2. *J Natl Cancer Inst* 83:775–779.

Al-Dewachi, H.S., Appleton, D.R., Watson, A.J., and Wright, N.A. 1979. Variation in the cell cycle time in the crypts of Lieberkuhn of the mouse. *Virchows Arch B* 31:37–44.

Al-Dewachi, H.S., Wright, N.A., Appleton, D.R., and Watson, A.J. 1974. The cell cycle time in the rat jejunal mucosa. *Cell Tissue Kinet* 7:587–594.

Al-Dewachi, H.S., Wright, N.A., Appleton, D.R., and Watson, A.J. 1975. Cell population kinetics in the mouse jejunal crypt. *Virchows Arch B* 18:225–242.

Al-Dewachi, H.S., Wright, N.A., Appleton, D.R., and Watson, A.J. 1980. The effect of a single injection of cytosine arabinoside on cell population kinetics in the mouse jejunal crypt. *Virchows Arch B* 34:299–309.

Alexander, J.P., Kudoh, S., Melsop, K.A., Hamilton, T.A., Edinger, M.G., Tubbs, R.R., Sica, D., Tuason, L., Klein, E., Bukowski, R.M., and Finke, J.H. 1993. T-cells infiltrating renal cell carcinoma display poor proliferative response even though they can produce interleukin 2 and express interleukin 2 receptors. *Cancer Res* 53:1380–1387.

Alexandrow, M.G., and Moses, H.L. 1995. Transforming growth factor beta and cell cycle regulation. *Cancer Res* 55:1452–1457.

Allegra, C.J. 1990. Antifolates. In *Cancer Chemotherapy,* B.A. Chabner and J.M. Collins, eds., pp. 110–153. Philadelphia: J.B. Lippincott.

Almoguera, C., Shibata, D., Forrester, K., Martin, J., Arnheim, N., and Perucho, M. 1988. Most human carcinomas of the exocrine pancreas contain mutant c-K-*ras* genes. *Cell* 53:549–554.

Alpert, M.E., Hutt, M.S., and Davidson, C.S. 1968. Hepatoma in Uganda. A study in geographic pathology. *Lancet* I:1265–1267.

Alvarez, O.A., Carmichael, D.F., and Declerck, Y.A. 1990. Inhibition of collagenolytic activity and metastasis by a recombinant human tissue inhibitor of metalloproteinases. *J Natl Cancer Inst* 82:589–595.

American Cancer Society 1996 Advisory Committee on Diet, Nutrition, and Cancer Prevention.

1996. Guidelines on diet, nutrition, and cancer prevention: Reducing the risk of cancer with healthy food choices and physical activity. *CA Cancer J Clin* 46:325–341.

Ames, B.N. 1987. What are the major carcinogens in the etiology of human cancer? Environmental pollution, natural carcinogens, and the causes of human cancer: Six errors. *Environ Health Perspect* 74:237–247.

Ames, B.N., Durston, W.E., Yamasaki, E., and Lee, F.D. 1973. Carcinogens are mutagens: A simple test system combining liver homogenates for activation and bacteria for detection. *Proc Natl Acad Sci USA* 70:2281–2285.

Ames, B.N., Magaw, R., and Gold, L.S. 1987. Ranking possible carcinogenic hazards. *Science* 236:271–280.

Anderson, T.J., Battersby, S., King, R.J.B., McPherson, K., and Going, J.J. 1989. Oral contraceptive use influences resting breast proliferation. *Human Pathol* 20:1139–1144.

Anderson, T.J., Ferguson, D.J., and Raab, G.M. 1982. Cell turnover in the "resting" human breast: Influence of parity, contraceptive pill, age and laterality. *Br J Cancer* 46:376–382.

Armstrong, B., and Doll, R. 1975. Environmental factors and cancer incidence and mortality in different countries, with special reference to dietary practices. *Int J Cancer* 15:617–631.

Armstrong, B.K. 1984. Melanoma of the skin. *Br Med Bull* 40:346–350.

Armstong, B.K. 1988. Epidemiology of malignant melanoma: Intermittent or total accumulated exposure to the sun? *J Dermatol Surg Oncol* 14:835–849.

Armstrong, P.B. 1984. Invasiveness of non-malignant cells. In *Invasion, Experimental and Clinical Implications,* M.M. Mareel and K.C. Calman, eds., pp. 126–167. Oxford: Oxford University Press.

Aronson, K.J., Siemiatychi, J., Dewar, R., and Gérin, M. 1996. Occupational risk factors for prostate cancer: Results from a case-control study in Montréal, Québec, Canada. *Am J Epidemiol* 143:363–373.

Ashby, J. 1994. Change the rules for food additives. *Nature* 368:582.

Ashworth, T.R. 1869. A case of cancer in which cells similar to those in the tumours were seen in the blood after death. *Aust Med J* 14:146–147.

Ausprunk, D.H., and Folkman, J. 1977. Migration and proliferation of endothelial cells during preformed and newly formed blood vessels during tumor angiogenesis. *Microvasc Res* 14:53–65.

Bader, J.P. 1972. Temperature-dependent transformation of cells infected with a mutant of Bryan Rous sarcoma virus. *J Virol* 10:267–276.

Bagshawe, K.D. 1992. Choriocarcinoma: A model for tumour markers. *Acta Oncol* 31:99–106.

Bal, D.G., and Foerster, S.B. 1991. Changing the American diet. Impact on cancer prevention policy recommendations and program implications for the American Cancer Society. *Cancer* 67:2671–2680.

Bal, D.G., Lloyd, J.C., and Manley, M.W. 1995. The role of the primary care physician in tobacco use prevention and cessation. *CA Cancer J Clin* 45:369–374.

Ballester, R., Marchuk, D., Boguski, M., Saulino, A., and Lechter, R. 1990. The NF1 locus encodes a protein functionally related to mammalian GAP and yeast IRA proteins. *Cell* 63:851–859.

Barber, H.R.K., and Brunschwig, A. 1968. Treatment and results of recurrent cancer of corpus uteri in patients receiving anterior and total pelvic exenteration. *Cancer* 22:949–955.

Bargmann, C.I., Hung, M.C., and Weinberg, R.A. 1986. Multiple independent activations of the *neu* oncogene by a point mutation altering the transmembrane domain of p185. *Cell* 45:649–657.

Barone, M.V., and Courtneidge, S.A. 1995. Myc but not Fos rescue of PDGF signalling block caused by kinase-inactive Src. *Nature* 378:509–512.

Barradell, L.B., and Faulds, D. 1994. Cyproterone. A review of its pharmacology and therapeutic efficacy in prostate cancer. *Drugs Aging* 5:59–80.

Barrett, J.C. 1993. Mechanisms of multistep carcinogenesis and carcinogen risk assessment. *Environ Health Perspect* 100:9–20.

Bar-Sagi, D., and Feramisco, J.R. 1985. Micro-injection of the *ras* oncogene protein into PC12 cells induces morphological differentiation. *Cell* 42:841–848.

Bartley,T. D., Bogenberger, J., Hunt. P., Li,Y.S., Lu, H.S., Martin, F., Chang, M.S., Samal, B., Nichol, J.L., Swift, S., Johnson, M.J., Hsu, R-Y, Parker, V.P., Suggs, S., Skrine, J.D., Merewether, L.A., Clogston, C., Hsu, E., Hokom, M.M., Hornkohl, A., Choi, E., Pangelinan, M., Sun, Y., Mar, V., McNinch, J., Simonet, L., Jacobsen, F., Xie, C., Shutter, J., Chute, H., Basu, R., Selander, L., Trollinger, D., Sieu, L., Padilla, D., Trail, G., Elliot, G., Izumi, R., Covey, T., Crouse, J., Garcia, A., Xu, W., Del Castillo, J., Biron, J., Cole, S., Hu, M.C-T., Pacifici, R., Ponting, I., Saris, C., Wen, D., Yung, Y.P., Lin, H., and Bosselman, R.A. 1994. Identification and cloning of a megakaryocyte growth and development factor that is a ligand for the cytokine receptor *Mpl Cell* 77:1117–1124.

Bartsch, H., Ohshima, H., Pignatelli, B., and Calmels, S. 1992. Endogenously formed N-nitroso compounds and nitrosating agents in human cancer etiology. *Pharmacogenetics* 2:272–277.

Baskaran, R., Dahmus, M.E., and Wang, J.Y.J. 1993. Tyrosine phosphorylation of mammalian RNA polymerase II carboxy-terminal domain. *Proc Natl Acad Sci USA* 90:11167–11171.

Bastida, E. 1988. The metastatic cascade: Potential approaches for the inhibition of metastasis. *Semin Thromb Hemost* 14:66–72.

Batzdorf, U., Black, K.L., and Selch, M.T. 1990. Neoplasms of the nervous system. In *Cancer Treatment*, 3rd ed., C.M. Haskell, ed., pp. 436–468. Philadelphia: W.B. Saunders.

Baum, M. 1976. The curability of breast cancer. *Br Med J* 1:439–442.

Bayer, M.H., Kaiser, H.E., and Micozzi, M.S. 1994. Abnormal growth processes in plants and animals: A comparison. *In Vivo* 8:3–16.

Bayreuther, K. 1960. Chromosomes in primary neoplastic growth. *Nature* 186:6–9.

Bell, B. 1794. *A Treatise on the Hydrocele or Sarcocele, or Cancer and Other Disease of the Testes.* Edinburgh.

Bender, R.A., Hamel, E., and Hande, K.R. 1990. Plant alkaloids. In *Cancer Chemotherapy*, B.A. Chabner and J.M. Collins, eds., pp. 253–275. Philadelphia: J.B. Lippincott.

Berchuk, A., Kamel, A., Whitaker, R., Kerns, B., Olt, G., Kinney, R., Soper, J.T., Dodge, R., Clarke-Pearson, D.L., Marks, P., McKenzie, S., Yin, S., and Bast, R.L., Jr. 1990. Overexpression of HER-2/neu is associated with poor survival in advanced epithelial ovarian cancer. *Cancer Res* 50:4087–4091.

Berek, J.S., and Hacker, N.F. 1990. Ovary and fallopian tubes. In *Cancer Treatment*, 3rd ed., C.M. Haskell, ed., pp. 295–325. Philadelphia: W.B. Saunders.

Berenblum, I., and Shubik, P. 1947. The role of croton oil application associated with a single painting of a carcinogen in tumour induction of the mouse's skin. *Br J Cancer* 1:379–383.

Berenson, J.R., and Gale, R.P. 1990. Acute lymphoblastic leukemia. In *Cancer Treatment*, 3rd ed., C.M. Haskell, ed., pp. 606–620. Philadelphia: W.B. Saunders.

Bergmann, L., Fenchel, K., Weidmann, E., Enzinger, H.M., Jahn, B., Jonas, D., and Mitrou, P.S. 1993. Daily alternating administration of high-dose apha-2b-interferon and interleukin-2 bolus infusion in metastatic renal cell cancer. *Cancer* 72:1733–1742.

Berkow, R.L., Schlabach, L., Dodson, R., Benjamin, W.H., Pettit, G.R., Rustagi, P., and Kraft, A.S. 1993. *In vivo* administration of the anticancer agent bryostatin 1 activates platelets and neutrophils and modulates protein kinase C activity. *Cancer Res* 53:2810–2815.

Berkowitz, R.S., and Goldstein, D.P. 1990. Gestational trophoblastic neoplasia. In *Cancer Treatment,* 3rd ed., C.M. Haskell, ed., pp. 366–372. Philadelphia: W.B. Saunders.

Berman, M.L., and Berek, J.S. 1990. Uterine corpus. In *Cancer Treatment,* 3rd ed., C.M. Haskell, ed., pp. 338–351. Philadelphia: W.B. Saunders.

Berrino, F., Panico, S., and Muti, P. 1989. Dietary fat, nutritional status, and endocrine-associated cancers. In *Diet and the Aetiology of Cancer,* A.B. Miller, ed., pp. 3–12. Berlin: Springer-Verlag.

Berx, G., Nollet, F., Strumane, K., and van Roy, F. 1997. An efficient and reliable multiplex PCR-SSCP mutation analysis test applied to the human E-cadherin gene. *Hum Mutat* 9:567–574.

Beukers, R., and Berends, W. 1960. Isolation and identification of the irradiation product of thymine. *Biochim Biophys Acta* 41:550–551.

Beutler, B., and Cerani, A. 1986. Cachectic and tumor necrosis factors as two sides of the same biological coin. *Nature* 320:584–588.

Bhanot, O.S., Grevatt, P.C., Donahue, J.M., Gabrielides, C.N., and Solomon, J.J. 1992. In vitro DNA replication implicates O^2-ethyldeoxythymidine in transversion mutagenesis by ethylating agents. *Nucleic Acids Res* 20:587–594.

Biggs, J., Hersperger, E., Steeg, P.S., Liotta, L.A., and Shearn, A. 1990. A Drosophila gene that is homologous to a mammalian gene associated with tumor metastasis codes for a nucleoside diphosphate kinase. *Cell* 63:933–940.

Billington, W.D. 1965. The invasiveness of transplanted mouse trophoblast and the influence of immunological factors. *J Reprod Fertil* 10:343–352.

Bingham, C.A. 1978. The cell cycle and cancer chemotherapy. *Am J Nurs* 78:1201–1205.

Binns, A.N., Wood, H.N., and Braun, A.C. 1981. Suppression of the tumorous state in crown gall teratomas of tobacco: A clonal analysis. *Differentiation* 19:97–102.

Birchmeier, W., Hülsken, J., and Behrens, J. 1995. E-cadherin as an invasion suppressor. In *Cell Adhesion and Human Disease,* Ciba Foundation Symposium 189, pp. 124–141. Chichester: Wiley.

Bister, K., and Jansen, H.W. 1986. Oncogenes in retroviruses and cells: Biochemistry and molecular genetics. *Adv Cancer Res* 47:99–188.

Bittner, J.J. 1936. Some possible effects of nursing on the mammary gland tumor incidence in mice. *Science* 84:162.

Bittner, J.J. 1937. Mammary tumors in mice in relation to nursing. *Am J Cancer* 30:530–538.

Bjorkoy, G., Overvatn, A., Diaz-Meco, M.T., Moscat, J., and Johansen, T. 1995. Evidence for a bifurcation of the mitogenic signaling pathway activated by Ras and phosphatidylcholine-hydrolyzing phospholipase C. *J Biol Chem* 270:21299–21306.

Blakeslee, A.F. 1934. New Jimson weeds from old chromosomes. *J Hered* 25:80–108.

Block, G. 1991. Vitamin C and cancer prevention: The epidemiological evidence. *Am J Clin Nutr* 53 (Suppl 1):270S–282S.

Blount, W.P. 1961. Turkey "X" disease. *J Br Turkey Fed* 9:522–528.

Boccardo, F., Bruzzi, P., Rubagotti, A., Nicolo, G.U., and Rosso, R. 1981. Estrogen-like action of tamoxifen on vaginal epithelium in breast cancer patients. *Oncology* 38:281–285.

Bodmer, W.F., Bailey, C.J., Bodmer, J., Bussey, H.J.R., Ellis, A., Gorman, P., Lucibello, F.C., Murday, V.A., Rider, S.H., Scambler, P., Sheer, D., Solomon, E., and Spurr, N.K. 1987. Localization of the gene for familial adenomatous polyposis on chromosome 5. *Nature* 328:614.

Boice, J.D., Jr., Greene, M.H., Killen, J.Y., Ellenberg, S.S., Keehn, R.J., McFadden, E., Chen, T.T., and Fraumeni, J.F., Jr. 1983. Leukemia and preleukemia after adjuvant treatment of gastrointestinal cancer with semustine methyl-CCNU. *N Engl J Med* 309:1079–1084.

Boone, C.W., Kelloff, G.J., and Malone, W.E. 1990. Identification of candidate cancer chemopreventive agents and their evaluation in animal models and human clinical trials: A review. *Cancer Res* 50:2–9.

Borthwick, N.J., Bofill, M., Hassan, I., Panayiotidis, P., Janossy, G., Salmon, M., and Akbar, A.N. 1996. Factors that influence activated CD8+ T-cell apoptosis in patients with acute herpesvirus infections: Loss of costimulatory molecules CD28, CD5 and CD6 but relative maintenance of Bax and Bcl-X expression. *Immunology* 88:508–515.

Bosari, S., Lee, A.K.C., DeLellis, R.A., Wiley, B.D., Heatley, G.J., and Silverman, M.L. 1992. Microvessel quantitation and prognosis in invasive breast carcinoma. *Hum Pathol* 23:755–761.

Bouroncle, B.A. 1994. Thirty-five years in the progress of hairy cell leukemia. *Leuk Lymphoma* 14 (Suppl 1):1–12.

Boutwell, J. 1982. Retinoids and inhibition of ornithine decarboxylase activity. *Am Acad Dermatol* 6:796–798.

Boveri, T. 1907. *Zellen-Studien. Heft 6. Die Entwicklung dispermer Seeigel-Eier. Ein Beitrag zur Befruchtungslehre und zur Theorie des Kerns.* Jena: Gustav Fischer.

Boveri, T. 1914. *Zur Frage der Erstehung Maligner Tumoren.* Jena: Gustav Fischer. (English translation by M. Boveri. 1929. *The Origin of Malignant Tumors.* Baltimore: Williams and Wilkins.)

Bowen, I.D., and Bowen, S.M. 1990. *Programmed Cell Death in Tumors and Tissues.* New York: Chapman and Hall.

Bowser, P.R., Wolfe, M.J., Forney, J.L., and Wooster, G.A. 1988. Seasonal prevalence of skin tumors from walleye *Stizostedion vitreum* from Oneida Lake, New York. *J Wildlife Dis* 24:292–298.

Boyar, A.P., Rose, D.P., Loughridge, J.R., Engle, A., Palgi, A., Laakso, K., Kinne, D., and Wynder, E.L. 1988. Response to a diet low in total fat in women with postmenopausal breast cancer: A pilot study. *Nutr Cancer* 11.93–99.

Boyle, P., Zaridze, D.G., and Smans, M. 1985. Descriptive epidemiology of colorectal cancer. *Int J Cancer* 36:9–18.

Brandom, W.F., Saccomanno, G., Archer, V.E., Archer, P.G., and Bloom, A.D. 1978. Chromosome aberrations as a biological dose-response indicator of radiation exposure in uranium miners. *Rad Res* 76:59–171.

Braun, A.C. 1943. Studies on tumor inception in crown gall disease. *Am J Bot* 30:674–677.

Braun, A.C. 1951. Recovery of crown gall tumor cells. *Cancer Res* 11:839–844.

Braun, A.C. 1956. The activation of two growth-substance systems accompanying the conversion of normal to tumor cells in crown gall. *Cancer Res* 16:53–56.

Braun, A.C. 1972. The usefulness of plant tumor systems for studying the basic cellular mechanisms that underlie neoplastic growth generally. In *Cell Differentiation,* R. Harris, P. Allin, and D. Viza, eds., pp. 115–118. Copenhagen: Munksgaard.

Braun, A.C. 1981. An epigenetic model for the origin of cancer. *Q Rev Biol* 56:33–60.

Briggs, R., and King, T.J. 1952. Transplantation of living nuclei from blastula cells into enucleated frogs' eggs. *Proc Natl Acad Sci USA* 38:455–463.

Brinster, R. 1993. Stem cells and transgenic mice in the study of development. *Int J Dev Biol* 37:89–99.

Brinster, R.L. 1974. Effects of cells transferred into the mouse blastocyst on subsequent development. *J Exp Med* 140:1049–1056.

Brinster, R.L., Chen, H.Y., Messing, A.A., Van Dyke, T., Levine, A.J., and Palmiter, R.D. 1984. Transgenic mice harboring SV40 T-antigen genes develop characteristic brain tumors. *Cell* 37:367–379.

Brown, J.E. 1990. *The Science of Human Nutrition.* San Diego: Harcourt, Brace, Jovanovich.

Bruyneel-Rapp, F., Dorsey, S.B., and Guin, J.D. 1988. The tanning salon: An area survey of equipment, procedures, and practices. *J Am Acad Dermatol* 18:1030–1038.

Bucher, N.L.R., and Malt, R.A. 1971. *Regeneration of the Liver and Kidney.* Boston: Little, Brown.

Buchkovich, K., Duffy, L.A., and Harlow, E. 1989. The retinoblastoma protein is phosphorylated during specific phases of the cell cycle. *Cell* 58:1097–1105.

Buday, L., and Downward, J. 1993. Epidermal growth factor regulates p21^{ras} through the formation of a complex of receptor, Grb2 adapter protein, and Sos nucleotide exchange factor. *Cell* 73:611–620.

Budillon, A., Cereseto, A., Kondrashin, A., Nesterova, M., Merlo, G., Clair, T., and Cho-Chung, Y.S. 1995. Point mutation of the autophosphorylation site or in the nuclear location signal causes protein kinase A RII beta regulatory subunit to lose its ability to revert transformed fibroblasts. *Proc Natl Acad Sci USA* 92:10634–10638.

Bueding, E., Ansher, S., and Dolan, P. 1985. Anticarcinogenic and other protective effects of dithiolthiones. In *Antimutagenesis and Anticarcinogenesis Mechanisms,* D.M. Shankel, P.E. Hartman, T. Kada, and A. Hollaender, eds., pp. 483–489. New York: Plenum.

Bullough, W.S. 1962. The control of mitotic activity in adult mammalian tissues. *Biol Rev* 37:301–342.

Bunn, P.A., Jr., and Ridgway, E.C. 1993. Paraneoplastic syndromes. In *Cancer Principles and Practice of Oncology,* 4th ed., V.T. DeVita, Jr., S. Hellman, and S.A. Rosenberg, eds., pp. 2026–2071. Philadelphia: J.B. Lippincott.

Burkitt, D. 1958. A sarcoma involving the jaws of African children. *Br J Surg* 46:218–225.

Burkitt, D.P. 1971. Epidemiology of cancer of the colon and rectum. *Cancer* 28:3–13.

Burkitt, D.P. 1975. Large-bowel cancer: An epidemiological jigsaw puzzle. *J Natl Cancer Inst* 54:3–6.

Burkitt, D.P., Walker, A.R.P., and Painter, N.S. 1974. Dietary fiber and disease. *J Am Med Assoc* 229:1068–1074.

Burmer, G.C., and Loeb, L.A. 1989. Mutations in the KRAS2 oncogene during progressive stages of human colon carcinoma. *Proc Natl Acad Sci USA* 86:2403–2407.

Bush, T.L., and Helzlsouer, K.J. 1993. Tamoxifen for the primary prevention of breast cancer: A review and critique of the concept and trial. *Epidemiol Rev* 15:233–243.

Butler, T.P., and Gullino, P.M. 1975. Quantitation of cell shedding into efferent blood of mammary adenocarcinoma. *Cancer Res* 35:512–516.

Butlin, H.J. 1892. Cancer of the scrotum in chimney-sweeps and others: II. Why foreign sweeps do not suffer from scrotal cancer. *Br Med J* 2:1–6.

Butrum, R.R., Clifford, C.K., and Lanza, E. 1988. NCI dietary guidelines: Rationale. *Am J Clin Nutr* 48:888–895.

Cain, K., Inayat-Hussain, S.H., Couet, C., and Cohen, G.M. 1996. A cleavage-site-directed inhibitor of interleukin-1 beta-converting enzyme-like protease inhibits apoptosis in primary cultures of rat hepatocytes. *Biochem J* 314:27–32.

Calabresi, P., and Schein, P.S., eds. 1993. *Medical Oncology: Basic Principles and Clinical Management of Cancer.* New York: McGraw-Hill.

Calabresi, P., and Parks, R.E., Jr. 1980. Antiproliferative agents and drugs used for immunosuppression. In *The Pharmacological Basis of Therapeutics,* A.G. Gilman, L.S. Goodman, and A. Gilman, eds., pp. 1256–1313. New York: Macmillan.

Calabresi, P., and Welch, A.D. 1962. Chemotherapy of neoplastic diseases. *Annu Rev Med* 13:147–202.

Call, K.M., Glaser, T., Ito, C.Y., Buckler, A.J., Pelletier, J., Haber, D.A., Rose, E.A., Kral, A., Yeger,

H., Lewis, W. H., Jones, C., and Housman, D.E. 1990. Isolation and characterization of a zinc finger polypeptide gene at the human chromosome 11 Wilms' tumor locus. *Cell* 60:509–520.

Calman, K.C. 1992. Cachexia. In *Oxford Textbook of Pathology*, vol. 1, J. O'D. McGee, P.G. Isaacson, and N.A. Wright, eds., pp. 715–717. Oxford: Oxford University Press.

Calnek, B.W. 1992. Chicken neoplasia – a model for cancer research. *Br Poult Sci* 33:3–16.

Cameron, I.L., Hardman, W.E., and Skehan, P. 1990. Regulation of growth in normal and neoplastic cell populations by a tissue sizer mechanism: Therapeutic implications. *Prog Clin Biol Res* 354A:61–79.

Campen, C.A., Jordan, V.C., and Gorski, J. 1985. Opposing biological actions of antiestrogens in vitro and in vivo: Induction of progesterone receptor in the rat and mouse uterus. *Endocrinology* 116:2327–2336.

Cannistra, S.A. 1990. Chronic myelogenous leukemia as a model for the genetic basis of cancer. *Hematol Oncol Clin North Am* 4:337–357.

Cano, E., and Mahadevan, L.C. 1995. Parallel signal processing among mammalian MAPKs. *Trends Biochem Sci* 20:117–122.

Carlson, D.L., Sauerbier, W., Rollins-Smith, L.A., and McKinnell, R.G. 1994a. The presence of DNA sequences of the Lucké herpesvirus in normal and neoplastic kidney tissue of *Rana pipiens*. *J Comp Pathol* 110:349–355.

Carlson, D.L., Sauerbier, W., Rollins-Smith, L.A., and McKinnell, R.G. 1994b. Fate of herpesvirus DNA in embryos and tadpoles cloned from Lucké renal carcinoma nuclei. *J Comp Pathol* 111:197–203.

Carlson, D.L., Williams, J.W., Rollins-Smith, L.A., Christ, C.G., John, J.C., Williams, C.S., and McKinnell, R.G. 1995. Pronephric carcinoma: Chromosomes of cells rescued from apoptosis by an oncogenic herpesvirus detected with a polymerase chain reaction. *J Comp Pathol* 113:277–286.

Carrel, A., and Burrows, M.T. 1911. Cultivation in vitro of malignant tumours. *J Exp Med* 13:571–575.

Carroll, K.K. 1986. Diet and carcinogenesis: Historical perspectives. *Adv Exp Med Biol* 206:45–53.

Carroll, K.K. 1991. Dietary fats and cancer. *Am J Clin Nutr* 53 (Suppl 4): 1064S -1067S.

Carroll, K.K., Gammal, E.B., and Plunkett, E.R. 1968. Dietary fat and mammary cancer. *Can Med Assoc J* 98:590–594.

Carroll, K.K., and Khor, H.T. 1975. Dietary fat in relation to tumorigenesis. *Prog Biochem Pharmacol* 10:308–353.

Carson, D.A., Wasson, D.B., Kaye, J., Ullman, B., Martin, D.W., Jr., Robins, R.K., and Montgomery, J.A. 1980. Deoxycytidine kinase-mediated toxicity of deoxyadenosine analogs toward malignant human lymphoblasts in vitro and toward murine L1210 leukemia in vivo. *Proc Natl Acad Sci USA* 77:6865–6869.

Case, R.A. 1969. Some environmental carcinogens. *Proc R Soc Med* 62:1061–1066.

Case, R.A.M., and Hosker, M.E. 1954. Tumours of the urinary bladder as an occupational disease in the rubber industry in England and Wales. *Br J Prev Soc Med* 8:39–50.

Case, R.A.M., Hosker, M.E., McDonald, D.B., and Pearson, J.T. 1954. Tumours of the urinary bladder in workmen engaged in the manufacture and use of certain dyestuff intermediates in the British chemical industry: Part I. The role of aniline, benzidine, alpha-naphthylamine and beta-naphthylamine. *Br J Ind Med* 11:75–104.

Castilla, L.H., Couch, F.J., Erdos, M.R., Hoskins, K.F., Calzone, K., Garber, J.E., Boyd, J., Lubin, M.B., Deshano, M.L., Brody, L.C., Collins, F.S., and Weber, B.L. 1994. Mutations in the *BRCA1* gene in families with early-onset breast and ovarian cancer. *Nat Genet* 8:387–391.

Cerutti, P. 1985. Prooxidant states and tumor promotion. *Science* 227:375–381.

Chabner, B.A. 1990a. Cytidine analogues. In *Cancer Chemotherapy*, B.A. Chabner and J.M. Collins, eds., pp. 154–179. Philadelphia: J.B. Lippincott.

Chabner, B.A. 1990b. Bleomycin. In *Cancer Chemotherapy*, B.A. Chabner and J.M. Collins, eds., pp. 341–355. Philadelphia: J.B. Lippincott.

Chabner, B.A. 1990c. Clinical strategies for cancer treatment: The role of drugs. In *Cancer Chemotherapy*, B.A. Chabner and J.M. Collins, eds., pp. 1–15. Philadelphia: J.B. Lippincott.

Chameaud, J., Masse, R., and Lafuma, J. 1984. Influence of radon daughter exposure at low doses on occurrence of lung cancer in rats. *Radiat Protect Dosim* 7:385–388.

Chang, R.L., Huang, M.T., Wood, A.W., Wong, C.Q., Newmark, H.L., Yagi, H., Sayer, J.M., Jerind, D.M., and Conney, A.H. 1985. Effect of ellagic acid and hydroxylated flavonoids on the tumorigenicity of benzo(a)pyrene and ±-7,8-dihydroxy-9,10-epoxy-7,8,9,10-tetrahydrobenzo(a)pyrene on mouse skin and in the newborn mouse. *Carcinogenesis* 6:1127–1133.

Chang, Y., Cesarman, E., Pessin, M.S., Lee, F., Culpepper, J., Knowles, D.M., and Moore, P.S. 1994. Identification of herpesvirus-like DNA sequences in AIDS-associated Kaposi's sarcoma. *Science* 266:1865–1869.

Chaplin, D.J., and Trotter, M.J. 1990. The nature of tumor hypoxia: Implications for therapy. *Prog Clin Biol Res* 354B:81–92.

Charbit, A., Malaise, E.P., and Tubiana, M. 1971. Relation between the pathological nature and the growth rate of human tumors. *Eur J Cancer* 7:307–315.

Chardin, P., Camonis, J.H., Gale, N.W., van Aelst, L., Schlessinger, J., Wigler, M.H., and Bar-Sagi, D. 1993. Human Sos1: A guanine nucleotide exchange factor for Ras that binds to GRB2. *Science* 260:1338–1343.

Charlson, M.E. 1985. Delay in the treatment of carcinoma of the breast. *Surg Gynecol Obstet* 160:393–399.

Chen, A., Licht, J.D., Wu, Y., Hellinger, N., Scher, W., and Waxman, S. 1994. Retinoic acid is required for and potentiates differentiation of acute promyelocytic leukemia cells by nonretinoid agents. *Blood* 84:2122–2129.

Chen, M.C., and Meguid, M.M. 1986. Postulated cancer prevention diets, a guide to food selections. *Surg Clin North Am* 66:931–945.

Chen, S.H., Shine, H.D., Goodman, J.C., Grossman, R.G., and Woo, S.L. 1994. Gene therapy for brain tumors: Regression of experimental gliomas by adenovirus-mediated gene transfer in vivo. *Proc Natl Acad Sci USA* 91:3054–3057.

Cheng, K.C., and Loeb, L.A. 1993. Genomic instability and tumor progression: Mechanistic considerations. *Adv Cancer Res* 60:121–157.

Chilton, M.D., Saiki, R.K., Yadav, N., Gordon, M.P., and Quetier, F. 1980. T-DNA from Agrobacterium tumefaciens Ti plasmid is in the nuclear fraction of crown gall tumor cells. *Proc Natl Acad Sci USA* 77:4060–4064.

Chin, K.V., Pastan, I., and Gottesman, M.M. 1993. Function and regulation of the human multidrug resistance gene. *Adv Cancer Res* 60:157–180.

Chiu, R., Boyle, W.J., Meek, J., Smeal, T., Hunter, T., and Karin, M. 1988. The c-fos protein interacts with c-jun/AP-1 to stimulate transcription of AP-1 responsive genes. *Cell* 54:541–552.

Chuang, T.Y., Popecsh, A., Su, W.P.D., and Chute, C.G. 1990. Basal cell carcinoma. *J Am Acad Dermatol* 22:413–417.

Clare, B.G., Kerr, A., and Jones, D.A. 1990. Characteristics of the nopaline catabolic plasmid in Agrobacterium strain K84 and K1026 used for biological control of crown gall disease. *Plasmid* 23:126–137.

Clark, L.C., Combs, G.F., Turnbull, B.W., Slate, E.H., Chalker, D.K., Chow, J., Davis, L.S., Glover, R.A., Graham, G.F., Gross, E.G., Krongrad, A., Lesher, J.L., Park, H.K., Sanders, B.B., Smith, C.L., and Taylor, J.R. 1996. Effects of selenium supplementation for cancer prevention in patients with carcinoma of the skin. *J Am Med Assoc* 276:1957–1963.

Clarke, A.R., Mandag, E.R., van Roon, M., van der Lugt, N.M.T., van der Valk, M., Hooper, M.L., Berns, A., and Te Riele, H. 1992. Requirement for a functional Rb-1 gene in murine development. *Nature* 359:328–330.

Clarke, A.R., Purdie, C.A., Harrison, D.J., Morris, R.G., Bird, C.C., Hooper, M.L., and Wyllie, A.H. 1993. Thymocyte apoptosis induced by p53-dependent and independent pathways. *Nature* 362:849–852.

Clarke, R., Lippman, M.E., and Dickson, R.B. 1990. Mechanisms of hormone and cytotoxic drug interactions in the development and treatment of breast cancer. *Prog Clin Biol Res* 322:243–278.

Cleary, M.L., Smith, S.D., and Sklar, J. 1986. Cloning and structural analysis of cDNAs for bcl-2 and a hybrid bcl-2/immunoglobulin transcript resulting from the t(14;18) translocation. *Cell* 47:19–28.

Cleaver, J.E. 1968. Defective repair replication of DNA in xeroderma pigmentosum. *Nature* 218:652–656.

Cleaver, J.E. 1994. It was a good year for DNA repair. *Cell* 76:1–4.

Cleaver, J.E., and Kraemer, K.H. 1989. Xeroderma pigmentosum. In *The Metabolic Basis of Inherited Disease,* 6th ed., C.R. Scriver, A.L. Beaudet, W.S. Sly, and D. Valle, eds., pp. 2949–2971. New York: McGraw-Hill.

Cohen, S.M., and Ellwein, L.B. 1991. Genetic errors, cell proliferation, and carcinogenesis. *Cancer Res* 51:6493–6505.

Coit, D.G. 1997. Cancer of the small intestine. In *Cancer, Principles and Practice of Oncology,* 5th ed., V.T. DeVita, S. Hellman, and S.A. Rosenberg, eds., pp. 1128–1143. Philadelpha: Lippincott-Raven.

Colombo, N. 1994. Controversial issues in the management of early epithelial ovarian cancer: Conservative surgery and the role of adjuvant therapy. *NIH Consensus Development Conference on Ovarian Cancer: Screening, Treatment, and Followup.* April 5–7, Bethesda, MD: NIH.

Colsher, P.L., and Wallace, R.B. 1989. Is modest alcohol consumption better than none at all? *Annu Rev Public Health* 10:203–219.

Colvin, M., and Chabner, B.A. 1990. Alkylating agents. In *Cancer Chemotherapy,* B.A. Chabner and J.M. Collins, eds., pp. 276–313. Philadelphia: J.B. Lippincott.

Coman, D.R. 1944. Decreased mutual adhesiveness, a property of cells from squamous cell carcinomas. *Cancer Res* 4:625–629.

Coman, D.R. 1953. Mechanisms responsible for origin and distribution of blood-borne tumor metastases; review. *Cancer Res* 13:397–404.

Conney, A.H. 1982. Induction of microsomal enzymes by foreign chemicals and carcinogenesis by polycyclic aromatic hydrocarbons. *Cancer Res* 42:4875–4917.

Conrad, R.A., Dobyns, B.M., and Sutow, W.W. 1970. Thyroid neoplasia as a late effect of exposure to radioactive iodine in fallout. *J Am Med Assoc* 214:316–324.

Cook, J.W., Hewett, C.L., and Hieger, I. 1933. The isolation of a cancer-producing hydrocarbon from coal tar. Parts I, II, and III. *J Chem Soc* Issue 395–405.

Cook, S.J., and McCormick, F. 1993. Inhibition by cAMP of Ras-dependent activation of Raf. *Science* 262:1069–1072.

Cooper, D.Y., Levin, S.S., Narrasimhulu, S., Rosenthal, O., and Estabrook, R.W. 1965. Photochemical action spectrum of the terminal oxidase of mixed function oxidase systems. *Science* 147:400–402.

Corliss, J. 1993. The Delaney clause: Too much of a good thing? *J Natl Cancer Inst* 85:600–603.

Cornwell, M.M., Tsuruo, T., Gottesman, M.M., and Pastan, I. 1987. ATP-binding properties of P-glycoprotein from multidrug-resistant KB cells. *Fed Am Soc Exp Biol J* 1:51–54.

Costantini, V., Zacharski, L.R., Memoli, V.A., Kudryk, B.J., Rousseau, S.M., and Stump, D.C. 1991. Occurrence of components of fibrinolysis pathways in situ in neoplastic and nonneoplastic human breast tissue. *Cancer Res* 51:354–358.

Cotran, R.S., Kumar, V., and Robbins, S.L. 1994. *Pathological Basis of Cancer,* 5th ed. Philadelphia: W.B. Saunders.

Courtneidge, S.A. 1994. Protein tyrosine kinases, with emphasis on the Src family. *Semin Cancer Biol* 5:239–246.

Coutard, H. 1932. Roentgenotherapy of epitheliomas of the tonsillar region, hypopharynx and larynx from 1920 to 1926. *Roentgenol Radium Ther* 28:313–331.

Cowell, J.K. 1989. One hundred years of retinoblastoma research. From the clinic to the gene and back again. *Ophthalmic Paediatr Genet* 10:75–88.

Cox, E.C. 1976. Bacterial mutator genes and the control of spontaneous mutation. *Annu Rev Genet* 10:135–156.

Croce, C.M., Thierfelder, W., Erikson, J., Nishikura, K., Finan, J., Lenoir, G., and Nowell, P.C. 1983. Transcriptional activation of an unrearranged and untranslocated c-*myc* oncogene by translocation of a Cl locus in Burkitt lymphoma. *Proc Natl Acad Sci USA* 80:6922–6926.

Croce, C.M., Tsujimoto, Y., Erikson, J., and Nowell, P. 1984. Chromosome translocations and B-cell neoplasia. *Lab Invest* 51:258–267.

Cross, F.T. 1988. Radon inhalation studies in animals. *Radiat Protect Dosim* 24:463–466.

Cross, F.T., Palmer, R.F., Filipy, R.E., Dagle, G.E., and Stuart, B.O. 1982. Carcinogenic effects of radon daughters, uranium ore dust and cigarette smoke in beagle dogs. *Health Phys* 42:233–252.

Culver, K., Cornetta, K., Morgan, R., Morecki, S., Aerbersold, T., Kasid, A., Lotve, M., Rosenberg, S.A., Anderson, W.F., and Blaese, R.M. 1991. Lymphocyte as cellular vehicles for gene therapy in mouse and man. *Proc Natl Acad Sci USA* 88:3155–3159.

Curran, W.J. 1988. Cancer-causing substances in food, drugs, and cosmetics: The de Minimis rule versus the Delaney clause. *N Engl J Med* 319:1262–1264.

Curren, R.D., Putman, D.L., Yang, L.L., Haworth, S.R., Lawlor, T.E., Plummer, S.M., and Harris, C.C. 1987. Genotoxicity of fecapentaene-12 in bacterial and mammalian cell assay systems. *Carcinogenesis* 8:349–352.

Cushing, H., and Wolback, S.B. 1927. The transformation of a malignant paravertebral neuroblastoma into a benign ganglioneuroma. *Am J Pathol* 3:203–216.

Daenen, S., Vellenga, E., van Dobbenburgh, O.A., and Halie, M.R. 1986. Retinoic acid as antileukemic therapy in a patient with acute promyelocytic leukemia and Aspergillus pneumonia. *Blood* 67:559–561.

Dalla Favera, R., Bregni, M., Erikson, J., Patterson, D., Gallo, R.C., and Croce, C.M. 1982. Assignment of the c-*myc* oncogene to the region of chromosome 8 which is translocated in Burkitt lymphoma cells. *Proc Natl Acad Sci USA* 79:7824–7827.

Damjanov, I. 1993. Teratocarcinoma: Neoplastic lessons about normal embryogenesis. *Int J Dev Biol* 37:39–46.

D'Amore, P.A. 1988. Antiangiogenesis as a strategy for antimetastasis. *Semin Thromb Hemost* 14:73–78.

Davidson, J.R. 1934. Attempt to inhibit development of tar-carcinoma in mice: Preliminary note. *Can Med Assoc J* 31:486–487.

Davis, J.M.G., Beckett, S.T., Bolton, R.E., Collins, P., and Middleton, A. P. 1978. Mass and number of fibres in the pathogenesis of asbestos-related lung disease in rats. *Br J Cancer* 37:673–688.

Davison, A.J. 1992. Channel catfish virus: A new type of herpesvirus. *Virology* 186:9–14.

Dawe, C.J. 1969. Phylogeny and oncogeny. In *Neoplasms and Related Disorders in Invertebrate and Lower Vertebrate Animals,* C.J. Dawe and J.C. Harshbarger, eds. Bethesda, MD: National Cancer Institute.

Dawe, C.J., Harshbarger, J.C., Kondo, S., Sugimura, T., and Takayama, S., eds. 1981. *Phyletic Approaches to Cancer.* Tokyo: Japan Scientific Societies Press.

Dean, C., Chetty, U., and Forrest, A.P.M. 1983. Effects of immediate breast reconstruction on psychosocial morbidity after mastectomy. *Lancet* 1:459–462.

DeCaprio, J.A., Ludlow, J.W., Figge, J., Shew, J.-Y., Huang, C.-M., Lee, W.-H., Marsillo, E., Paucha, E., and Livingston, D.M. 1988. SV40 large tumor antigen forms a specific complex with the product of the retinoblastoma susceptibility gene. *Cell* 54:275–283.

De Guise, S., Lagace, A., and Beland, P. 1994. Tumors in St. Lawrence beluga whales Delphinapterus leucas. *Vet Pathol* 31:444–449.

De Jong, P.J., Gorsovsky, A.J., and Glickman, B.W. 1988. Spectrum of spontaneous mutation at the APRT locus of Chinese hamster ovary cells: An analysis at the DNA sequence level. *Proc Natl Acad Sci USA* 85:3499–3503.

deKernion, J.B., and Berry, D. 1980. The diagnosis and treatment of renal cell carcinoma. *Cancer* 45:1947–1956.

De Klein, A., Van Kessel, A.G., Grosveld, G., Bartram, C.R., Hagemijer, A., Bootsma, D., Spurr, N.E., Heisterkamp, N., Groffen, J., and Stephenson, J.R. 1982. A cellular oncogene is translocated to the Philadelphia chromosome in chronic myelocytic leukemia. *Nature* 300:765–767.

de Vita, V.T., Jr., Hellman, S., and Rosenberg, S.A., eds. 1997. *Cancer: Principles and Practice of Oncology,* 5th ed., Philadelphia: Lippincott-Raven.

Deng, T., and Karin, M. 1994. C-Fos transcriptional activity stimulated by H-ras-activated protein kinase distinct from JNK and ERK. *Nature* 371:171–175.

Denker, H.W. 1980. Embryo implantation and trophoblast invasion. In *Cell Movement and Neoplasia,* M. DeBrabander, ed., pp. 151–162. Oxford: Pergamon.

Der, C.J., Finkel, T., and Cooper, G.M. 1986. Biological and biochemical properties of human ras^H genes mutated at codon 61. *Cell* 44:167–176.

Dérijard, B., Hibi, M., Wu, I.-H., Barret, T., Su, B., Deng, T., Karin, M., and Davis, R.J. 1994. JNK1: A protein kinase stimulated by UV light and Ha-Ras that binds and phosphorylates the c-Jun activation domain. *Cell* 76:1025–1037.

de Sauvage, F.J., Hass, P.E., Spencer, S.D., Malloy, B.E., Gurney, A.L., Spencer, S.A., Darbonne, W.C., Henzel, W.J., Wong, S.C., Kuang, W.J, Oles, K.J., Hultgren, D., Solberg, L.A., Jr., Goeddel, D.V., and Eaton, D.L. 1994. Stimulation of megakaryocytopoiesis and thrombopoiesis by the c-Mpl ligand. *Nature* 369:533–538.

DeVilliers, E.-M. 1989. Heterogeneity of the human papillomavirus group. *J Virol* 53:4898–4903.

DiBerardino, M.A. 1962. The karyotype of *Rana pipiens* and investigation of its stability during embryonic differentiation. *Dev Biol* 5:101–126.

DiBerardino, M.A. 1979. Nuclear and chromosomal behavior in amphibian nuclear transplants. *Int Rev Cytol* (Suppl 9):129–160.

DiBerardino, M.A. 1987. Genomic potential of differentiated cells analyzed by nuclear transplantation. *Am Zool* 27:623–644.

DiBerardino, M.A. 1997. *Genomic Potential of Differentiated Cells.* New York: Columbia University Press.

DiBerardino, M.A., King, T.J., and McKinnell, R.G. 1963. Chromosome studies of a frog renal adenocarcinoma line carried by intraocular transplantation. *J Natl Cancer Inst* 31:769–789.

DiBerardino, M.A., and McKinnell, R.G. 1997. Backward compatible. *The Sciences* 37(5):32–37.

Diffey, B.L. 1987. Cosmetic solaria and malignancies of the skin. *Photodermatology* 4:273–276.

Di Fiore, P.P., Pierce, J.H., Kraus, M.H., Segatto, O., King, C.R., and Aaronson, S.A. 1987. *erb*B-2 is a potent oncogene when overexpressed in NIH3T3 cells. *Science* 237:178–182.

Diwan, B.A., Ohshima, M., and Rice, J.M. 1989. Promotion by sodium barbital of renal cortical and transitional cell tumors, but not intestinal tumors, in F344 rats given methylacetoxy-methylnitrosamine, and lack of effect of phenobarbital, amobarbital, or barbituric acid on development of either renal or intestinal tumors. *Carcinogenesis* 10:183–188.

Dixon, F.J., and Moore, R.A. 1952. Tumors of the male sex organs. Atlas of tumor pathology section viii, Fascicle 31b and 32. Washington, DC. Armed Forces Institute of Pathology.

Doi, S., Goldstein, B., Hug, H., and Weinstein, I.B. 1993. Expression of multiple isoforms of protein kinase C in normal human colon mucosa and colon tumor and decreased levels of protein kinase C beta and eta mRNAs in the tumors. *Mol Carcinog* 11:197–203.

Doll, R. 1992. Health and the environment in the 1990s. *Am J Public Health* 82:933–941.

Doll, R., and Bradford-Hill, A. 1950. Smoking and carcinoma of the lung. *Br Med J* 2:739–748.

Doll, R., and Peto, R. 1981. The causes of cancer. *J Natl Cancer Inst* 66:1191–1308.

Donaldson, K.L., Goolsby, G.L., and Wahl, A.F. 1994. Cytotoxicity of the anticancer agents cisplatin and Taxol during cell proliferation and the cell cycle. *Int J Cancer* 57:847–855.

Donehower, L.A., Harvey, M., Slagle, B.L., McArthur, M.J., Montgomery, Jr., C.A., Butel, J.S., and Bradley, A. 1992. Mice deficient for p53 are developmentally normal but susceptible to spontaneous tumours. *Nature* 356:215–221.

Dong, J-T, Lamb, P.W., Rinker-Schaeffer, C.W., Vukanovic, J., Ichikawa, T., Isaacs, J.T., and Barrett, J.C. 1995. KAI1, a metastasis suppressor gene for prostate cancer on human chromosome 11p11.2. *Science* 268:884–886.

Doniach, I., and Pelc, S.R. 1950. Autoradiographic techniques. *Br J Radiol* 23:184–192.

Doniger, J., Jacobson, E.D., Krell, K., and DiPaolo, J.A. 1981. Ultraviolet light action spectra for neoplastic transformation and lethality of Syrian hamster embryo cells correlate with spectrum for pyrimidine dimer formation in cellular DNA. *Proc Natl Acad Sci USA* 78:2378–2382.

Doonan, J., and Hunt, T. 1996. Why don't plants get cancer? *Nature* 380:481–482.

Dorr, R.T., and Fritz, W.L. 1980. *Cancer Chemotherapy Handbook.* New York: Elsevier.

Dorr, R.T., and Von Hoff, D. 1994. *Cancer Chemotherapy Handbook.* Norwalk, CT: Appleton and Lange.

Dotto, G.P., Parada, L.F., and Weinberg, R.A. 1985. Specific growth response of *ras*-transformed fibroblasts to tumor promoters. *Nature* 318:472–475.

Downward, J. 1990. The ras superfamily of small GTP-binding proteins. *Trends Biochem Sci* 15:469–472.

Drapkin, R., Sancar, A., and Reinberg, D. 1994. Where transcription meets repair. *Cell* 77:9–12.

Drenou, B., Fardel, O., Amiot, L., and Fauchet, R. 1993. Detection of P glycoprotein activity on normal and leukemic CD34+ cells. *Leuk Res* 17:1031–1035.

Drets, M.E., and Shaw, M.W. 1971. Specific banding patterns of human chromosomes. *Proc Natl Acad Sci USA* 68:2073–2077.

Driessens, M.H.E., Stroeken, P.J.M., Erena, N.F.R., van der Valk, M.A., van Rijthoven, E.A.M., and Roos, E. 1995. Targeted disruption of CD44 in MDAY-D2 lymphosarcoma cells has no effect on subcutaneous growth or metastatic capacity. *J Cell Biol* 131:1849–1855.

Drinkwater, N.R. 1990. Experimental models and biological mechanisms for tumor promotion. *Cancer Cells* 2:8–14.

Druckrey, H., Ivankovic, S., and Preussmann, R. 1966. Teratogenic and carcinogenic effects in the offspring after single injection of ethylnitrosourea to pregnant rats. *Nature* 210:1378–1379.

Duan, D.R., Pause, A., Burgess, W.H., Aso, T., Chen, D.Y.T., Garrett, K.P., Conaway, R.C., Conaway, J.W., Linehan, W.M., and Klausner, R.D. 1995. Inhibition of transcription elongation by the VHL tumor suppressor protein. *Science* 269:1402–1406.

Dulic, V., Kaufmann, W.K., Wilson, S.J., Tlsty, T.D., Lees, E., Harper, J.W., Elledge, S.J., and Reed, S.I. 1994. p53-dependent inhibition of cyclin-dependent kinase activities in human fibroblasts during radiation-induced G1 arrest. *Cell* 76:1013–1023.

Dunn, J.E. 1975. Cancer epidemiology in populations of the United States – with emphasis on Hawaii and California – and Japan. *Cancer Res* 35:3240–3245.

Dutreix, J., Tubiana, M., Wambersie, A., and Malaise, E. 1971. The influence of cell proliferation in tumours and normal tissues during fractionated radiotherapy. *Eur J Cancer* 7:205–213.

Dworkin, H.J., Meier, D.A., and Kaplan, M. 1995. Advances in the management of patients with thyroid disease. *Semin Nucl Med* 25:205–220.

Eardley, A.M., Heller, G., and Warrell, R.P., Jr. 1994. Morbidity and costs of remission induction therapy with all-trans retinoic acid compared with standard chemotherapy in acute promyelocytic leukemia. *Leukemia* 8:934–939.

Earle, J. 1808. *Chirurgical works of Percivall Pott.* A new edition with his last corrections to which are added a short account of the life of the author, a method of curing the hydrocele by injection and occasional notes and observations. London: J. Johnson.

Eichner, E.R. 1987. Exercise, lymphokines, calories, and cancer. *Physician Sportsmed* 15:109–116.

Elion, G.B., and Hitchings, G.H. 1965. Metabolic basis for the actions of analogs of purines and pyrimidines. *Adv Chemother* 2:91–177.

Ellinger, M.S., King, D.R., and McKinnell, R.G. 1975. Androgenetic haploid development produced by ruby laser irradiation of anuran ova. *Rad Res* 62:117–122.

Elston, R.A., Kent, M.L., and Drum, A.S. 1988. Transmission of hemic neoplasia in the bay mussel, *Mytilus edulis,* using whole cells and cell homogenate. *Dev Comp Immunol* 12:719–727.

Elwood, J.M., Lee, J.A.H., Walter, S.D., Mo, T., and Green, A.E.S. 1974. Relationship of melanoma and other skin cancer mortality to latitude and ultraviolet radiation in the United States and Canada. *Int J Epidemiol* 3:325–332.

Encarnacion, C.A., Ciocca, D.R., McGuire, W.L., Clark, G.M., Fuqua, S.A., and Osborne, C.K. 1993. Measurement of steroid hormone receptors in breast cancer patients on tamoxifen. *Breast Cancer Res Treat* 26:237–246.

Engell, H.C. 1955. Cancer cells in the circulating blood. *Acta Chir Scand* (Suppl 201):1–70.

English, H.F., Santner, S.J., Levine, H.B., and Santen, R.J. 1986. Inhibition of testosterone production with ketoconazole alone and in combination with a gonadotropin releasing hormone analogue in the rat. *Cancer Res* 46:38–42.

Ennemoser, O., Ambach, W., Auer, T., Brunner, P., Schneider, P., Oberaigner, W., Purtscheller, F., and Stingl, V. 1994. High indoor radon concentrations in an alpine region of western Tyrol. *Health Phys* 67:151–154.

Ennemoser, O., Ambach, W., Brunner, P., Schneider, P., and Oberaigner, W. 1993. High domestic and occupational radon exposures: A comparison. *Lancet* 342:47.

Ensoli, B., Barillari, G., and Gallo, R.C. 1991. Pathogenesis of AIDS-associated Kaposi's sarcoma. *Hematol Oncol Clin North Am* 5:281–295.

Enterline, H.T., and Coman, D.R. 1950. The amoeboid motility of human and animal neoplastic cells. *Cancer* 3:1033–1038.

Epner, D.E., Partin, A.W., Schalken, J.A., Isaacs, J.T., and Coffey, D.S. 1993. Association glyceraldehyde-3-phosphate dehydrogenase expression with cell motility and metastatic potential of rat prostatic adenocarcinoma. *Cancer Res* 53:1995–1997.

Erikson, J., Nishikura, K., ar-Rushdi, A., Finan, J., Emanuel, B., Lenoir, G., Nowell, P.C., and Croce, C.M. 1983. Translocation of an immunoglobulin kappa locus to a region 3' of an unrearranged c-*myc* oncogene enhances c-*myc* transcription. *Proc Natl Acad Sci USA* 80:7581–7585.

Essigmann, J.M., Croy, R.G., Bennett, R.A., and Wogan, G.N. 1982. Metabolic activation of aflatoxin B$_1$: Patterns of DNA adduct formation, removal, excretion in relation to carcinogenesis. *Drug Metab Rev* 13:581–602.

Ettinghausen, S.E., and Rosenberg, S.A. 1995. Immunotherapy and gene therapy of cancer. *Adv Surg* 28:223–254.

Evans, M.J., and Kaufman, M.H. 1981. Establishment in culture of pluripotential cells from mouse embryos cultured in medium conditioned by teratocarcinoma cells. *Nature* (Lond.) 292:154–156.

Ewen, M.E. 1994. The cell cycle and the retinoblastoma protein family. *Cancer Metastasis Rev* 13:45–66.

Ewen, M.E., Sluss, H.K., Sherr, C.J., Matsushime, H., Kato, J., and Livingston, D.M. 1993. Functional interaction of the retinoblastoma protein with mammalian D-type cyclins. *Cell* 73:487–497.

Ewing, J. 1916. Pathological aspects of some problems of experimental cancer research. *J. Cancer Res* 1:71–86.

Ewing, J. 1928. *Neoplastic Diseases,* 3rd ed. Philadelphia: W.B. Saunders.

Fabre, I., Fabre, G., and Goldman, I.D. 1984. Polyglutamylation, an important element in methotrexate cytotoxicity and selectivity in tumor versus murine granulocytic progenitor cells in vitro. *Cancer Res* 44:3190–3195.

Faisst, S., and Meyer, S. 1992. Compilation of vertebrate-encoded transcription factors. *Nucleic Acids Res* 20:3–26.

Falk, A., and Sachs, L. 1980. Clonal regulation of the induction of macrophage- and granulocyte-inducing proteins for normal and leukemic myeloid cells. *Int J Cancer* 26:595–601.

Fankhauser, G. 1945. The effects of changes in chromosome number on amphibian development. *Q Rev Biol* 20:20–78.

Fantl, W.J., Escobedo, J.A., Martin, G.A., Turck, C.W., del Rosario, M., McCormick, F., and Williams, L.T. 1992. Distinct phosphoproteins on a growth factor receptor bind to specific molecules that mediate different signaling pathways. *Cell* 69:413–423.

Fausto, N., and Webber, E.M. 1993. Control of liver growth. *Crit Rev Eukaryot Gene Expr* 3:316–373.

Fearon, E.R., and Vogelstein, B. 1990. A genetic model for colorectal tumorigenesis. *Cell* 61:759–767.

Feig, D.I., Reid, T.M., and Loeb, L.A. 1994. Reactive oxygen species in tumorigenesis. *Cancer Res* (Suppl 54):1890s–1894s.

Feinberg, A.P., and Coffey, D.S. 1982. Organ site specificity for cancer in chromosomal instability disorders. *Cancer Res* 42:3252–3254.

Felt, E.P. 1965. *Plant Galls and Gall Makers.* New York: Hafner.

Fenaux, P. 1993. The role of all-trans-retinoic acid in the treatment of acute promyelocytic leukemia. *Acta Haematol* 89(Suppl 1):22–27.

Fenoglio-Preiser, C.M., and R.V.P. Hutter. 1985. Colorectal polyps: Pathologic diagnosis and clinical significance. *CA Cancer J Clin* 35:322–344.

Ferguson, D.J., and Anderson, T.J. 1981a. Ultrastructural observations on cell death by apoptosis in the "resting" human breast. *Virchows Arch Pathol Anat* 393:193–203.

Ferguson, D.J., and Anderson, T.J. 1981b. Morphological evaluation of cell turnover in relation to the menstrual cycle in the "resting" human breast. *Br J Cancer* 442:177–181.

Fidler, I.J. 1978. Tumor heterogeneity and the biology of cancer invasion and metastasis. *Cancer Res* 38:2651–2660.

Fidler, I.J. 1984. The evolution of biological heterogeneity in metastatic neoplasms. In *Cancer Invasion and Metastasis: Biologic and Therapeutic Aspects,* pp. 5–30. New York: Raven Press.

Fidler, I.J. 1990. Critical factors in the biology of human cancer metastasis: Twenty-eighth G.H.A. Clowes memorial award lecture. *Cancer Res* 50:6130–6138.

Fidler, I.J. 1997. Molecular biology of cancer: Invasion and metastasis. In *Cancer: Principles and Practice of Oncology,* 5th ed., V.T. DeVita, Jr., S. Hellman, and S.A. Rosenberg, eds., pp. 135–152. Philadelphia: Lippincott-Raven.

Fidler, I.J., and Hart, I.R. 1982. Biological diversity in metastatic neoplasms: Origins and implications. *Science* 217:998–1003.

Fidler, I.J., and Kripke, M.L. 1977. Metastasis results from pre-existing variant cells within a malignant tumor. *Science* 197:893–895.

Fieser, L.F., and Newman, M.S. 1935. Methycholanthrene from cholic acid. *J Am Chem Soc* 57:961.

Figlin, R.A., Holmes, E.C., Petrovich, Z., and Sarna, G.P. 1990. Lung cancer. In *Cancer Treatment,* 3rd ed., C.M. Haskell, ed., pp. 165–188. Philadelphia: W.B. Saunders.

Findlay, G.M. 1928. Ultra-violet light and skin cancer. *Lancet* 2:1070–1073.

Fisher, B., Bauer, M., Margolese, R., Poisson, R., Pilch, Y., Redmond, C., Fisher, E.R., Wolmark, N., Deutsch, M., Montague, E., Saffer, E., Wickerham, D.L., Lerner, H., Glass, A., Shibata, H., Deckers, P., Ketcham, A., Oishi, R., and Russell, I. 1985. Five-year results of a randomized clinical trial comparing total mastectomy and segmental mastectomy with or without radiation in the treatment of breast cancer. *N Engl J Med* 312:665–673.

Fisher, B., and Fisher, E.R. 1967. Experimental evidence in support of the dormant tumor cell. *Science* 130:918–919.

Fisher, B., Gunduz, N., and Saffer, E.A. 1983. Influence of the interval between primary tumor removal and chemotherapy on kinetics and growth of metastases. *Cancer Res* 43:1488–1492.

Fisher, B., Montague, E., Redmond, C., Barton, B., Borland, D., Fisher, E.R., Deutsch, M., Schwarz, G., Margolese, R., Donegan, W., Volk, H., Konvolinka, C., Gardner, B., Cohn, I., Jr., Lesnick, G., Cruz, A.B., Lawrence, W., Nealon, T., Butcher, H., and Lawton, R. 1977. Comparison of radical mastectomy with alternative treatments for primary breast cancer. *Cancer* 39:2827–2839.

Fisher, B., Osborne, C.K., Margolese, R., and Bloomer, W. 1993. Neoplasms of the breast. In *Cancer Medicine,* vol. 2, J.F. Holland, E. Frei III, R.C. Bast, Jr., D.W. Kufe, D.L. Morton, and R.R. Weichselbaum, eds., pp. 1706–1774. Philadelphia: Lea and Febiger.

Fisher, B., Redmond, C., Fisher, E.R., Bauer, M., Wolmark, N., Wickerham, D.L., Deutsch, M., Montague, E., Margoleses, R., and Foster, R. 1985. Ten-year results of a randomized clinical trial comparing radical mastectomy and total mastectomy with or without radiation. *N Engl J Med* 312:674–681.

Fisher, B., Redmond, C., Poisson, R., Capian, R., Wickerham, D.L., Wolmark, N., Fisher, E.R., Deutsch, M., Margolese, R., Pitch, Y., Glass, A., Shibata, H., Lerner, H., Terz, J., and Sidorovich, L. 1989. Eight-year results of a randomized clinical trial comparing total mastectomy and lumpectomy with or without radiation in the treatment of breast cancer. *N Engl J Med* 320:822–828.

Fisher, B., Saffer, E.A., Rudock, C., Coyle, J., and Gunduz, N. 1989a. Presence of a growth stimulating factor in serum following primary tumor removal in mice. *Cancer Res* 49:1996–2001.

Fisher, B., Saffer, E.A., Rudock, C., Coyle, J., and Gunduz, N. 1989b. Effect of local or systemic
treatment prior to primary tumor removal on the production and response to a serum growth
stimulating factor in mice. *Cancer Res* 49:2002–2004.

Fisher, B., Wolmark, N., Fisher, E.R., and Deutsch, M. 1985. Lumpectomy and axillary dissection
for breast cancer: Surgical, pathological, and radiation considerations. *World J Surg* 9:692–698.

FitzGerald, M.G., MacDonald, D.J., Krainer, M., Hoover, I., O'Neil, E., Unsal, H., Silva-Arrieto,
S., Finkelstein, D.M., Beer-Romero, P., Englert, C., Sgroi, D.C., Smith, B.L., Younger, J.W.,
Garber, J.E., Duda, R.B., Mayzel, K.A., Isselbacher, K.J., Friend, S.H., and Haber, D.A. 1996.
Germ-line *BRCA1* mutations in Jewish and non-Jewish women with early-onset breast cancer. *N
Engl J Med* 334:143–149.

Flynn, P.J., Miller, W.J., Weisdorf, D.J., Arthur, D.C., Brunning, R., and Branda, R.F. 1983.
Retinoic acid treatment of acute promyelocytic leukemia: In vitro and in vivo observations. *Blood*
62:1211–1217.

Folkman, J. 1985. Tumor angiogenesis. *Adv Cancer Res* 43:175–203.

Folkman, J. 1993. Tumor angiogenesis. In *Cancer Medicine,* 3rd ed., J.F. Holland, E. Frei, R.C.
Bast, D.W. Kufe, D.L. Morton, and R.R. Weichselbaum, eds., pp. 153–170. Melbourne, PA:
Lea and Febiger.

Fontham, E.T., Pickle, L.W., Haenszel, W., Correa, P., Lin, Y.P., and Falk, R.T. 1988. Dietary vitamins A and C and lung cancer risk in Louisiana. *Cancer* 62:2267–2273.

Food and Drug Administration. 1986. Listing of D and C Orange No. 17 for use in externally
applied drugs and cosmetics. *Fed Regist* 51:28331.

Forbes, J.F. 1990. Surgery, kinetics and biological considerations in planning adjuvant therapy protocols. *Prog Clin Biol Res* 354A:133–146.

Formica, J.V. 1989. Crown gall neoplasms. In *The Pathobiology of Neoplasia,* A.E. Sirica, ed., pp.
497–512. New York: Plenum.

Fornander, T., Rutqvist, L.E., and Wilking, N. 1991. Effects of tamoxifen on the female genital
tract. *Ann NY Acad Sci* 622:469–476.

Forrester, K., Almoguera, C., Han, K., Grizzle, W.E., and Perucho, M. 1987. Detection of high
incidence of K-*ras* oncogenes during human colon tumorigenesis. *Nature* 327:298–303.

Fosså, S.D., Gunderson, R., and Moe, B. 1990. Recombinant interferon-alpha combined with
prednisone in metastatic renal cell carcinoma. *Cancer* 65:2451–2454.

Foulds, L. 1965. Multiple etiologic factors in neoplastic development. *Cancer Res* 25:1339–1347.

Foulds, L. 1969. *Neoplastic Development,* vol. 1. New York: Academic Press.

Fowler, R.G., Schaaper, R.M., and Glickman, B.W. 1986. Characterization of mutational specificity
within the lacI gene for a mutD5 mutator strain of Escherichia coli defective in 3′-5′ exonuclease
proofreading activity. *J Bacteriol* 167:130–137.

Fox, H., and Buckley, C.H. 1992. The female genital tract and ovaries. In *Oxford Textbook of Pathology,* J.O'D. McGee, P.G. Isaacson, and N.A. Wright, eds., pp. 1563–1639. Oxford: Oxford University Press.

Francke, U., Holmes, L.B., Atkins, L., and Riccardi, V.M. 1979. Aniridia-Wilms' tumor association: Evidence for specific deletion of 11p13. *Cytogenet Cell Genet* 24:185–192.

Fraumeni, J.F., Hoover, R.N., DeVesa, S.S., and Kinlen, L.J. 1989. Epidemiology of cancer. In *Cancer: Principles and Practice of Oncology,* 4th ed., V.T. DeVita, S. Hellman, and S.A. Rosenberg,
eds., pp. 150–181. Philadelphia: J.B. Lippincott.

Fredriksson, M., Bengtsson, N.O., Hardell, L., and Axelson, O. 1989. Colon cancer, physical activity, and occupational exposures. *Cancer* 63:1838–1842.

Frieben, A. 1902. Cancroid des rechten Handruckens nach lang dauerden Einwirkung von Roentgenstrahlen. *Fortschr Geb Rontgenstr* 6:106–108.

Friedman, L.S., Ostermeyer, E.A., Szabo, C.I., Dowd, P., Lynch, E.D., Rowell, S.E., and King, M-C. 1994. Confirmation of *BRCA1* by analysis of germline mutations linked to breast and ovarian cancer in ten families. *Nat Genet* 8:399–404.

Friedman, R.J., Rigel, D.S., Silverman, M.K., Kopf, A.W., and Vossaert, K.A. 1991. Malignant melanoma in the 1990s. *CA Cancer J Clin* 41:201–226.

Friedman, R.M., Grimley, P., and Baron, S. 1996. Biological effects of the interferons and other cytokines. *Biotherapy* 8:189–198.

Friedwald, W.F., and Rous, P. 1950. The pathogenesis of deferred cancer: A study of the after effects of methylcholanthrene upon the rabbit skin. *J Exp Med* 91:459–484.

Friend, S.H., Bernards, R., Rogelj, S., Weinberg, R.A., Rapaport, J.M., Albert, D.M., and Dryja, T.P. 1986. A human DNA segment with properties of the gene that predisposes to retinoblastoma and osteosarcoma. *Nature* 323:643–646.

Friend, S.H., Horowitz, J.M., Gerber, M.R., Wang, X.-F., Bogenmann, E., Li, F.P., and Weinberg, R.A. 1987. Deletions of a DNA sequence in restinoblastomas and mesenchymal tumors: Organization of the sequence and its encoded protein. *Proc Natl Acad Sci USA* 84:9059–9063.

Frisch, R.E., Wyshak, G., Albright, N.L., Albright, T.E., Schiff, I., Jones, K.P., Witschi, J., Shiang, E., Koff, E., and Marguglio, M. 1985. Lower prevalence of breast cancer and cancers of the reproductive system among former college athletes compared to non-athletes. *Br J Cancer* 52:885–891.

Frisch, R.E., Wyshak, G., Albright, N.L., Albright, F.E., Schiff, I., Witschi, J., and Margullo, M. 1987. Lower lifetime occurrence of breast cancer and cancers of the reproductive system among former college athletes. *Am J Clin Nutr* 45:328–335.

Frisch, S.M., and Francis, H. 1994. Disruption of epithelial cell-matrix interactions induces apoptosis. *J Cell Biol* 124:619–626.

Frixen, U.H., Behrens, J., Sachs, M., Eberle, G., Voss, B., Warda, A., Lochner, D., and Birchmeier, W. 1991. E-Cadherin-mediated cell-cell adhesion prevents invasiveness of human carcinoma cells. *J Cell Biol* 113:173–185.

Frost, P., and Levin B. 1992. Clinical implications of metastatic process. *Lancet* 339:1458–1461.

Frye, F.L., Gillespie, D.S., and Maruska, E. 1991. Multifocal fibrosarcoma in a goliath frog *Gigantorana goliath*. In *4. Internationales Colloquium für Pathologie und Therapie der Reptilien und Amphibien*, K. Gabrisch, B. Schildger, and P. Zwart, eds., pp. 177–178. Bad Nauheim: Deutsche Veterinärmedizinische Gesellschaft e.V.

Fuchs, D.A., and Johnson, R.K. 1978. Cytologic evidence that Taxol, an antineoplastic agent from *Taxus brevifolia*, acts as a mitotic spindle poison. *Cancer Treat Rep* 62:1219–1222.

Fujimoto, J., Ichigo, S., Hirose, R., Sakaguchi, H., and Tamaya, T. 1997. Expression of E-cadherin and alpha- and beta-catenin mRNAs in uterine cervical cancers. *Tumor Biol* 18:206–212.

Furth, J. 1953. Conditioned and autonomous neoplasms: A review. *Cancer Res* 13:477–492.

Gaidano, G., and Dalla-Favera, R. 1997. Lymphomas. In *Cancer: Principles and Practice of Oncology*, 5th ed., V.T. DeVita, Jr., S. Hellman, and S.A. Rosenberg, eds., pp. 2131–2145. Philadelphia: Lippincott-Raven.

Gale, N.W., Kaplan, S., Lowenstein, E.J., Schlessinger, J., and Bar-Sagi, D. 1993. Grb2 mediates the EGF-dependent activation of guanine nucleotide exchange on Ras. *Nature* 363:88–92.

Gallagher, R.E., Said, F., Pua, I., Papenhausen, P.R., Paietta, E., and Wiernik, P.H. 1989. Expression of retinoic acid receptor-alpha mRNA in human leukemia cells with variable responsiveness to retinoic acid. *Leukemia* 3:789–795.

Gambarotta, G., Pistoi, S., Giordano, S., Comoglio, P.M., and Santoro, C. 1994. Structure and inducible regulation of the human MET promoter. *J Biol Chem* 269:12852–12857.

Garabrant, D.H., Peters, J.M., Mack, T.M., and Bernstein, L. 1984. Job activity and colon cancer risk. *Am J Epidemiol* 119:1005–1014.

Garbisa, S., Pozzatti, R., Muschel, R.J., Saffiotti, U., Ballin, M., Goldfarb, R.H., Khoury, G., and Liotta, L.A. 1987. Secretion of Type IV collagenolytic protease and metastatic phenotype: Induction by transfection with c-Ha-*ras* but not c-Ha-*ras* plus Ad2-Ela. *Cancer Res* 47:1523–1528.

Garfinkel, D. 1958. Studies on pig liver microsomes: 1. Enzymic and pigment composition of different microsomal fractions. *Arch Biochem Biophys* 77:493–509.

Garland, C., Barrett-Conner, E., Rossof, A.H., Shekelle, R.B., Criqui, M.H., and Paul, O. 1985. Dietary vitamin D and calcium and risks of colorectal cancers: A 19-year prospective study in men. *Lancet* 1:307–309.

Garland, F.C., White, M.R., Garland, C.F., Shaw, E., and Gorham, E.D. 1990. Occupational sunlight exposure and melanoma in the U.S. Navy. *Arch Environ Health* 45:261–267.

Garnick, M. B. 1994. The dilemmas of prostate cancer. *Scientific American* 1994 (April):72–81.

Garro, A.J., and Lieber, C.S. 1990. Alcohol and cancer. *Annu Rev Pharmacol Toxicol* 30:219–249.

Gateff, E. 1981. Malignancies of genetic origin in *Drosophila*. In *Phyletic Approaches to Cancer,* C.J. Dawe, J.C. Harshbarger, S. Kondo, T. Sugimura, and S. Takayama, eds., pp. 311–318. Tokyo: Japan Scientific Societies Press.

Gateff, E. 1994. Tumor-suppressor genes, hematopoietic malignancies and other hematopoietic disorders of *Drosophila melanogaster. Ann NY Acad Sci* 712:260–279.

Gateff, E., Wismar, J., Habtemichael, N., Löffler, T., Dreschers, S., Kaiser, S., and Protin, U. 1996. Functional analysis of *Drosophila* developmental genes instrumental in tumor suppression. *In Vivo* 10:211–216.

Gehring, W. 1968. The stability of the differentiated state in cultures of imaginal discs in *Drosophila*. In *The Stability of the Differentiated State,* H. Ursprung, ed., p. 136. Berlin: Springer-Verlag.

Gelfand, M.J. 1993. Meta-iodobenzylguanidine in children. *Semin Nucl Med* 23:231–242.

Gerhardsson, M., Norell, S.E., Kiviranta, H., Pederson, N.L., and Ahlbom, A. 1986. Sedentary jobs and colon cancer. *Am J Epidemiol* 123:775–780.

Gerwitz, G., and Yallow, R.S. 1974. Ectopic ACTH production in carcinoma of the lung. *J Clin Invest* 53:1022–1032.

Ghelelovitch, S. 1969. Melanotic tumors in *Drosophila melanogaster.* In *Natl Cancer Inst Monogr* 31:263–275.

Ghosh, D. 1993. Status of the transcription factor database (TFD). *Nucleic Acids Res* 21:3117–3118.

Gilbert, S.F. 1991. *Developmental Biology,* 3rd ed. Sunderland, MA: Sinauer.

Gilman, A., and Philips, F.S. 1946. The biological actions and therapeutic applications of the β-chloroethylamines and sulfides. *Science* 103:409–415.

Giovannucci, E., Ascherio, A., Rimm, E.B., Stampfer, M.J., Colditz, G.A., and Willett, W.C. 1995. Intake of carotenoids and retinol in relation to risk of prostate cancer. *J Natl Cancer Inst* 87:1767–1776.

Girard, F., Strausfeld, U., Fernandez, A., and Lamb, N.J. 1991. Cyclin A is required for the onset of DNA replication in mammalian fibroblasts. *Cell* 67:1169–1179.

Glazer, R.I., and Rohlff, C. 1994. Transcriptional regulation of multidrug resistance in breast cancer. *Breast Cancer Res Treat* 31:263–271.

Gnarra, J.R, Tory, K., Weng, Y., Schmidt, L., Wei, M.H., Li, H., Latif, F., Liu, S., Chen, F., Duh,

F.-M., Lubensky, I., Duan, D.R., Florence, C., Pozzatti, R., Walther, M.M., Bander, N.H., Grossman, H.B., Brauch, H., Pomer, S., Brooks, J.D., Isaacs, W.B., Lerman, M.I., Zbar, B., and Linehan, W.M. 1994. Mutations of the VHL tumour suppressor gene in renal carcinoma. *Nat Genet* 7:85–90.

Godin, I., Wylie, C., and Heasman, J. 1990. Genital ridges exert long-range effects on mouse primordial germ cell numbers and direction of migration in culture. *Development* 108:357–363.

Goel, S.C. 1983. Role of cell death in the morphogenesis of the amniote limbs. In *Limb Development and Regeneration, Part A,* J.F. Fallon and A.L. Caplan, eds., pp. 175–182. New York: Alan R. Liss.

Going, J.J., Anderson, T.J., Battersby, S., and Macintyre, C.C.A. 1988. Proliferative and secretory activity in human breast during natural and artificial menstrual cycles. *Am J Pathol* 130:193–204.

Gold, L.S., Slone, T.H., Stern, B.R., Manley, N.B., and Ames, B.N. 1992. Rodent carcinogens: Setting priorities. *Science* 258:261–265.

Goldfarb, M., Shimizu, K., Perucho, M., and Wigler, M. 1982. Isolation and preliminary characterization of a human transforming gene from T24 bladder carcinoma cells. *Nature* 296:404–409.

Goldhirsch, A., and Gelber, R.D. 1996. Endocrine therapies of breast cancer. *Semin Oncol* 23:494–505.

Goldie, J.H., and Coldman, A.J. 1979. A mathematical model for relating the drug sensitivity of tumors to their spontaneous mutation rate. *Cancer Treat Rep* 63:1727–1733.

Goldin, A., and Schabel, F.M. 1981. Cancer concepts derived from animal chemotherapy studies. *Cancer Treat Rep* 65(Suppl):11–19.

Gong, Y.L., Koplan, J.P., Feng, W., Chen, C.H.C., Zheng, P., and Harris, J.R. 1995. Cigarette smoking in China. *J Am Med Assoc* 274:1232–1234.

Goodman, L.S., Wintrobe, M.M., Dameshek, W., Goodman, M.J., Gilman, A., and McLennan, M. 1946. Nitrogen mustard therapy: Use of methylbis (β-chloroethyl) amino hydrochloride for Hodgkin's disease, lymphosarcoma, leukemia, and certain allied and miscellaneous disorders. *J Am Med Assoc* 132:126–132.

Gootwine, E., Webb, C.G., and Sachs, L. 1982. Participation of myeloid leukemia cells injected into embryos in haematopoietic differentiation in adult mice. *Nature* 299:63–65.

Gotoh, N., Tojo, A., Muroya, K., Hashimoto, Y., Hattori, S., Nakamura, S., Takenawa, T., Yazaki, Y., and Shibuya, M. 1994. Epidermal growth factor-receptor mutant lacking the auto-phosphorylation sites induces phosphorylation of Shc protein and Shc-Grb2/ASH association and retains mitogenic activity. *Proc Natl Acad Sci USA* 91:167–171.

Gottesman, M.M., Hrycyna, C.A., Schoenlein, P.V., Germann, U.A., and Pastan, I. 1995. Genetic analysis of the multidrug transporter. *Ann Rev Genet* 29:607–649.

Graff, J.R., Herman, J.G., Lapidus, R.G., Chopra, H., Xu, R., Jarrard, D.F., Isaacs, W.B., Pitha, P.M., Davidson, N.E., and Baylin, S.B. 1995. E-Cadherin expression is silenced by DNA hypermethylation in human breast and prostate carcinomas. *Cancer Res* 55:5195–5199.

Graham, C.F., and Wareing, P.F. 1976. *The Developmental Biology of Plants and Animals.* Philadelphia: W.B. Saunders.

Graham, S., Marshall, J., Haughay, B., Mittelman, A., Swanson, M., Zielezny, M., Byers, T., Wilkinson, G., and West, D. 1988. Dietary epidemiology of cancer of the colon in western New York. *Am J Epidemiol* 128:490–503.

Gramzinski, R.A., Parchment, R.E., and Pierce, G.B. 1990. Evidence linking programmed cell death in the blastocyst to polyamine oxidation. *Differentiation* 43:59–65.

Greaves, M.H., Hariri, G., Newman, R.A., Sutherland, D.R., Ritter, M.A., and Ritz, J. 1983. Selective expression of the common acute lymphoblastic leukemia gp100 antigen on immature lymphoid cells and their malignant counterparts. *Blood* 61:628–639.

Greenberg, E.R., Baron, J.A., Tosteson, T.D., Freeman, D.H., Jr., Beck, G.J., Bond, J.H., Colacchio, T.A., Coller, J.A., Frankl, H.D., Haile, R.W., Mandel, J.S., Nierenberg, D.W., Rothstein, R., Snover, D.C., Stevens, M.M., Summers, R.W., and van Stolk, R.U. 1994. A clinical trial of antioxidant vitamins to prevent colorectal adenoma. *N Engl J Med* 331:141–147.

Greenberger, L.M., Cohen, D., and Horwitz, S.B. 1994. In vitro models of multiple drug resistance. *Cancer Treat Res* 73:69–106.

Greene, H.S.N. 1957. Heterotransplantation of tumors. *Ann NY Acad Sci* 69:818–829.

Greenwald, P., Lanza, E., and Eddy, G.A. 1987. Dietary fiber in the reduction of colon cancer risk. *J Am Diet Assoc* 87:1178–1188.

Grem, J.L. 1990. Fluorinated pyrimidines. In *Cancer Chemotherapy,* B.A. Chabner and J.M. Collins, eds., pp. 180–224. Philadelphia: J.B. Lippincott.

Grenman, S.E., Roberts, J.A., England, B.G., Gronroos, M., and Carey, T.E. 1988. In vitro growth regulation of endometrial carcinoma cells by tamoxifen and medroxyprogesterone acetate. *Gynecol Oncol* 30:239–250.

Grever, M.R., and Grieshaber, C.K. 1993. Toxicology by organ system. In *Cancer Medicine,* vol. 1, J.F. Holland, E. Frei III, R.C. Bast, Jr., D.W. Kufe, D.L. Morton, and R.R. Weichselbaum, eds., pp. 683–697. Philadelphia: Lea and Febiger.

Griffiths, J.D., McKinna, J.A., Rowbotham, H.D., Tsolakidis, P., and Salsbury, A.J. 1973. Carcinoma of the colon and rectum: Circulating malignant cells and five-year survival. *Cancer* 31:226–236.

Grimmer, G., and Misfeld, J. 1983. Environmental carcinogens: A risk for man? Concept and strategy of the identification of carcinogens in the environment. In *Environmental Carcinogens: Polycyclic Aromatic Hydrocarbons,* G. Grimmer, ed., pp. 1–26. Boca Raton: CRC Press.

Griswold, D.P., Jr., Schabel, F.M., Jr., Wilcox, W.S., Simpson-Herren, L., and Skipper, H.E. 1968. Success and failure in the treatment of solid tumors: I. Effects of cyclophosphamide (NSC-26271) on primary and metastatic plasmacytoma in the hamster. *Cancer Chemother Rep* 52:345–387.

Gritz, E.R. 1988. Cigarette smoking: The need for action by health professionals. *CA Cancer J Clin* 38:194–212.

Grobstein, C., and Zwilling, E. 1953. Modification of growth and differentiation of chorio-allantoic grafts of chick blastoderm pieces after culture of a glass slot interface. *J Exp Zool* 122:259–284.

Gronroos, M., Maenpaa, J., Kangas, L., Erkkola, R., Paul, R., and Grenman, S. 1987. Steroid receptors and response of endometrial cancer to hormones in vitro. *Ann Chir Gynaecol Suppl* 202:76–79.

Gross, L. 1970. *Oncogenic Viruses.* Oxford: Pergamon Press.

Gross, L. 1988. Inhibition of the development of tumors or leukemia in mice and rats after reduction of food intake. *Cancer* 62:1463–1465.

Grossman, L. 1997. Epidemiology of ultraviolet-DNA repair capacity and human cancer. *Environ Health Perspect* 105 (Suppl 4):927–930.

Gunduz, N., Fisher, B., and Saffer, E.A. 1979. Effect of surgical removal on the growth and kinetics of residual tumor. *Cancer Res* 39:3861–3865.

Gurney, A.L., Carver Moore, K., de Sauvage, F.J., and Moore, M.W. 1994. Thrombocytopenia in c-mpl-deficient mice. *Science* 265:1445–1447.

Gutman, A., and Wasylyk, B. 1991. Nuclear targets for transcription regulation by oncogenes. *Trends Genet* 7:49–54.

Gutterman, J.U. 1994. Cytokine therapeutics: Lessons from interferon alpha. *Proc Natl Acad Sci USA* 91:1198–1205.

Haas, G.P., and Sakr, W.A. 1997. Epidemiology of prostate cancer. *CA Cancer J Clin* 47:273–287.

Haas-Kogan, D.A., Kogan, S.C., Levi, D., Dazin, P., T'Ang, A., Fung, Y.-K., and Israel, M.A. 1995. Inhibition of apoptosis by the retinoblastoma gene product. *Eur Mol Biol Organ J* 14:461–472.

Haenszel, W. 1963. Variations in skin cancer incidence within the United States. *Natl Cancer Inst Monogr* 10:225–243.

Haffner, R., and Oren, M. 1995. Biochemical properties and biological effects of p53. *Curr Opin Genet Dev* 5:84–90.

Hagag, N., Halegua, S., and Viola, M. 1986. Inhibition of growth factor induced differentiation of PC12 cells by micro-injection of antibody to *ras* p21. *Nature* 319:680–682.

Hahn, K.A., Bravo, L., Adams, W.H., and Frazier, D.L. 1994. Naturally occurring tumors of dogs as comparative models for cancer therapy research. *In Vivo* 8:133–144.

Hai, T., and Curran, T. 1991. Cross-family dimerization of transcription factors fos/jun and ATF/CREB alters DNA-binding specificity. *Proc Natl Acad Sci USA* 88:3720–3724.

Hait, W.N., and Aftab, D.T. 1992. Rational design and preclinical pharmacology of drugs for reversing multidrug resistance. *Biochem Pharmacol* 43:103–107.

Hall, J.M., Lee, M.K., Newman, B., Morrow, J.E., Anderson, L.A., Huey, B., and King, M-C. 1990. Linkage of early-onset familial breast cancer to chromosome 17q21. *Science* 250:1684–1689.

Halsted, W.S. 1894–1895. The results of operations for the cure of cancer of the breast at the Johns Hopkins Hospital from 1889–1894. *Johns Hopkins Hosp Rep* 4:297–350.

Hanks, G.E., Myers, C.E., and Scardino, P.T. 1993. Cancer of the prostate. In *Cancer: Principles and Practice of Oncology*, 4th ed., V.T. DeVita, Jr., S. Hellman, and S.A. Rosenberg, eds., pp. 1073–1113. Philadelphia: J.B. Lippincott.

Hannon, G.J., and Beach, D. 1994. p15INK4 is a potential effector of TGF-beta-induced cell cycle arrest. *Nature* 371:257–261.

Haque, A., and Kanz, M.F. 1988. Asbestos bodies in children's lungs. *Arch Pathol Lab Med* 112:514–518.

Harbour, J.W., Lai, S.-L., Whang-Peng, J., Gazdar, A.F., Minna, J.D., and Kaye, F.J. 1988. Abnormalities in structure and expression of the human retinoblastoma gene in SCLC. *Science* 241:353–357.

Harris, C.C., Hirohashi, S., Ito, N., Pitot, H.C., Sugimura, T., Terada, M., and Yokota, J. 1992. Multistage carcinogenesis: The twenty-second international symposium of the Princess Takamatsu Cancer Research Fund. *Cancer Res* 52:4837–4840.

Harris, H. 1988. The analysis of malignancy in cell fusion: The position in 1988. *Cancer Res* 48:3302–3306.

Harris, R.W.C., Key, T.J.A., Silcocks, P.B., Bull, D., and Wald, N.J. 1991. A case control study of dietary carotene in men with lung cancer and men with other epithelial cancers. *Nutr Cancer* 15:63–68.

Harrisson, F., Andries, L., and Vakaet, L. 1988. The chick blastoderm: Current views on cell biological events guiding intercellular communication. *Cell Diff* 22:83–106.

Harshbarger, J.C., Charles, A.M., and Spero, P.M. 1981. Collection and analysis of neoplasms in sub-homeothermic animals from a phyletic point of view. In *Phyletic Approaches to Cancer*, C.J. Dawe, J.C. Harshbarger, S. Kondo, T. Sugimura, and S. Takayama, eds., pp. 357–384. Tokyo: Japan Scientific Societies Press.

Hart, B.L., Mettler, F.A., Jr., and Harley, N.H. 1989. Radon: Is it a problem? *Radiology* 172:593–599.

Hartwell, L. 1992. Defects in a cell cycle checkpoint may be responsible for the genomic instability of cancer cells. *Cell* 71:543–546.

Haseltine, W.A. 1983. Ultraviolet light repair and mutagenesis revisited. *Cell* 33:13–17.

Haskell, C.M., ed. 1990. *Cancer Treatment,* 3rd ed. Philadelphia: W.B. Saunders.

Haskell, C.M. 1990. Principles and practice of cancer chemotherapy. In *Cancer Treatment,* 3rd ed., C.M. Haskell, ed., pp. 21–43. Philadelphia: W.B. Saunders.

Haskell, C.M., Giuliano, A.E., Thompson, R.W., and Zarem, H.A. 1990. Breast cancer. In *Cancer Treatment,* 3rd ed., C.M. Haskell, ed., pp. 123–164. Philadelphia: W.B. Saunders.

Haskell, C.M., Selch, M.T., and Ramming, K.P. 1990a. Colon and rectum. In *Cancer Treatment,* 3rd ed., C.M. Haskell, ed., pp. 232–254. Philadelphia: W.B. Saunders.

Haskell, C.M., Selch, M.T., and Ramming, K.P. 1990b. Stomach. In *Cancer Treatment,* 3rd ed., C.M. Haskell, ed., pp. 217–231. Philadelphia: W.B. Saunders.

Hassan, H.T., and Zander, A. 1996. Stem cell factor as a survival and growth factor in human normal and malignant hematopoiesis. *Acta Haematologica* 95:257–262.

Haut, M., Steeg, P.S., Willson, J.K.V., and Markowitz, S.D. 1991. Induction of *nm23* gene expression in human colonic neoplasms and equal expression in colon tumors of high and low metastatic potential. *J Natl Cancer Inst* 83:712–716.

Hawley-Nelson, P., Vousden, K.H., Hubbert, N.L., Lowry, D.R., and Schiller, J.T. 1989. HPV16 E6 and E7 proteins cooperate to immortalize human foreskin keratinocytes. *EMBO J* 8:3905.

Hayakawa, K., Morita, T., Augustus, L.B., von Eschenbach, A.C., and Itoh, K. 1992. Human renal-cell carcinoma cells are able to activate natural killer cells. *Int J Cancer* 51:290–295.

Hayashi, Y., Nagao, M., Sugimura, T., Takayama, S., Tomatis, L., Wattenberg, L.W., and Wogan, G.N., eds. 1986. Preface. *Int Symp Princess Takamatsu Cancer Res Fund* 16:vii.

Hayle, A.J., Darling, D.L., Taylor, A.R., and Tarin, D. 1993. Transfection of metastatic capability with total genomic DNA from human and mouse metastatic tumours. *Differentiation* 54:177–189.

Hebert, J.R., and Kabat, G.C. 1991. Implications for cancer epidemiology of differences in dietary intake associated with alcohol consumption. *Nutr Cancer* 15:107–119.

Hecker, E. 1967. Phorbol esters from croton oil: Chemical nature and biological activities. *Naturwissenschaften* 54:282–284.

Heckler, M.M. 1984. Quoted in a news article in the *NIH Record* 36:1–12.

Hedrick, R.P., and Sano, T. 1989. Herpesviruses of fishes. In *Viruses of Lower Vertebrates,* W. Ahne and E. Kurstak, eds., pp. 161–170. Berlin: Springer-Verlag.

Hei, T.K., Piao, C.Q., He, Z.Y., Vannais, D., and Waldren, C.A. 1992. Chrysotile fiber is a strong mutagen in mammalian cells. *Cancer Res* 52:6305–6309.

Heidelberger, C. 1973. Pyrimidine and pyrimidine nucleosides. In *Cancer Medicine,* vol. 1, J.F. Holland, E. Frei III, R.C. Bast, Jr., D.W. Kufe, D.L. Morton, and R.R. Weichselbaum, eds., pp. 768–791. Philadelphia: Lea and Febiger.

Heisterkamp, N., Jenster, G., ten Hoeve, J., Zovich, D., Pattengale, P.K., and Groffen, J. 1990. Acute leukemia in *bcr/abl* transgenic mice. *Nature* 344:251–253.

Held, W., Acha-Orbea, H., MacDonald, H.R., and Waanders, G.A. 1994. Superantigens and retroviral infection: Insights from mouse mammary tumor virus. *Immunol Today* 15:184–190.

Helgason, S., Wilking, N., Carlstrom, K., Damber, M.G., and von Schoultz, B. 1982. A comparative study of the estrogenic effects of tamoxifen and 17 beta-estradiol in postmenopausal women. *J Clin Endocrinol Metab* 54:404–408.

Heller, I. 1930. Occupational cancers. *J Ind Hyg* 12:169–197.

Hemminki, K., ed. 1994. DNA adducts: Identification and biological significance. *IARC Scientific Publication 125,* International Agency for Research on Cancer, Lyon.

Henderson, B.E., Bernstein, L., and Ross, R. 1997. Etiology of cancer: Hormonal factors. In *Cancer, Principles and Practice of Oncology,* 5th ed., V.T. DeVita, Jr., S. Hellman, and S.A. Rosenberg, eds., pp. 219–229. Philadelphia: Lippincott-Raven.

Hengartner, M.O., and Horvitz, H.R. 1994. Programmed cell death in *Caenorhabditis elegans. Curr Opin Genet Dev* 4:581–586.

Hennekens, C.H., Buring, J.E., Manson, J.E., Stampfer, M., Rosner, B., Cook, N.R., and Belanger, C. 1996. Lack of effect of long-term supplementation with beta carotene on the incidence of malignant neoplasms and cardiovascular disease. *N Engl J Med* 334:1145–1149.

Hennessy, C., Henry, J.A., May, F.E.B., Westley, B.R., Angus, B., and Lennard, T.W.J. 1991. Expression of the antimetastatic gene *nm*23 in human breast cancer: An association with good prognosis. *J Natl Cancer Inst* 83:281–285.

Henson, D.E., Block, G., and Levine, M. 1991. Ascorbic acid: Biologic functions and relation to cancer. *J Natl Cancer Inst* 83:547–550.

Heppner, G.H. 1982. Tumor subpopulation interactions. In *Tumor Cell Heterogeneity: Origins and Implications,* A.H. Owens, D.S. Coffey, and S.B. Baylin, eds., pp. 225–236. New York: Academic Press.

Heppner, G.H. 1993. Cancer cell societies and tumor progression. *Stem Cells* 11:199–203.

Heppner, G.H., Dexter, D.L., DeNucci, T., Miller, F.R., and Calabresi, P. 1978. Heterogeneity in drug sensitivity among tumor cell subpopulations of a single mouse mammary tumor. *Cancer Res* 38:3758–3763.

Herman, J.R., Bhartia, P.K., Ziemke, J., Ahmad, J., and Larko, D. 1996. UV-B increases 1979–1992 from decreases in total ozone. *Geophys Res Lett* 23:2117–2120.

Hermann, G.G., Geertsen, P.F., von der Maase, H., and Zeuthen, J. 1991. Interleukin-2 dose, blood monocyte and CD25+ lymphocyte counts as predictors of clinical response to interleukin-2 therapy in patients with renal cell carcinoma. *Cancer Immunol Immunother* 34:111–114.

Herrero, R., Brinton, L.A., Reeves, W.C., Brenes, M.M., Tenorio, F., de Britton, R.C., Gaitan, E., Garcia, M., and Rawles, W.E. 1990. Sexual behavior, venereal diseases, hygiene practices, and invasive cervical cancer in a high-risk population. *Cancer* 65:380–386.

Higginson, J., Muir, C.S., and Muñoz, N. 1992. *Human Cancer: Epidemiology and Environmental Causes.* Cambridge: Cambridge University Press.

Higinbotham, K.G., Rice, J.M., Reed, C.D., Watatani, M., Enomoto, T., Anderson, L.M., and Perantoni, A.O. 1996. Variant mutational activation of the K-ras oncogene in renal mesenchymal tumors induced in newborn F344 rats by methyl(methoxymethyl)nitrosamine. *Carcinogenesis,* in press.

Hill, M.J. 1974. Bacteria and the etiology of colonic cancer. *Cancer* 34(Suppl):815–818.

Hinds, P.W. 1995. The retinoblastoma tumor suppressor protein. *Curr Opin Genet Dev* 5:79–83.

Hirano, T., Manabe, T., and Takeuchi, S. 1993. Serum cathepsin B levels and urinary excretion of cathepsin B in the cancer patients with remote metastasis. *Cancer Lett* 70:41–44.

Hirayama, T. 1979. Diet and cancer. *Nutr Cancer* 1:67–81.

Hocman, G. 1988. Chemoprevention of cancer: Selenium. *Int J Biochem* 20:123–132.

Hoffman, E.C., Reyes, H., Chu, F.F., Sander, F., Conley, L.H., Brooks, B.A., and Hankinson, O. 1991. Cloning of a factor required for activity of the Ah dioxin receptor. *Science* 252:954–958.

Hoffman, S.A., Pashkis, K.E., DeBias, D.A., Cantarow, A., and Williams, T.L. 1962. The influence of exercise on the growth of transplanted rat tumors. *Cancer Res* 22:597–599.

Holaday, D.A. 1969. History of the exposure of miners to radon. *Health Phys* 16:547–552.

Holland, J.F., Frei, E., III, Bast, R.C., Jr., Kufe, D.W., Morton, D.L., and Weichselbaum, R.R., eds. 1993. *Cancer Medicine*. Philadelphia: Lea and Febiger.

Holman, C.D.J., Armstrong, B.K., and Heenan, P.J. 1986. Relationship of cutaneous malignant melanoma to individual sunlight exposure habits. *J Natl Cancer Inst* 76:403–414.

Holmes, E.C., Livingston, R., and Turrisi, A., III. 1993. Neoplasms of the thorax. In *Cancer Medicine,* vol. 2, J.F. Holland, E. Frei III, R.C. Bast, Jr., D.W. Kufe, D.L. Morton, and R.R. Weichselbaum, eds., pp. 1285–1337. Philadelphia: Lea and Febiger.

Holt, J.T., Thompson, M.E., Szabo, C., Robinson-Benion, C., Arteaga, C.L., King, M-C, and Jensen, R.A. 1996. Growth retardation and tumour inhibition by *BRCA1*. *Nat Genet* 12:296–302.

Honegger, A.M., Schmidt, A., Ullrich, A., and Schlessinger, J. 1990. Evidence for epidermal growth factor (EGF)-induced intermolecular autophosphorylation of the EGF receptors in living cells. *Mol Cell Biol* 10:4035–4044.

Hopwood, D., and Levison, D.A. 1976. Atrophy and apoptosis in the cyclical human endometrium. *J Pathol* 119:159–166.

Horowitz, J.M., Yandell, D.W., Park, S.H., Canning, S., Whyte, P., Buchkovich, K., Harlow, E., Weinberg, R.A., and Dryja, T.P. 1989. Point mutational activation of the retinoblastoma antioncogene. *Science* 243:937–940.

Horwitz, K.B., Koseki, Y., and McGuire, W.L. 1978. Estrogen control of progesterone receptor in human breast cancer: Role of estradiol and antiestrogen. *Endocrinology* 103:1742–1751.

Hostetler, L.W., Romerdahl, C.A., and Kripke, M.L. 1989. Specificity of antigens on UV radiation-induced antigenic tumor cell variants measured *in vitro* and *in vivo*. *Cancer Res* 49:1207–1213.

Houlston, R.S., McCarter, E., Parbhoo, S., Scurr, J.H., and Slack, J. 1992. Family history and risk of breast cancer. *J Med Genet* 29:154–157.

Howard, E.B., Britt, J.O., Jr., and Simpson, J.G. 1983. Neoplasms in marine mammals. In *Pathobiology of Marine Mammal Diseases,* vol. 2, E.B. Howard, ed., p. 146. Boca Raton: CRC Press.

Howe, G., Rohan, T., Decarli, A., Iscovich, J., Kaldor, J., Katsoyanni, K., Marubini, E., Miller, A., Riboli, E., Toniolo, P., and Trichopoulos, D. 1991. The association between alcohol and breast cancer risk: Evidence from combined analysis of six dietary case-control studies. *Int J Cancer* 47:707–710.

Howley, P.M., Ganem, D., and Kieff, E. 1997. Etiology of cancer: Viruses. Section 2, DNA viruses. In *Cancer: Principles and Practice of Oncology,* 5th ed., V.T. DeVita, Jr., S. Hellman, and S.A. Rosenberg, eds., pp. 168–184. Philadelphia: Lippincott-Raven.

Hsu, I.C., Metcalf, R.A., Sun, T., Welsh, J., Wang, N.J., and Harris, C.C. 1991. P53 gene mutational hotspot in human hepatocellular carcinomas from Qidong, China. *Nature* 350:427–428.

Hsu, T.C. 1979. *Human and Mammalian Cytogenetics*. New York: Springer-Verlag.

Huang, H.J.S., Lee, J.K., Shew, J.Y., Chen, P.L., Bookstein, R., Friedmann, T., Lee, E.Y.-H.P., and Lee, W.-H. 1988. Suppression of the neoplastic phenotype by replacement of the RB gene in human cancer cells. *Science* 242:1563–1566.

Huang, J.C., Svoboda, D.L., Reardon, J.T., and Sancar, A. 1992. Human nucleotide excision nuclease removes thymine dimers from DNA by incising the 22nd phosphodiester bond 5′ and the 6th phosphodiester bond 3′ to the photodimer. *Proc Natl Acad Sci USA* 89:3664–3668.

Huang, M.E., Ye, Y.C., Chen, S.R., Chai, J.R., Lu, J.X., Zhoa, L., Gu, L.J., and Wang, Z.Y. 1988. Use of all-trans retinoic acid in the treatment of acute promyelocytic leukemia. *Blood* 72:567–572.

Huff, J.E., McConnell, E.E., Haseman, J.K., Boorman, G.A., Eustis, S.L., Schwetz, B.A., Rao, G.N., Jameson, C.W., Hart, L.G., and Rall, D.P. 1988. Carcinogenesis studies: Results of 398 experiments on 104 chemicals from the U.S. National Toxicology Program. *Ann NY Acad Sci* 534:1–30.

Huggins, C.B., Grand, L.C., and Brillantes, F.P. 1961. Mammary cancer induced by a single feeding of polynuclear hydrocarbons and its suppression. *Nature* 189:204–207.

Huggins, C.B., and Hodges, C.V. 1941. Studies on prostate cancer: I. The effect of castration, of estrogen and of androgen injection on serum phosphatases in metastatic carcinoma of the prostate. *Cancer Res* 1:293–297.

Hunter, B.R., Tweedell, K., and McKinnell, R.G. 1990. PNKT4B cells exhibit motility at invasion-restrictive temperatures. *Proc Am Assoc Cancer Res* 31:66.

Hunter, T. 1991. Cooperation between oncogenes. *Cell* 64:249–270.

Hunter, T., and Sefton, B.M. 1980. The transforming gene product of Rous sarcoma virus phosphorylates tyrosine. *Proc Natl Acad Sci USA* 77:1311–1315.

Huxley, J. 1958. *Biological Aspects of Cancer.* New York: Harcourt, Brace.

Hyde, J.N. 1906. On the influence of light in the production of cancer of the skin. *Am J Med Sci* 131:1–22.

IARC Monographs on the Evaluation of Carcinogenic Risks to Humans: Solar and Ultraviolet Radiation. 1992. United Kingdom, IARC Vol. 55.

Ikeda, K., Sasaki, K., Tasaka, T., Nagai, M., Kawanishi, K., Takahara, J., and Irino, S. 1994. PML-RAR alpha fusion transcripts by RNA PCR in acute promyelocytic leukemia in remission and its correlation with clinical outcome. *Int J Hematol* 60:197–205.

Ikenaga, M. 1994. Cytotoxic action of alkylating agents in human tumor cells and its relationship to apoptosis. *Gann Tokyo Kagaku Ryoho* 21:596–601.

Isaacs, J.T., and Coffey, D.S. 1981. Adaptation versus selection as the mechanism responsible for the relapse of prostatic cancer to androgen ablation therapy as studied in the Dunning R-3327-H adenocarcinoma. *Cancer Res* 41:5070–5075.

Isaacs, J.T., Heston, W.D., Weissman, R.M., and Coffey, D.S. 1978. Animal models of the hormone-sensitive and -insensitive prostatic adenocarcinomas, Dunning R-3327-H, R-3327-HI, and R-3327-AT. *Cancer Res* 38:4353–4359.

Isaacs, J.T., Wake, N., Coffey, D.S., and Sandberg, A.A. 1982. Genetic instability coupled to clonal selection as a mechanism for tumor progression in the Dunning R-3327 rat prostatic adenocarcinoma system. *Cancer Res* 42:2353–2371.

Isobe, M., Emanuel, B.S., Givol, D., Oren, M., and Croce, C.M. 1986. Localization of gene for human p53 tumour antigen to band 17p13. *Nature* 320:84–85.

Issemann, I., and Green, S. 1990. Activation of a member of the steroid receptor superfamily by peroxisome proliferators. *Nature* 347:645–650.

Jackson, P., and Baltimore, D. 1989. N-terminal mutations activate the leukemogenic potential of the myristoylated form of c-*abl*. *Eur Mol Biol Organ J* 8:6649–6653.

Jacob, M., Christ, B., and Jacob, H.J. 1978. On the migration of myogenic stem cells into the prospective wing region of chick embryos. *Anat Embryol* 153:179–193.

Jacob, M., Christ, B., and Jacob, H.J. 1979. The migration of myogenic cells from the somites into the leg region of avian embryos. *Anat Embryol* 157:219–309.

Jacobs, K., Shoemaker, C., Rudersdorf, R., Neill, S.D., Kaufman, R.J., Musson, A., Seehra, J., Jones, S.S., Hewick, R., Fritsch, E.F., Kawakita, M., Shimivu, T., and Miyake, T. 1985. Isolation and characterization of genomic and cDNA clones of human erythropoietin. *Nature* 313:806–810.

Janick-Buckner, D., Barua, A.B., and Olson, J.A. 1991. Induction of HL-60 cell differentiation by water soluble and nitrogen-containing conjugates of retinoic acid and retinol. *Fed Am Soc Exp Biol J* 5:320–325.

Jarrett, W.F.H., Crawford, E.M., Martin, W.B., and Davie, F. 1964. A virus-like particle associated with leukemia lymphosarcoma. *Nature* 202:567–568.

Jaurand, M.C. 1989. Particulate-state carcinogenesis: A survey of recent studies on the mechanisms of action of fibres. In *Nonoccupational Exposure to Mineral Fibres,* J. Bignon, J. Peto, and R. Saraccik, eds., pp. 54–73. IARC Scientific Publication 90. Lyon: International Agency for Research on Cancer.

Jelkmann, W., and Metzen, E. 1996. Erythropoietin in the control of red cell production. *Anatomischer Anzeiger* 178:391–403.

Jensen, O.M. 1989. Dietary fiber carbohydrate and cancer: Epidemiological evidence. In *Diet and the Aetiology of Cancer,* A.B. Miller, ed., pp. 21–29. Berlin: Springer-Verlag.

Jerina, D.M., Chadha, A., Cheh, A.M., Schurdak, M.E., Wood, A.W., and Sayer, J.M. 1991. Covalent bonding of bay-region diol epoxides to nucleic acids. *Adv Exp Med Biol* 283:533–553.

Jewett, H.J. 1970. The case for radical perineal prostatectomy. *J Urol* 103:195–199.

Jhappan, C., Stahle, C., Harkins, R.N., Fausto, N., Smith, G.H., and Merlino, G.T. 1990. TGFα overexpression in transgenic mice induces liver neoplasia and abnormal development of the mammary gland and pancreas. *Cell* 61:1137–1146.

Jiang, R., Kato, M., Bernfield, M., and Grabel, L.B. 1995. Expression of syndecan-1 changes during the differentiation of visceral and parietal endoderm from murine F9 teratocarcinoma cells. *Differentiation* 59:225–233.

Jiang, W., Kahn, S.M., Tomita, N., Zhang, Y.-J., Lu, S.-H., and Weinstein, I.B. 1992. Amplification and expression of the human cyclin D gene in esophageal cancer. *Cancer Res* 52:2980–2983.

Jimenez, J.J., and Yunis, A.A. 1987. Tumor cell rejection through terminal cell differentiation. *Science* 238:1278–1280.

Johnson, I.S., Armstrong, J.G., and Gorman, M. 1963. The vinca alkaloids: A new class of oncolytic agents. *Cancer Res* 23:1390–1427.

Johnson, I.S., Wright, H.F., and Svoboda, G.H. 1960. Antitumor principles derived from *Vinca rosea* Linn: I. Vincaleukoblastine and leurosine. *Cancer Res* 20:1016–1022.

Johnson, R.L., Rothman, A.L., Xie, J., Goodrich, L.V., Bare, J.W., Bonifas, J.M., Quinn, A.G., Myers, R.M., Cox, D.R., Epstein, E.H., Jr., and Scott, M.P. 1996. Human homolog of patched, a candidate gene for the basal cell nevus syndrome. *Science* 272:1668–1671.

Johnston,T.P., McCaleb, G.S., and Montgomery, J.A. 1963. The synthesis of antineoplastic agents: XXXII. N-nitroso-ureas. *J Med Chem* 6:669–681.

Jonasson, J., Povey, S., and Harris, H. 1977. The analysis of malignancy by cell fusion: VII. Cytogenetic analysis of hybrids between malignant and diploid cells and of tumours derived from them. *J Cell Sci* 24:217–254.

Jordan, V.C. 1982. Metabolites of tamoxifen in animals and man: Identification, pharmacology, and significance. *Breast Cancer Res Treat* 2:123–138.

Jordan, V.C. 1988. Long-term tamoxifen therapy to control or to prevent breast cancer: Laboratory concept to clinical trials. *Prog Clin Biol Res* 262:105–123.

Jordan, V.C., Gottardis, M.M., and Satyaswaroop, P.G. 1991. Tamoxifen-stimulated growth of human endometrial carcinoma. *Ann NY Acad Sci* 622:439–446.

Jordan, V.C., and Robinson, S.P. 1987. Species-specific pharmacology of antiestrogens: Role of metabolism. *Fed Proc* 46:1870–1874.

Jostes, R.F. 1996. Genetic, cytogenetic, and carcinogenic effects of radon: A review. *Mutat Res* 340:125–139.

Kadlubar, F.F., Miller, J.A., and Miller, C.E. 1977. Hepatic microsomal N-glucuronidation and nucleic acid binding of N-hydroxy arylamines in relation to urinary bladder carcinogenesis. *Cancer Res* 37:805–814.

Kalil, M., and Hildebrandt, A.C. 1981. Pathology and distribution of plant tumors. In *Neoplasms – Comparative Pathology of Growth in Animals, Plants, and Man,* H.E. Kaiser, ed., pp. 813–821. Baltimore: Williams and Wilkins.

Kamimura, M., Kotani, M., and Yamagata, K. 1980. The migration of presumptive germ cells through the endodermal cell mass in *Xenopus laevis:* A light and electron microscopic study. *J Embryol Exp Morph* 59:1–17.

Kaplan, H.S. 1981. Progress in radiation therapy of cancer. In *Cancer: Achievements, Challenges, and Prospects for the 1980s,* J.H. Burchenal and H.F. Oettgen, eds., vol. 2, pp. 829–857. New York: Grune and Stratton.

Kaplan, K.B., Bibbins, K.B., Swedlow, J.R., Arnaud, M., Morgan, D.O., and Varmus, H.E. 1994. Association of the amino-terminal half of c-Src with focal adhesions alters their properties and is regulated by phosphorylation of tyrosine 527. *Eur Mol Biol Organ J* 13:4745–4756.

Kasai, H., Crain, P.F., Kuchino, Y., Nishimura, S., Ootsuyama, A., and Tanooka, H. 1986. Formation of 8-hydroxyguanine moiety in cellular DNA by agents producing oxygen radicals and evidence for its repair. *Carcinogenesis* 7:1849–1851.

Kashishian, A., Kazlauskas, A., and Cooper, J.A. 1992. Phosphorylation sites in the PDGF receptor with different specificities for binding GAP and PI3 kinase in vivo. *Eur Mol Biol Organ J* 11:1373–1382.

Kastan, M.B., Onyekwere, O., Sidransky, D., Vogelstein, B., and Craig, R.W. 1991. Participation of p53 protein in the cellular response to DNA damage. *Cancer Res* 51:6301–6311.

Kastan, M.B., Zhan, Q., El-Deiry, W.S., Carrier, F., Jacks, T., Walsh, W.V., Plunkett, B.S., Vogelstein, B., and Fornace, A.J., Jr. 1992. A mammalian cell cycle checkpoint pathway utilizing p53 and GADD45 is defective in ataxia-telangiectasia. *Cell* 71:587–597.

Kato, K., Yoshida, J., Mizuno, M., Sugita, K., and Emi, N. 1994. Retroviral transfer of herpes simplex thymidine kinase gene into glioma cells causes targeting of gancyclovir cytotoxic effect. *Neurol Med Chir Tokyo* 34:339–344.

Katzenellenbogen, B.S., Norman, M.J., Eckert, R.L., Peltz, S.W., and Mangel, W.F. 1984. Bioactivities, estrogen receptor interactions, and plasminogen activator-inducing activities of tamoxifen and hydroxy-tamoxifen isomers in MCF-7 human breast cancer cells. *Cancer Res* 44:112–119.

Kaushansky, K., Lok, S., Holly, R.D., Broudy, V.C., Lin, N., Bailey, M.C., Forstrom, J.W., Buddle, M.M., Oort, P.J., Hagen, F.S., Roth, G.J., Papayannopoulou, T., and Foster, D.C. 1994. Promotion of megakaryocyte progenitor expansion and differentiation by the c-Mpl ligand thrombopoietin. *Nature* 369:568–571.

Kawai, S., and Hanafusa, H. 1971. The effects of reciprocal changes in temperature on the transformed state of cells infected with a Rous sarcoma virus mutant. *Virology* 46:470–479.

Kedes, D.H., Operskalski, E., Busch, M., Kohn, R., Flood, J., and Ganem, D. 1996. The seroepidemiology of human herpesvirus 8 (Kaposi's sarcoma-associated herpesvirus): Distribution of infection in KS risk groups and evidence for sexual transmission. *Nature Med* 2:918–924.

Kellie, S., Horvath, A.R., Felice, G., Anand, R., Murphy, C., and Westwick, J. 1993. The interaction of the tyrosine kinase pp60[src] with membrane and cytoskeletal components. *Symp Soc Exp Biol* 47:267–282.

Kelly, M.G., O'Gara, R.W., Adamson, R.H., and Gadekar, K. 1966. Induction of hepatic cell carcinomas in monkeys with N-nitrosodiethylamine. *J Natl Cancer Inst* 36:323–351.

Kennaway, E.L. 1924. On the cancer-producing factor in tar. *Br Med J* 1:564–567.

Kern, S.E., and Vogelstein, B. 1991. Genetic alterations in colorectal tumors. In *Origins of Human Cancer,* J. Brugge, T. Curran, E. Harlow, and F. McCormick, eds., pp. 577–585. Plainview, NY: Cold Spring Harbor Press.

Kerr, A. 1980. Biological control of crown gall through production of agrocin 84. *Plant Dis* 4:24–30.

Kerr, J.B., and McElroy, C.T. 1993. Evidence for large upward trends of ultraviolet-B radiation linked to ozone depletion. *Science* 262:1032–1034.

Kerr, J.F., and Searle, J. 1973. Deletion of cells by apoptosis during castration-induced involution of the rat prostate. *Virchows Arch B* 13:87–102.

Kerr, J.F.R., Wyllie, A.H., and Currie, A.R. 1972. Apoptosis: A basic biological phenomenon with wide-ranging implications in tissue kinetics. *Br J Cancer* 26:239–257.

Kessler, I.I. 1976. Human cervical cancer as a venereal disease. *Cancer Res* 36:783–791.

Kim, B. 1996. Biological therapy: Interferons, interleukins, and monoclonal antibodies. In *Cancer Management: A Multidisciplinary Approach,* R. Pazdur, L.R. Coia, W.J. Hoskins, and L.D. Wagman, eds., pp. 581–592. Huntington, NY: PRR.

Kimbrough, R.D. 1983. Determining exposure and biochemical effects in human population studies. *Environ Health Perspect* 48:77–79.

Kimura, N., Shimazada, N., Nomura, K., and Watanabe, K. 1990. Isolation and characterization of cDNA clone encoding rat nucleoside diphosphate kinase. *J Biol Chem* 265:15744–15749.

Kimura, T., and Yoshimizu, M. 1988. Salmon herpesvirus: OMV, *Oncorhynchus masou virus.* From an international symposium of tumors of lower vertebrates, Munich, August 22–25, 1988.

King, D.R. 1979. *Banding studies of chromosomes from cultured leopard frog blood.* Thesis. University of Minnesota, Minneapolis.

King, M.C., Rowell, S., and Love, S.M. 1993. Inherited breast and ovarian cancer. What are the risks? What are the choices? *J Am Med Assoc* 269:1975–1980.

King, T.J., and McKinnell, R.G. 1960. An attempt to determine the developmental potentialities of the cancer cell nucleus by means of transplantation. *Cell Physiology of Neoplasia,* pp. 591–617. Austin: University of Texas Press.

Kinzler, K.W., and Vogelstein, B. 1996. Lessons from hereditary colorectal cancer. *Cell* 87:159–170.

Kirkby, D.R.S. 1965. The "invasiveness" of the trophoblast. In *The Early Conceptus: Normal and Abnormal,* pp. 68–73. London: E and S Livingstone.

Kiyohara, C., Hirohata, T., and Inutsuka, S. 1996. The relationship between aryl hydrocarbon hydroxylase and polymorphisms of the CYP1A1 gene. *Jpn J Cancer Res* 87:18–24.

Kizaka-Kondoh, S., Matsuda, M., and Okayama, H. 1996. CrkII signals from epidermal growth factor to Ras. *Proc Natl Acad Sci USA* 93:12177–12182.

Klein, E.A. 1996. Hormone therapy for prostate cancer: A topical perspective. *Urology* 47(1A Suppl):3–12.

Kleinsmith, L.J., and Pierce, G.B. 1964. Multipotentiality of single embryonal carcinoma cells. *Cancer Res* 24:1544–1551.

Klingenberg, M. 1958. Pigments of rat liver microsomes. *Arch Biochem Biophys* 75:376–386.

Knudson, A.G. 1971. Mutation and cancer: Statistical study of retinoblastoma. *Proc Natl Acad Sci USA* 68:820–823.

Knudson, A.G., Jr. 1985. Hereditary cancer, oncogenes, and antioncogenes. *Cancer Res* 45:1437–1443.

Knudson, A.G. 1993. Antioncogenes and human cancer. *Proc Natl Acad Sci USA* 90:10914–10921.

Knudson, A.G., and Strong, L.C. 1972. Mutation and cancer: A model for Wilms' tumor of the kidney. *J Natl Cancer Inst* 48:313–324.

Knudson, C.M., Tung, K., Brown, G., and Korsmeyer, S. 1995. Bax deficient mice demonstrate lymphoid hyperplasia but male germ cell death. *Science* 270:96–99.

Kohn, E.C., and Liotta, L.A. 1995. Molecular insights into cancer invasion: Strategies for prevention and intervention. *Cancer Res* 55:1856–1862.

Koizumi, S., Curt, G.A., Fine, R.L., Griffin, J.D., and Chabner, B.A. 1985. Formation of methotrexate polyglutamates in purified myeloid precursor cells from normal human bone marrow. *J Clin Invest* 75:1008–1011.

Kolata, G. 1987. Dietary fat – breast cancer link questioned. *Science* 235:436.

Konishi, N., Hiasa, Y., Matsuda, H., Tao, M., Tsuzuki, T., Hayashi, I., Kitahori, Y., Shiraishi, T., Yatani, R., Shimazaki, J., and Lin, J.C. 1995. Intratumor cellular heterogeneity and alterations in ras oncogene and p53 tumor suppressor gene in human prostate carcinoma. *Am J Pathol* 147:1112–1122.

Kononov, M.E., Bassuner, B., and Gelvin, S.B. 1997. Integration of T-DNA binary vector 'backbone' sequences into the tobacco genome: Evidence for multiple complex patterns of integration. *Plant J* 11:945–957.

Kottler, M.J. 1974. From 48 to 46: Cytological technique, preconception, and the counting of human chromosomes. *Bull Hist Med* 48:465–502.

Kouri, R.E., McKinney, C.E., Slomiany, D.J., and Snodgrass, D.R. 1982. Positive correlation between high aryl hydrocarbon hydroxylase activity and primary lung cancer as analyzed in cryopreserved lymphocytes. *Cancer Res* 42:5030–5037.

Koury, S.T., Bondurant, M.C., and Koury, M.J. 1988. Localization of erythropoietin synthesizing cells in murine kidneys by in situ hybridization. *Blood* 71:524–527.

Kraemer, K.H., Lee, M.M., and Scotto, J. 1987. Xeroderma pigmentosum. Cutaneous, ocular, and neurologic abnormalities in 830 published cases. *Arch Dermatol* 123:241–250.

Kraljic, I., Kovacic, K., and Tarle, M. 1994. Serum TPS, PSA, and PAP values in relapsing stage D2 adenocarcinoma of the prostate. *Urol Res* 22:329–332.

Kraus, R.S., Daston, D.L., Caspary, W.J., and Eling, T.E. 1986. Peroxidase-mediated metabolic activation of the thyroid carcinogen amitrole. *Proc Am Assoc Cancer Res* 27:112.

Krek, W., Xu, G., and Livingston, D.M. 1995. Cyclin A-kinase regulation of E2F-1 DNA binding function underlies suppression of an S phase checkpoint. *Cell* 83:1149–1158.

Kripke, M.L. 1986. Photoimmunology: The first decade. *Curr Probl Dermatol* 15:164–175.

Kripke, M.L. 1990. Effects of UV radiation on tumor immunity (Review). *J Natl Cancer Inst* 82:1392–1396.

Kripke, M.L., and Yarosh, D.B. 1994. Repair of UV-damaged genes by topical application of liposomes containing T4 endonuclease. V: Implications for skin cancer. *Proceedings of the Eighth International Conference of the International Society of Differentiation ISD,* Hiroshima, Japan, pp. 9–12.

Kripke, M.L., Cox, P.A., Alas, L.G., and Yarosh, D.B. 1992. Pyrimidine dimers in DNA initiate systemic immunosuppression in UV-irradiated mice. *Proc Natl Acad Sci USA* 98:7516–7520.

Kritchevsky, D. 1990. Influence of caloric restriction and exercise on tumorigenesis in mice. *Proc Soc Exp Biol Med* 193:35–38.

Krontiris, T.G., and Cooper, G.M. 1981. Transforming activity of human tumor DNAs. *Proc Natl Acad Sci USA* 78:1181–1184.

Kuchino, Y., Mori, F., Kasai, H., Inoue, H., Iwai, S., Miura, K., Ohtsuka, E., and Nishimura, S.

1987. Misreading of DNA templates containing 8-hydroxydeoxyguanosine at the modified base and at adjacent residues. *Nature* 327:77–79.

Kumar, R., Sukumar, S., and Barbacid, M. 1990. Activation of *ras* oncogenes preceding the onset of neoplasia. *Science* 248:1101–1104.

Kumar, V., Cotran, R.S., and Robbins, S.L. 1992. *Robbins Pathologic Basis of Disease,* 5th ed. Philadelphia: W.B. Saunders.

Kune, S., Kune, G.A., and Watson, L.F. 1987. Case-control study of dietary etiological factors: The Melbourne colorectal cancer study. *Nutr Cancer* 9:21–42.

Kurzik-Dumke, U., Phannavong, B., Gundacker, D., and Gateff, E. 1992. Genetic, cytogenetic and developmental analysis of the *Drosophila melanogaster* tumor suppressor gene *lethal(2)tumorous imaginal discs (1(2)tid). Differentiation* 51:91–104.

Kurzrock, R., Gutterman, J.U., and Talpaz, M. 1988. The molecular genetics of Philadelphia chromosome-positive leukemias. *N Engl J Med* 319:990–998.

Kyprianou, N., English, H.F., Davidson, N.E., and Isaacs, J.T. 1991. Programmed cell death during regression of the MCF-7 human breast cancer following estrogen ablation. *Cancer Res* 51:162–166.

Kyriakis, J.M., App, H., Zhang, X.F., Banerjee, P., Brautigan, D.L., Rapp, U.R., and Avruch, J. 1992. Raf-1 activates MAP kinase-kinase. *Nature* 358:417–421.

Lacombe, C., DaSilva, L., Bruneval, P., Fournier, J-G, Wendling, S., Casadevall, N., Camilleri, J-P, Bariety, J., Varet, B., and Tambourin, P. 1988. Peritubular cells are the site of erythropoietin synthesis in the murine hypoxic kidney. *J Clin Invest* 81:620–623.

Lam, W.C., Delikatny, E.J., Orr, F.W., Wass, J., Varani, J., and Ward, P.A. 1981. The chemotactic response of tumor cells. *Am J Pathol* 104:69.

Lambert, A.R. 1916. Tissue cultures in the investigation of cancer. *J Cancer Res* 1:169–182.

Land, H., Parada, L.F., and Weinberg, R.A. 1983. Tumorigenic conversion of primary embryo fibroblasts requires at least two cooperating oncogenes. *Nature* 304:596–602.

Landis, S.H., Murray, T., Bolden, S., and Wingo, P.A. 1998. Cancer statistics, 1998. *CA Cancer J Clin* 48:6–29.

Landrigan, P.J. 1996. The prevention of occupational cancer. *CA Cancer J Clin* 46:67–69.

Lane, P., Vichi, P., Bain, D.L., and Tritton, T.R. 1987. Temperature dependence studies of adriamycin uptake and cytotoxicity. *Cancer Res* 47:4038–4042.

Langen, P. 1985. Chalones and other endogenous inhibitors of cell proliferation. In *Biological Response Modifiers, New Approaches to Disease Intervention,* P.F. Torrence, ed., pp. 265–291. Orlando, FL: Academic Press.

Langston, A.A., Malone, K.E., Thompson, J.D., Daling, J.R., and Ostrander, E.A. 1996. *BRCA1* mutations in a population-based sample of young women with breast cancer. *N Engl J Med* 334:137–142.

Lapidot, T., Pflumio, F., Doedens, M., Murdoch, B., Williams, D.E., and Dick, J.E. 1992. Cytokine stimulation of multilineage hematopoiesis from immature human cells engrafted in SCID mice. *Science* 255:1137–1141.

Laster, W.R., Jr., Mayo, J.G., Simpson-Herren, L., Griswold, D.P., Jr., Lloyd, H.H., Schabel, F.M., Jr., and Skipper, H.E. 1969. Success and failure in the treatment of solid tumors. II: Kinetic parameters and "cell cure" of moderately advanced carcinoma 755. *Cancer Chemother Rep* 53:169–188.

Lea, A.J. 1966. Dietary factors associated with death-rates from certain neoplasms in man. *Lancet* 2:332–333.

Leav, I., Merk, F.B., Ofner, P., Goodrich, G., Kwan, P.W., Stein, B.M., Sar, M., and Stumpf, W.E.

1978. Bipotentiality of response to sex hormones by the prostate of castrated or hypophysectomized dogs. Direct effects of estrogen. *Am J Pathol* 93:69–92.

Le Beau, M.M. 1997. Molecular biology of cancer: Cytogenetics. In *Cancer: Principles and Practice of Oncology,* 5th ed., V.T. DeVita, Jr., S. Hellman, and S.A. Rosenberg, eds., pp. 103–119. Philadelphia: Lippincott-Raven.

LeDoux, S.P., Thangada, M., Bohr, V.A., and Wilson, G.L. 1991. Heterogeneous repair of methyl-nitrosourea-induced alkali-labile sites in different DNA sequences. *Cancer Res* 51:775–779.

Lee, E.Y.-H.P., To, H., Shew, J.-Y., Bookstein, R., Scully, P., and Lee, W.-H. 1988. Inactivation of the retinoblastoma susceptibility gene in human breast cancers. *Science* 241:218–221.

Lee, J.A.H., and Strickland, D. 1980. Malignant melanoma: Social status and outdoor work. *Br J Cancer* 41:757–763.

Leevers, S.J., Paterson, H.F., and Marshall, C.J. 1994. Requirement for Ras and Raf activation is overcome by targeting Raf to the plasma membrane. *Nature* 369:411–414.

Lehvaslaiho, H., Lehtola, L., Sistonen, L., and Alitalo, K. 1989. A chimeric EGF-R-neu proto-oncogene allows EGF to regulate neu tyrosine kinase and cell transformation. *Eur Mol Biol Organ J* 8:159–166.

Lemen, R.A., Dement, J.M., and Wagoner, J.K. 1980. Epidemiology of asbestos-related diseases. *Environ Health Perspect* 20:1–21.

Lemieux, N., Milot, J., Barsoum-Homsy, M., Michaud, J., Leund, T.K., and Righer, C.L. 1989. First cytogenetic evidence of homozygosity for the retinoblastoma deletion in chromosome 13. *Cancer Genet Cytogenet* 43:73–78.

Lenihan, J. 1993. *The Good News About Radiation,* pp. 1–173. Madison: Cogito Books.

Lennon, S.V., Martin, S.J., and Cotter, T.G. 1991. Dose-dependent induction of apoptosis in human tumour cell lines by widely divergent stimuli. *Cell Prolif* 24:203–214.

Lenormand, P., Sardet, C., Pages, G., L'Allemain, G., Brunet, A., and Pouyssegur, J. 1993. Growth factors induce nuclear translocation of MAP kinases (p42mapk) but not of their activator MPA kinase kinase (p45mapkk) in fibroblasts. *J Cell Biol* 122:1079–1088.

Leone, A., Flatow, U., Richterr-King, C., Sandeen, M.A., Margulies, I.M.K., Liotta, L.A., and Steeg, P.S. 1991. Reduced tumor incidence, metastatic potential, and cytokine responsiveness of nm23-transfected melanoma cells. *Cell* 65:25–35.

Lewis, M.G., Lafrado, L.J., Haffer, K., Gerber, J., Sharpee, R.L., and Olsen, R.G. 1988. Feline leukemia virus vaccine: New developments. *Vet Microbiol* 17:297–308.

Lewison, E.E., ed. 1976. Conference on Spontaneous Regression of Cancer. *Natl Cancer Inst Monogr* 44:99–102.

Li, F.P. 1988. Cancer families: Human models of susceptibility to neoplasia – the Richard and Hinda Rosenthal Foundation Award Lecture. *Cancer Res* 48:5381–5386.

Li, F.P., and Fraumeni, J.F. 1969. Soft-tissue sarcomas, breast cancer, and other neoplasms. A familial syndrome. *Ann Intern Med* 71:747–752.

Li, J.J., and Deshaies, R.J. 1993. Exercising self-restraint: Discouraging illicit acts of S and M in eukaryotes. *Cell* 74:223–226.

Li, Y., Mohammad, R.M., al-Katib, A., Varterasian, M.L., and Chen, B. 1997. Bryostatin 1 (bryol)-induced monocytic differentiation in THP-1 human leukemia cells is associated with enhanced c-fyn tyrosine kinase and M-CSF receptors. *Leuk Res* 21:391–397.

Liang, R., Chow, W.S., Chiu, E., Chan, T.K., Lie, A., Kwong, Y.L., and Chan, L.C. 1993. Effective salvage therapy using all-trans retinoic acid for relapsed and resistant acute promyelocytic leukemia. *Anti-Cancer Drugs* 4:339–340.

Liebermann, D., Hoffman-Liebermann, B., and Sachs, L.1982. Regulation and role of different macrophage-and granulocyte-inducing proteins in normal and leukemic myeloid cells. *Int J Cancer* 29:159–161.

Liebl, E.C., and Martin, G.S. 1992. Intracellular targeting of pp60[src] expression: Localization of v-src to adhesion plaques is sufficient to transform chicken embryo fibroblasts. *Oncogene* 7:2417–2428.

Lijinsky, W., and Shubik, P. 1964. Benzo[a]pyrene and other polynuclear hydrocarbons in charcoal-broiled meat. *Science* 145:53–55.

Lin, T.S., Luo, M.Z., Liu, M.C., Pai, S.B., Dutschman, G.E., and Cheng, Y.C. 1994. Synthesis and biological evaluation of 2′,3′-dideoxy-L-pyrimidine nucleosides as potential antiviral agents against human immunodeficiency virus (HIV) and hepatitis B virus (HBV). *J Med Chem* 37:798–803.

Lindahl, T., and Nyberg, B. 1972. Rate of depurination of native deoxyribonucleic acid. *Biochemistry* 11:3610–3618.

Linehan, W.M., Lerman, M.I., and Zbar, B. 1995. Identification of the von Hippel-Lindau (VHL) gene: Its role in renal cancer. *J Am Med Assoc* 273:564–570.

Liotta, L.A., Abe, S., Robey, P.G., and Martin, G.R. 1979. Preferential digestion of basement membrane collagen by an enzyme derived from a metastatic murine tumor. *Proc Natl Acad Sci USA* 76:2268–2272.

Liotta, L.A., and Rao, C.N. 1985. Role of the extracellular matrix in cancer. *Ann NY Acad Sci* 460:333–344.

Liotta, L.A., Rao, C.N., and Barsky, S.H. 1983. Tumor invasion and the extracellular matrix. *Lab Invest* 49:636–649.

Livant, D.L., Linn, S., Markwart, S., and Shuster, J. 1995. Invasion of selectively permeable sea urchin embryo basement membrane by metastatic tumor cells, but not by their normal counterparts. *Cancer Res* 55:5085–5093.

Loeb, L.A. 1989. Endogenous carcinogenesis: Molecular oncology into the twenty-first century. *Cancer Res* 49:5489–5496.

Lofberg, J., Ahlbors, K., and Fallstrom, C. 1980. Neural crest cell migration in relation to extracellular matrix organization in the embryonic axolotl trunk. *Dev Biol* 75:148–167.

Lofroth, G. 1989. Environmental tobacco smoke: Overview of chemical composition and genotoxic components. *Mutation Res* 222:73–80.

Lok, S., Kaushansky, K., Holly, R.D., Kuijper, J.L., Lofton-Day, C.E., Oort, P.J., Grant, F.J., Heipel, M.D., Burkhead, S.K., Kramer, J.M., Bell, L.A., Sprecher, C.A., Blumberg, H., Johnson, R., Prunkard, D., Ching, A.S.T., Matthewes, S.L., Bailey, M.C., Forstrom, J.W., Buddle, M.M., Osborn, S.G., Evans, S.J., Sheppard, P.O., Presnell, S.R., O'Hara, P.J., Hagen, F.S., Roth, G.J., and Foster D.L. 1994. Cloning and expression of murine thrombopoietin cDNA and stimulation of platelet production in vivo. *Nature* 369:565–568.

Longnecker, D.S., and Curphey, T.J. 1975. Adenocarcinoma of the pancreas in azaserine-treated rats. *Cancer Res* 35:2249–2258.

Loser, R., Seibel, K., and Eppenberger, U. 1985. No loss of estrogenic or anti-estrogenic activity after demethylation of roloxifene 3-OH-tamoxifen. *Int J Cancer* 36:701–703.

Loser, R., Seibel, K., Roos, W., and Eppenberger, U. 1985. In vivo and in vitro antiestrogenic action of 3-hydroxytamoxifen, tamoxifen and 4-hydroxytamoxifen. *Eur J Cancer Clin Oncol* 21:985–990.

Lotem, J., and Sachs, L. 1982. Mechanisms that uncouple growth and differentiation in myeloid leukemia cells: Restoration of requirement for normal growth-inducing protein without restoring induction of differentiation-inducing protein. *Proc Natl Acad Sci USA* 79:4347–4351.

Loury, D.L., Goldsworthy, T.L., and Butterworth, B.E. 1987. The value of measuring cell replication as a predictive index of tissue-specific tumorigenic potential. In Banbury Report 25: *Nongenotoxic mechanisms in carcinogenesis,* B.E. Butterworth and T.J. Slaga, eds., pp. 119–136. New York: Cold Spring Harbor Press.

Lovec, H., Sewing, A., Lucibello, F.C., Muller, R., and Moroy, T. 1994. Oncogenic activity of cyclin D1 revealed through cooperation with Ha-*ras:* Link between cell cycle control and malignant transformation. *Oncogene* 9:323–326.

Lowenstein, E.J., Daly, A.G., Batzer, A.G., Li, W., Margolis, B., Lammers, R., Ullrich, A., Skolnik, E.Y., Bar-Sagi, D., and Schlessinger, J. 1992. The SH2 and SH3 domain-containing protein GRB2 links receptor tyrosine kinases to ras signaling. *Cell* 70:431–442.

Luande, J., Henschke, C.I., and Mohammed, N. 1985. The Tanzanian human albino skin. *Cancer* 55:1823–1828.

Lucké, B. 1934. A neoplastic disease of the kidney of the frog, *Rana pipiens. Am J Cancer* 20:352–379.

Lucké, B. 1938. Carcinoma in the leopard frog: Its probable causation by a virus. *J Exp Med* 68:457–468.

Lucké, B. 1939. Characteristics of frog carcinoma in tissue culture. *J Exp Med* 70:269–276.

Lucké, B., and Schlumberger, H. 1949. Induction of metastasis of frog carcinoma by increase of environmental temperature. *J Exp Med* 89:269–278.

Luer, C.A. 1986. Inhibitors of angiogenesis from shark cartilage. *Fed Proc* 45:949.

Lust, J.M., Carlson, D.L., Kowles, R., Rollins-Smith, L., Williams, J.W., III, and McKinnell, R.G. 1991. Allografts of tumor nuclear transplantation embryos: Differentiation competence. *Proc Natl Acad Sci USA* 88:6883–6887.

Luttrell, D.K., Lee, A., Lansing, T.J., Crosby, R.M., Jung, K.D., Willard, D., Luther, M., Rodriguez, M., Berman, J., and Gilmer, T.M. 1994. Involvement of pp60[c-src] with two major signaling pathways in human breast cancer. *Proc Natl Acad Sci USA* 91:83–87.

Lyman, S.D., and Jordan, V.C. 1985. Metabolism of tamoxifen and its uterotrophic activity. *Biochem Pharmacol* 34:2787–2794.

Lynch, H.T. 1985a. Classics in oncology. Alfred Scott Warthin, M.D., Ph.D. (1866–1931). *CA Cancer J Clin* 35:345–347.

Lynch, H.T. 1985b. Hereditary colon cancer: Polyposis and nonpolyposis variants. *CA Cancer J Clin* 35:95–114.

Lynch, H.T., Lynch, P.M., Albano, W.A., and Lynch, J.F. 1981. The cancer syndrome: A status report. *Dis Colon Rectum* 24:311–322.

Macchiarini, P., Fontanini, G., Hardin, M.J., Squartini, F., and Angeletti, C.A. 1992. Relation of neovascularisation to metastasis of non-small-cell lung cancer. *Lancet* 340:145–146.

MacDonald, A.D., and MacDonald, J.C. 1987. *Asbestos-Related Malignancy,* K.H. Antman and J. Aisner, eds., pp. 31–55. Orlando: Grune and Stratton.

Maclure, M., Katz, R.B.A., Bryant, M.S., Skipper, P.L., and Tannenbaum, S.R. 1989. Elevated blood levels of carcinogens in passive smokers. *Am J Public Health* 79:1381–1384.

Madigan, M.P., Ziegler, R.G., Benichou, J., Byrne, C., and Hoover, R.N. 1995. Proportion of breast cancer cases in the United States explained by well-established risk factors. *J Natl Cancer Inst* 87:1681–1685.

Magee, P.N., and Barnes, J.M. 1956. The production of malignant primary hepatic tumours in the rat by feeding dimethylnitrosamine. *Br J Cancer* 10:114–122.

Magrath, I. 1990. The pathogenesis of Burkitt's lymphoma. *Adv Cancer Res* 55:133–270.

Maheswaran, S., Park, S., Bernard, A., Morris, J.F., Rauscher, F.J., III, Hill, D.E., and Haber, D.A.

1993. Physical and functional interaction between WT1 and p53 proteins. *Proc Natl Acad Sci USA* 90:5100–5104.

Malaise, E.P., Chavaudra, N., and Tubiana, M. 1973. The relationship between growth rate, labelling index and histological type of human solid tumours. *Eur J Cancer* 9:305–312.

Malkin, D. 1993. p53 and Li-Fraumeni syndrome. *Cancer Genet Cytogenet* 66:83–92.

Malkin, D., Li, F.P., Strong, L.C., Fraumeni, J.F., Nelson, C.E., Kim, D.H., Kassel, J., Gryka, M.A., Bischoff, F.Z., Tainsky, M.A., and Friend, S.H. 1990. Germ line p53 mutations in a familial syndrome of breast cancer, sarcomas, and other neoplasms. *Science* 250:1233–1238.

Mannens, M., Slater, R.M., Heyting, C., Bliek, J., Hoovers, J., Bleeker-Wagemakers, E.M., Voûte, P.A., Coad, N., Frants, R.R., and Pearson, P.L. 1987. Chromosome 11, Wilms' tumour and associated diseases. *Cytogenet Cell Genet* 46:655.

Manni, A., and Arafah, B.M. 1981. Tamoxifen-induced remission in breast cancer by escalating the dose to 40 mg daily after progression on 20 mg daily: A case report and review of the literature. *Cancer* 48:873–875.

Mareel, M., Bracke, M., Van Roy, F., and Vakaet, L. 1993. Expression of E-Cadherin in embryonic ingression and cancer invasion. *Int J Dev Biol* 37:227–235.

Mareel, M.M., Bracke, M.E., and Storme, G.A. 1985. Mechanisms of tumor spread: A brief overview. In *The Cancer Patient: Illness and Recovery*, E. Grundmann, ed., pp. 59–64. Stuttgart: Gustav Fischer Verlag.

Mareel, M.M., DeBaetselier, P., and Van Roy, F.M. 1991. *Mechanisms of Invasion and Metastasis.* Boca Raton: CRC Press.

Mareel, M.M., Storme, G.A., DeBruyne, G.K., and Van Cauwenberge, R.M. 1982. Vinblastine, vincristine, and vindesine: Antiinvasive effect on MO4 mouse fibrosarcoma cells in vitro. *Eur J Cancer Clin Oncol* 18:199–210.

Mareel, M.M., Van Roy, F.M., and Bracke, M.E. 1993. How and when do tumor cells metastasize. *Crit Rev Oncogenesis* 4:559–594.

Margolese, R., Poisson, R., Shibata, H., Pilch, Y., Lerner, H., and Fisher, B. 1987. The technique of segmental mastectomy (lumpectomy) and axillary dissection. *Surgery* 102:828–834.

Marks, R. 1996a. Prevention and control of melanoma: The public health approach. *CA Cancer J Clin* 46:199–216.

Marks, R. 1996b. Squamous cell carcinoma. *Lancet* 347:735–738.

Marmot, M.G., Rose, G., Shipley, M.J., and Thomas, B.J. 1981. Alcohol and mortality: A U-shaped curve. *Lancet* 1:580–583.

Marshall, M. 1995. Interactions between Ras and Raf: Key regulatory proteins in cellular transformation. *Mol Reprod Dev* 42:493–499.

Martin, G.R. 1981. Isolation of a pluripotential cell line from early mouse embryos cultured in medium conditioned by teratocarcinoma stem cells. *Proc Natl Acad Sci USA* 78:7634–7638.

Martineau, D., Lagacé, A., Massé, R., Morin, M., and Béland, P. 1985. Transitional cell carcinoma of the urinary bladder in a beluga whale *Delphinapterus leucas. Can Vet J* 26:297–302.

Martland, H.S. 1931. The occurrence of malignancy in radioactive persons. *Am J Cancer* 15:2435–2516.

Marx, J. 1996. A second breast cancer susceptibility gene is found. *Science* 271:30–31.

Masahito, Prince, Ishikawa, T., and Sugana, H. 1988. Fish tumors and their importance to cancer research. *Jpn J Cancer Res (Gann)* 79:545–555.

Masahito, Prince, Nishioka, M., Ueda, H., Kato, Y., Yamazaki, I., Nomura, K., Sugano, H., and Kitagawa, T. 1994. Frequent development of pancreatic carcinomas in the *Rana nigromaculata* group. *Proceedings 8th Int. Conf. Int. Soc. Diff.*, pp. 183–186. Hiroshima, Japan: International Society of Differentiation.

Masahito, Prince, Nishioka, M., Ueda, H., Kato, Y., Yamazaki, I., Nomura, K., Sugano, H., and Kitagawa, T. 1995. Frequent development of pancreatic carcinomas in the *Rana nigromaculata* group. *Cancer Res* 55:3781–3784.

Matsumura, Y., and Tarin, D. 1992. Significance of CD44 gene products for cancer diagnosis and disease evaluation. *Lancet* 340:1053–1058.

Matsumoto, J., Akiyama, T., Nemoto, N., Masahito, Prince, and Ishikawa, T. 1993. Appearance of tumorous phenotypes in goldfish erythrophores transfected with ras, src, and myc oncogenes and spontaneous differentiation of the transformants in vitro. *J Invest Dermatol* 100:214S–221S.

Matz, G. 1969. Histology and transmission of Locusta migratoria (L.) tumors (Insecta, Orthoptera). *Natl Cancer Inst Monogr* 31:465–473.

Maurer, H.S., Pendergrass, T.W., Borges, W., and Honig, G.R. 1979. The role of genetic factors in the etiology of Wilms' tumor. *Cancer* 43:205–208.

Mayo, J.G., Laster, W.R., Jr., Andrews, C.M., and Schabel, F.M., Jr. 1972. Success and failure in the treatment of solid tumors. 3: "Cure" of metastatic Lewis lung carcinoma with methyl-CCNU (NSC-95442) and surgery-chemotherapy. *Cancer Chemother Rep* Part 1, 56:183–195.

McBurney, M.W. 1993. P19 embryonal carcinoma cells. *Int J Dev Biol* 37:135–140.

McCarthy, J.B., Basara, M.L., Palm, S.L., Sas, D.F., and Furcht, L.T. 1985. The role of cell adhesion proteins – laminin and fibronectin – in the movement of malignant and metastatic cells. *Cancer Metastasis Rev* 4:125–152.

McCarthy, N.J., Smith, C.A., and Williams, G.T. 1992. Apoptosis in the development of the immune system: Growth factors, clonal selection and bcl-2. *Cancer Metastasis Rev* 11:157–178.

McCormack, J.J., and Johns, D.G. 1990. Purine and purine nucleoside antimetabolites. In *Cancer Chemotherapy*, B.A. Chabner and J.M. Collins, eds., pp. 234–252. Philadelphia: J.B. Lippincott.

McCutcheon, M., Coman, D.R., and Moore, F.B. 1948. Studies on invasiveness of cancer. Adhesiveness of malignant cells in various human adenocarcinomas. *Cancer* 1:460–467.

McDonnell, T.J., Deane, N., Platt, F.M., Nunez, G., Jaeger, U., McKearn, J.P., and Korsmeyer, S.J. 1989. Bcl-2-immunoglobulin transgenic mice demonstrate extended B cell survival and follicular lymphoproliferation. *Cell* 57:79–88.

McGinnis, J.M., and Foege, W.H. 1993. Actual causes of death in the United States. *J Am Med Assoc* 270:2207–2212.

McKinnell, R.G. 1973. Nuclear transplantation. In *Seventh National Cancer Conference Proceedings*, pp. 65–72. Philadelphia: J.B. Lippincott.

McKinnell, R.G. 1989a. Expression of differentiated function in neoplasms. In *The Pathobiology of Neoplasia*, A.E. Sirica, ed., pp. 435–460. New York: Plenum.

McKinnell, R.G. 1989b. Neoplastic cells: Modulation of the differentiated state. In *Genomic Adaptability in Somatic Cell Specialization*, M.A. DiBerardino and L.D. Etkin, eds., pp. 199–236. New York: Plenum.

McKinnell, R.G. 1994. Reduced oncogenic potential associated with differentiation of the Lucké renal adenocarcinoma. *In Vivo* 8:65–70.

McKinnell, R.G., Bruyneel, E.A., Mareel, M.M., Tweedell, K.S., and Mekela, P. 1988. Temperature-dependent malignant invasion in vitro by frog renal carcinoma-derived PNKT4B cells. *Clin Exp Metastasis* 6:49–59.

McKinnell, R.G., and Carlson, D.L. 1997. The Lucké renal adenocarcinoma, an anuran neoplasm:

Studies at the interface of pathology, virology and differentiation competence. *J Cell Physiol* 173:115–118.

McKinnell, R.G., Carlson, D.L., Christ, C.G., and John, J.C. 1995. Detection of Lucké tumor herpesvirus DNA sequences in normal and neoplastic tissue obtained from the northern leopard frog, Rana pipiens. *Comptes Rendus du Cinquième Colloque International de Pathologie des Reptiles et des Amphibiens,* NRG Repro Facility BV 's-Hertogensbosch, Pays-Bas, pp. 13–16.

McKinnell, R.G., and Cunningham, W.P. 1982. Herpesviruses in metastatic Lucké renal adenocarcinoma. *Differentiation* 22:41–46.

McKinnell, R.G., DeBruyne, G.K., Mareel, M.M., Tarin, D., and Tweedell, K.S. 1984. Cytoplasmic microtubules of normal and tumor cells of the leopard frog. Temperature effects. *Differentiation* 26:231–234.

McKinnell, R.G., Deggins, B.A., and Labat, D.D. 1969. Transplantation of pluripotential nuclei from triploid frog tumors. *Science* 165:394–396.

McKinnell, R.G., Kren, B.T., Bergad, R., Schultheis, M., Byrne, T., and Schaad, J.W., IV. 1980. Dominant lethality in *Xenopus laevis* induced with triethylene melamine. *Teratog Carcinog Mutagen* 1:283–294.

McKinnell, R.G., Lust, J.M., Sauerbier, W., Rollins-Smith, L., Williams, J.W., III, Williams, C.S., and Carlson, D.L. 1993. Genomic plasticity of the Lucké carcinoma: A review. *Int J Develop Biol* 37:213–219.

McKinnell, R.G., Mareel, M.M., Bruyneel, E.A., Seppanen, E.D., and Mekala, P.R. 1986. Invasion in vitro by explants of Lucké renal carcinomas cocultured with normal tissue is temperature dependent. *Clin Exp Metastasis* 4:237–243.

McKinnell, R.G., Picciano, D.J., and Schaad, J.W., IV. 1979. Dominant lethality in frog embryos after paternal treatment with triethylenemelamine. *Environ Mutagen* 1:221–231.

McKinnell, R.G., Sauerbier, W., Lust, J.M., Williams, J.W., III, Williams, C.S., Rollins-Smith, L., and Carlson, D.L. 1991. The Lucké renal adenocarcinoma and its herpesvirus. In *4. Internationales Colloquium für Pathologie und Therapie der Reptilien und Amphibien,* K. Gabrisch, B. Schildger, and P. Zwart, eds., pp. 204–218. Bad Nauheim: Deutsche Veterinärmedizinische Gesellschaft e.V.

McKinnell, R.G., and Tarin, D. 1984. Temperature dependent metastasis of the Lucké renal carcinoma and its significance for studies on mechanisms of metastasis. *Cancer Metastasis Rev* 3:373–386.

McKinnell, R.G., and Tweedell, K.S. 1970. Induction of renal tumors in triploid leopard frogs. *J Natl Cancer Inst* 44:1161–1166.

McNally, W.D. 1932. The tar in cigarette smoke and its possible effects. *Am J Cancer* 16:1502–1514.

McTiernan, A., and Thomas, D.B. 1986. Evidence for a protective effect of lactation on risk of breast cancer in young women: Results from a case-control study. *Am J Epidemiol* 124:353–358.

Mellon, I., Spivak, G., and Hanawalt, P.C. 1987. Selective removal of transcription-blocking DNA damage from the transcribed strand of the mammalian DHFR gene. *Cell* 51:241–249.

Melo, J.V. 1996. The molecular biology of chronic myeloid leukemia. *Leukemia* 10:751–756.

Mendelsohn, M.L. 1960. The growth fraction: A new concept applied to tumors. *Science* 132:1496.

Mercep, M., Weissman, A.M., Frank, S.J., Klausner, R.D., and Ashwell, J.D. 1989. Activation-driven programmed cell death and T cell receptor zeta eta expression. *Science* 246:1162–1165.

Mertens, V.E. 1930. Zigaretterauch eine Ursache des Lungenkrebses? *Z Krebsforsch* 32:82–91.

Methia, N., Louache, F., Vainchenker, W., and Wendling, F. 1993. Oligodeoxynucleotides antisense to the proto-oncogene c-mpl specifically inhibit in vitro megakaryocytopoiesis. *Blood* 82:1395–1401.

Meyn, M.S. 1995. Ataxia-telangiectasia and cellular responses to DNA damage. *Cancer Res* 55:5991–6001.

Michaels, L. 1964. Cancer incidence and mortality in patients having anticoagulant therapy. *Lancet* 2:832–835.

Miki, Y., Swensen, J., Shattuck-Eidens, D., Futreal, P.A., Harshman, K., Tavtigian, S., Liu, Q., Cochran, C., Bennett, L.M., Ding, W., Bell, R., Rosenthal, J., Hussey, C., Tran,T., McClure, M., Frye, C., Hattier, T., Phelps, R., Haugen-Strano, A., Katcher, H., Yakumo, K., Gholami, Z., Shaffer, D., Stone, S., Bayer, S., Wray, C., Bogden, R., Dayananth, P., Ward, J., Tonin, P., Narod, S., Bristow, P.K., Norris, F.H., Helvering, L., Morrison, P., Rosteck, P., Lai, M., Barrett, J.C., Lewis, C., Neuhausen, S., Cannon-Albright, L., Goldgar, D., Wiseman, R., Kamb, A., and Skolnick, M.H. 1994. A strong candidate for the breast and ovarian cancer susceptibility gene *BRCA1*. *Science* 266:66–71.

Milburn, M.V., Tong, L., deVoss, A.M., Brunger, A., Yamaizumi, Z., Nishimura, S., and Kim, S.-H. 1990. Molecular switch for signal transduction: Structural differences between active and inactive forms of protooncogenic ras proteins. *Science* 247:939–945.

Miles, S.A., Mitsuyasu, R.I., and Aboulafia, D.M. 1997. AIDS-related malignancies. In *Cancer: Principles and Practice of Oncology*, 5th ed., V.T. DeVita, S. Hellman, and S.A. Rosenberg, eds., pp. 2445–2467. Philadelphia: Lippincott-Raven.

Miller, A.B. 1989. Vitamins, minerals, and other dietary factors. In *Diet and the Aetiology of Cancer*, A.B. Miller, ed., pp. 39–54. Berlin: Springer-Verlag.

Miller, E.C. 1978. Some current perspectives on chemical carcinogenesis in humans and experimental animals. *Cancer Res* 38:1479–1496.

Miller, J.A. 1970. Carcinogenesis by chemicals: An overview. *Cancer Res* 30:559–576.

Miller, R.W. 1966. Delayed radiation effects in atomic bomb survivors. *Science* 166:569–574.

Mintz, B., and Illmensee, K. 1975. Normal genetically mosaic mice produced from malignant teratocarcinoma cells. *Proc Natl Acad Sci USA* 72:3585–3589.

Miranda, A., Janssen, G., Hodges, L., Peralta, E.G., and Ream, W. 1992. Agrobacterium tumefaciens transfers extremely long T-DNA by a unidirectional mechanism. *J Bacteriol* 174:2288–2297.

Misra, R.P., Bonni, A., Miranti, C.K., Rivera, V.M., and Sheng, Z.H. 1994. L-type voltage-sensitive calcium channel activation stimulates gene expression by a serum response factor-dependent pathway. *J Biol Chem* 269:25483–25493.

Mitelman, F. 1991. *Catalog of Chromosome Aberrations in Cancer*, 4th ed. (2 vols.). New York: Wiley-Liss.

Miura, M., Zhu, H., Rotello, R., Hartwieg, E.A., and Yuan, J. 1993. Induction of apoptosis in fibroblasts by IL-1β-converting enzyme, a mammalian homolog of the *C. elegans* cell death gene *ced-3*. *Cell* 75:653–660.

Miyashita, T., and Reed, J.C. 1995. Tumor suppressor p53 is a direct transcriptional activator of the human bax gene. *Cell* 80:293–299.

Molineux, G., Migdalska, A., Haley, J., Evans, G.S., and Dexter, T.M. 1994. Total marrow failure induced by pegylated stem-cell factor administered before 5-fluorouracil. *Blood* 3:3491–3499.

Moodie, S.A., and Wolfman, A. 1994. The 3Rs of life: Ras, Raf and growth regulation. *Trends Genet* 10:44–48.

Moore, K.L., and Persaud, T.V.N. 1993. *The Developing Human*, 5th ed. Philadelphia: W.B. Saunders.

Moore, M.A.S., and Owen, J.J.T. 1967. Stem cell migration in developing myeloid and lymphoid systems. *Lancet* 2:658–659.

Morison, W.L. 1988. Skin cancer and artificial sources of UV radiation. *J Dermatol Surg Oncol* 14:893–896.

Mørk, S., Laerum, O.D., and deRidder, L. 1984. Invasiveness of tumours of the central nervous system. In *Invasion, Experimental and Clinical Implications,* M.M. Mareel and K.C. Calman, eds., pp. 79–125. Oxford: Oxford University Press.

Morris, C.M., Reeve, A.E., Fitzgerald, P.H., Hollings, P.E., Beard, M.E.J., and Heaton, D.C. 1986. Genomic diversity correlates with clinical variation in Ph^1-negative chronic myeloid leukaemia. *Nature* 320:281–283.

Morrison, V.A. 1994. Chronic leukemias. *CA Cancer J Clin* 44:353–377.

Morse, D., Dailey, R.C., and Bunn, J. 1976. Prehistoric multiple myeloma. In *Essays on the History of Medicine,* S. Jarcho, ed., pp. 413–424. New York: Science History Publications/USA.

Morton, D.L., Antman, K.H., and Tepper, J. 1993. Soft tissue sarcoma. In *Cancer Medicine,* vol. 2 J.F. Holland, E. Frei III, R.C. Bast, Jr., D.W. Kufe, D.L. Morton, and R.R. Weichselbaum, eds., pp. 1858–1887. Philadelphia: Lea and Febiger.

Morton, D.L., Cochran A.J., and Lazar, G. 1990. Melanoma. In *Cancer Treatment,* 3rd ed., C.M. Haskell, ed., pp. 500–512. Philadelphia: W.B. Saunders.

Mossman, B.T., Bignon, J., Corn, M., Seaton, A., and Gee, J.B.L. 1990. Asbestos: Scientific developments and implications for public policy. *Science* 247:294–301.

Mossman, B.T., Kamp, D.W., and Weitzman, S.A. 1996. Mechanisms of carcinogenesis and clinical features of asbestos-associated cancers. *Cancer Invest* 14:466–480.

Moulton, J.E. 1978. *Tumors in Domestic Animals.* Berkeley: University of California Press.

Mourney, R.J., and Dixon, J.E. 1994. Protein tyrosin phosphatases: Characterization of extracellular and intracellular domains. *Curr Opin Genet Dev* 4:31–39.

Mueller, C.B., and Jeffries, W. 1975. Cancer of the breast: Its outcome as measured by the rate of dying and causes of death. *Ann Surg* 182:334–341.

Mugrauer, G., Alt, F.W., and Ekblom, P. 1988. N-myc proto-oncogene expression during organogenesis in the developing mouse as revealed by in situ hybridization. *J Cell Biol* 107:1325–1335.

Muir, C., Waterhouse, J., Mack, T., Powell, J., and Whelan, S., eds. 1987. *Cancer Incidence in Five Continents,* Vol. V. IARC Scientific Publications No. 88. Lyon: IARC.

Muller, A.J., Young, J.C., Pendergast, A.-M., Pondel, M., Littman, D.R., and Witte, O.N. 1991. BCR first exon sequences specifically activate the BCR/ABL tyrsoine kinase oncogene of Philadelphia chromosome-positive human leukemias. *Mol Cell Biol* 11:1785–1792.

Mundoz-Dorado, J., Inouye, M., and Inouye, S. 1990. Nucleoside diphosphate kinase from *Myxococcus xanthus:* I. Cloning and sequencing of the gene. *J Biol Chem* 265:2702–2706.

Munger, K., Werness, B.A., Dyson, N., Phelps, W.C., and Howley, P.M. 1989. Complex formation of human papillomavirus E7 proteins with the retinoblastoma tumor suppressor gene product. *Eur Mol Biol Organ J* 8:4099–4105.

Munson, L. 1987. Carcinoma of the mammary gland in a mare. *J Am Vet Med Assoc* 191:71–72.

Murphy, C.S., Langan-Fahey, S.M., McCague, R., and Jordan, V.C. 1990. Structure-function relationships of hydroxylated metabolites of tamoxifen that control the proliferation of estrogen-responsive T47D breast cancer cells in vitro. *Mol Pharmacol* 38:737–743.

Murray, L., Chen, B., Galy, A., Chen, S., Tushinski, R., Uchida, N., Negrin, R., Tricot, G., Jagannath, S., Vesole, D., Barlogie, B., Hoffman, R., and Tsukamoto, A. 1995. Enrichment of human hematopoietic stem cell activity in the CD34+Thy-1+Lin-subpopulation from mobilized peripheral blood. *Blood* 85:368–378.

Murray, L., DiGusto, D., Chen, B., Chen, S., Combs, J., Conti, A., Galy, A., Negrin, R., Tricot, G., and Tsukamoto, A. 1994. Analysis of human hematopoietic stem cell populations. *Blood Cells* 20:364–369.

Muschel, R.J., Williams, J.E., Lowy, D.R., and Liotta, L.A. 1985. Harvey *ras* induction of metastatic potential depends upon oncogene activation and the type of recipient cell. *Am J Pathol* 121:1–8.

Muss, H.B., Smith, L.R., and Cooper, M.R. 1987. Tamoxifen rechallenge: Response to tamoxifen following relapse after adjuvant chemohormonal therapy for breast cancer. *J Clin Oncol* 5:1556–1558.

Myers, C.E., and Chabner, B.A. 1990. Anthracyclines. In *Cancer Chemotherapy*, B.A. Chabner and J.M. Collins, eds., pp. 356–381. Philadelphia: J.B. Lippincott.

Myers, G.H., Jr., Fehrenbaker, L.G., and Kelalis, P.P. 1968. Prognostic significance of renal vein invasion by hypernephroma. *J Urol* 100:420–423.

Naegele, R.F., Granoff, A., and Darlington, R.W. 1974. The presence of the Lucké herpesvirus genome in induced tadpole tumors and its oncogenicity: Koch-Henle postulates fulfilled. *Proc Natl Acad Sci USA* 71:830–834.

Naghshineh, R., Hagdoost, I.S., and Mokhber-Dezfuli, M.R. 1991. A retrospective study of the incidence of bovine neoplasms in Iran. *J Comp Pathol* 105:235–239.

Nakachi, K., Imai, K., Hayashi, S., Watanabe, J., and Kawajiri, K. 1991. Genetic susceptibilty to squamous cell carcinoma of the lung in relation to cigarette smoking dose. *Cancer Res* 51:5177–5180.

Nakamori, S., Ishikawa, O., Ohhigashi, H., Kameyama, M., Furukawa, H., Sasaki, Y., Inaji, H., Higashiyama, M., Imaoka, S., Iwanaga, T., Funai, H., Wada, A., and Kimura, N. 1993. Expression of nucleoside diphosphate kinase/*nm*23 gene product in human pancreatic cancer: An association with lymph node metastasis and tumor invasion. *Clin Exp Metastasis* 11:151–158.

Nakazawa, H., English, D., Randell, P.L., Nakazawa, K., Martel, N., Armstrong, B.K., and Yamasaki, H. 1994. UV and skin cancer: Specific p53 gene mutation in normal skin as a biologically relevant exposure measurement. *Proc Natl Acad Sci USA* 91:360–364.

Nass, S.J., Li, M., Amundadottir, L.T., Furth, P.A., and Dickson, R.B. 1996. Role for Bcl-xL in the regulation of apoptosis by EGF and TGF beta 1 in c-myc overexpressing mammary epithelial cells. *Biochem Biophys Res Commun* 227:248–256.

National Council on Radiation Protection and Measurements. 1987. *Ionizing radiation exposure of the population of the United States.* NCRP Report 93. Bethesda, MD: National Council on Radiation Protection and Measurements.

National Research Council. 1982. *Diet, Nutrition and Cancer.* Washington, DC: National Academy Press.

Needham, J. 1942. *Biochemistry and Morphogenesis.* Cambridge: Cambridge University Press.

Neel, J.V., Schull, W.J., Awa, A.A., Satoh, C., Kato, H., Otake, M., and Yoshimoto, Y. 1990. The children of parents exposed to atomic bombs: Estimates of the genetic doubling dose of radiation for humans. *Am J Hum Genet* 46:1053–1072.

Nelson, D.R., Kamataki, T., Waxman, D.J., Guengerich, F.P., Estabrook, R.W., Feyereisen, R., Gonzalez, F.J., Coon, M.J., Gunsalus, I.C., Gotoh, O., Okuda, K., and Nebert, D.W. 1993. The P450 superfamily: Update on new sequences, gene mapping, accession numbers, early trivial names of enzymes and nomenclature. *DNA Cell Biol* 12:1–51.

Nelson, D.R., and Strobel, H.W. 1987. Evolution of cytochrome P-450 protein. *Mol Biol Evol* 4:572–593.

Nelson, M.A., Futscher, B.W., Kinsella, T., Wymer, J., and Bowden, G.T. 1992. Detection of mutant Ha-*ras* genes in chemically initiated mouse skin epidermis before the development of benign tumors. *Proc Natl Acad Sci USA* 89:6398–6402.

Neuberger, J.S. 1992. Residential radon exposure and lung cancer: An overview of ongoing studies. *Health Phys* 63:503–509.

Neuman, E., Flemington, E.K., Sellers, W.R., and Kaelin, W.G., Jr. 1994. Transcription of the E2F-1 gene is rendered cell cycle dependent by E2F DNA-binding sites within its promoter. *Mol Cell Biol* 14:6607–6615.

Neuwirth, H., and de Kernion, J.B. 1990. In *Prostate Cancer Treatment,* 3rd ed., C.M. Haskell, ed., pp. 737–749, Philadelphia: W.B. Saunders.

Neuwirth, H., Figlin, R.A., and deKernion, J.B. 1990. Kidney. In *Cancer Treatment,* 3rd ed., C.M. Haskell, ed., pp. 769–778. Philadelphia: W.B. Saunders.

Neve, J. 1991. Physiological and nutritional importance of selenium. *Experientia* 47:187–193.

Newberne, P.M., and Conner, M.W. 1988. Dietary modifiers of cancer. *Prog Clin Biol Res* 259:105–129.

Newcomb, P.A., Storer, B.E., and Marcus, P.M. 1995. Cigarette smoking in relation to risk of large bowel cancer in women. *Cancer Res* 55:4906–4909.

Newton, A.C. 1995. Protein kinase C: Structure, function, and regulation. *J Biol Chem* 270:28495–28498.

Nicolson, G.L. 1982. Cancer metastasis: Organ colonization and the cell-surface properties of malignant cells. *Biochim Biophys Acta* 695:113–176.

Nikitin, A.Y., Ballering, L.A.P., and Rajewsky, M.F. 1991. Early mutatins of the *neu* (*erb*B-2) gene during ethylnitrosourea-induced oncogenesis in the rat Schwann cell lineage. *Proc Natl Acad Sci USA* 88:9939–9943.

Nishisho, I., Nakamura, Y., Miyoshi, Y., Miki, Y., Ando, H., Horii, A., Koyama, K., Utsunomiya, J., Baba, S., Hedge, P., Markham, A., Krush, A.J., Petersen, G., Hamilton, S.R., Nilbert, M.C., Levy, D.B., Bryan, T.M., Preisinger, A.C., Smith, K.J., Su, L-K, Kinzler, K.W., and Vogelstein, B. 1991. Mutations of chromosome 5q21 genes in FAP and colorectal cancer patients. *Science* 253:665–669.

Nishizuka, Y. 1992. Intracellular signaling by hydrolysis of phospholipids and activation of protein kinase C. *Science* 258:607–614.

Noble, R.L., Beer, C.T., and Cutts, J.H. 1958. Further biological activities of vincaleukoblastine – an alkaloid isolated from *Vinca rosea* (L.). *Biochem Pharmacol* 1:347–348.

Noel, A.C., Calle, A., Emonard, H.P., Nusgens, B.V., Simar, L., Foidart, J., Lapiere, C.M., and Foidart, J.M. 1991. Invasion of reconstituted basement membrane matrix is not correlated to the malignant metastatic cell phenotype. *Cancer Res* 51:405–414.

Noguchi, S., Miyauchi, K., Nishizawa, Y., and Koyama, H. 1988. Induction of progesterone receptor with tamoxifen in human breast cancer with special reference to its behavior over time. *Cancer* 61:1345–1349.

Norton, L. 1979. Thoughts on a role for cell kinetics in cancer chemotherapy. In *Controversies in Cancer: Design of Trials and Treatment,* H.J. Tagnon and M.J. Staquet, eds., pp. 105–115. New York: Masson.

Norton, L. 1990. Biology of residual breast cancer after therapy: A kinetic interpretation. *Prog Clin Biol Res* 354A:109–132.

Norton, L., and Simon, R. 1986. The Norton-Simon hypothesis revisited. *Cancer Treat Rep* 70:163–169.

Novick, D., Cohen, B., and Rubinstein, M. 1992. Soluble interferon-a receptor molecules are present in body fluids. *Fed Exp Biol Soc Lett* 314:445–448.

Nowell, P.C. 1976. The clonal evolution of tumor cell populations. *Science* 194:23–25.

Nowell, P.C. 1992. Biology of disease: Cancer, chromosomes, and genetics. *Lab Invest* 66:407–417.

Nowell, P.C., and Hungerford, D.A. 1960a. A minute chromosome in human granulocytic leukemia. *Science* 132:1497.

Nowell, P.C., and Hungerford, D.A. 1960b. Chromosome studies on normal and leukemic human leukocytes. *J Natl Cancer Inst* 25:85–109.

Ochsner, A., and DeBakey, M. 1939. Primary pulmonary malignancy: Treatment of total pneumonectomy: Analysis of seventy-nine collected cases and presentation of seven personal cases. *Surg Gynecol Obstet* 68:435–451.

Ochsner, A., and DeBakey, M. 1940. Surgical considerations of primary carcinoma of the lung. *Surgery* 8:992–1023.

Ochsner, A., and DeBakey, M. 1941. Carcinoma of the lung. *Arch Surg* 42:209–258.

Ogilvie, D., McKinnell, R.G., and Tarin, D. 1984. Temperature-dependent elaboration of collagenase by the renal adenocarcinoma of the leopard frog, *Rana pipiens*. *Cancer Res* 44:3438–3441.

Okey, A.B. 1990. Enzyme induction in the cytochrome P-450 system. *Pharmacol Ther* 45:241–298.

Okey, A.B., Roberts, E.A., Harper, P.A., and Denison, M.S. 1986. Induction of drug-metabolizing enzymes: Mechanisms and consequences. *Clin Biochem* 19:132–141.

Omenn, G.S., Goodman, G.E., Thornquist, M.D., Balmes, J., Cullen, M.R., Glass, A., Keogh, J.P., Meyskens, F.L., Valanis, B., Williams, J.H., Barnhart, S., and Hammar, S. 1996. Effects of a combination of beta carotene and vitamin A on lung cancer and cardiovascular disease. *N Engl J Med* 334:1150–1155.

Ortner, D.J., and Putschar, W.G.J. 1981. *Identification of pathological conditions in human skeletal remains*. Washington, DC: Smithsonian Institution Press.

Orzechowski, A., Schrenk, D., and Bock, K.W. 1992. Metabolism of 1- and 2-naphthylamine in isolated rat hepatocytes. *Carcinogenesis* 13:2227–2232.

Osborne, C.K., Coronado-Heinsohn, E.B., Hilsenbeck, S.G., McCue, B.L., Wakeling, A.E., McClelland, R.A., Manning, D.L., and Nicholson, R.I. 1995. Comparison of the effects of a pure steroidal antiestrogen with those of tamoxifen in a model of human breast cancer. *J Natl Cancer Inst* 87:746–750.

Osborne, C.K., Yochmowitz, M.G., Knight, W.A., and McGuire, W.L. 1980. The value of estrogen and progesterone receptors in the treatment of breast cancer. *Cancer* 46 (Suppl 12):2884–2888.

O'Sullivan, C., and Lewis, C.E. 1994. Tumour-associated leucocytes: Friends or foes in breast carcinoma. *J Pathol* 172:229–235.

Ozes, O.N., Klein, S.B., Reiter, Z., and Taylor, M.W. 1993. An interferon resistant variant of the hairy cell leukemic cell line, Eskol: Biochemical and immunological characterization. *Leuk Res* 17:983–990.

Paffenbarger, R.S., Hyde, R.T., and Wing, A.L. 1987. Physical activity and incidence of cancer in diverse populations: A preliminary report. *Am J Clin Nutr* 45:312–317.

Paffenbarger, R.S., Hyde, R.T., Wing, A.L., and Hsieh, C.C. 1986. Physical activity, all-cause mortality, and longevity of college alumni. *N Engl J Med* 314:605–613.

Paget, S. 1889. The distribution of secondary growths in cancer of the breast. *Lancet* 1:571–573.

Palmer, S. 1986. Dietary considerations for risk reduction. *Cancer* 58(Suppl 8):1949–1953.

Palmer, S., and Bakshi, K. 1983. National Academy of Sciences – Report of the committee on diet, nutrition, and cancer. *J Natl Cancer Inst* 70:1151–1170.

Papadopoulos, N., Nicolaides, N.C., Wei, Y-F., Ruben, S.M., Carter, K.C., Rosen, C.A., Haseltine, W.A., Fleischmann, R.D., Fraser, C.M., Adams, M.D., Venter, J.C., Hamilton, S.R., Petersen, G.M., Watson, P., Lynch, H.T., Peltomäki, P., Mecklin, J-P., de la Chapelle, A., Kinzler, K.W., and Vogelstein, B. 1994. Mutation of a *mutL* homolog in hereditary colon cancer. *Science* 263:1625–1629.

Parchment, R.E. 1993. The implications of a unified theory of programmed cell death, polyamines, oxyradicals and histogenesis in the embryo. *Int J Dev Biol* 37:75–83.

Parchment, R.E., and Pierce, G.B. 1989. Polyamine oxidation, programmed cell death, and regulation of melanoma in the murine limb. *Cancer Res* 49:6680–6686.

Parczyk, K., and Schneider, M.R. 1996. The future of antihormone therapy: Innovations based on an established principle. *J Cancer Res Clin Oncol* 122:383–396.

Parker, R.G. 1990. Principles of radiation oncology. In *Cancer Treatment,* 3rd ed., C.M. Haskell, ed., pp. 15–21. Philadelphia: W.B. Saunders.

Parker, S.L., Tong, T., Bolden, S., and Wingo, P.A. 1997. Cancer statistics, 1997. *CA Cancer J Clin* 47:5–27.

Parry, S.L., Hasbold, J., Holman, M., and Klaus, G.G. 1994. Hypercross-linking surface IgM or IgD receptors on mature B cells induces apoptosis that is reversed by costimulation with IL-4 and anti-CD40. *J Immunol* 152:2821–2829.

Parry, S.L., Holman, M.J., Hasbold, J., and Klaus, G.G. 1994. Plastic-immobilized anti-mu or anti-delta antibodies induce apoptosis in mature murine B lymphocytes. *Eur J Immunol* 24:974–979.

Pastorino, U., Chiesa, G., Infante, M., Soresi, E., Clerica, M., Valente, M., Belloni, P.A., and Ravasi, G. 1991. Safety of high-dose vitamin A. Randomized trial on lung cancer chemoprevention. *Oncology* 48:131–137.

Patel, N.H., and Rothenberg, M.L. 1994. Multidrug resistance in cancer chemotherapy. *Invest New Drugs* 12:1–13.

Patil, S., Merrick, S., and Lubs, H.A. 1971. Identification of each human chromosome with a modified Giemsa stain. *Science* 173:821–822.

Patterson, B.H., and Block, G. 1988. Food choices and the cancer guidelines. *Am J Public Health* 78:282–286.

Pause, A., Lee, S., Worrell, R.A., Chen, D.Y., Burgess, W.H., Linehan, W.M., and Klausner, R.D. 1997. The von Hippel-Lindau tumor-suppressor gene product forms a stable complex with human CUL-2, a member of the Cdc53 family of proteins. *Proc Natl Acad Sci USA* 94:2156–2161.

Payne, D.M., Rossomando, A.J., Martino, P., Erickson, A.K., Her, J.-H., Shananowitz, J., Hunt, D.F., Weber, M.J., and Sturgill, T.W. 1991. Identification of the regulatory phosphorylation sites in pp42/mitogen-activated protein kinase (MAP kinases). *Eur Mol Biol Organ J* 10:885–892.

Payne, L.N. 1994. Marek's disease. In *Encyclopedia of Virology,* vol. 2, pp. 832–837. London: Academic Press.

Pearl, R. 1922. New data on the influence of alcohol on the expectation of life in man. *Am J Hygiene* 2:463–466.

Pecorelli, S. 1994. Management of the symptomatic patient: Surgery. *NIH Consensus Development Conference on Ovarian Cancer.* Bethesda, MD: National Institutes of Health.

Pelicci, G., Lanfrancone, L., Grignani, F., McGlade, J., Cavallo, F., Forni, G., Nicoletti, I., Pawson, T., and Pelicci, P.G. 1992. A novel transforming protein (SHC) with an SH2 domain is implicated in mitogenic signal transduction. *Cell* 70:93–104.

Peltomäki, P., Aaltonen, L.A., Sistonen, P., Pylkkänen, L., Mecklin, J-P., Järvinen, H., Green, J.S., Jass, J.R., Weber, J.L., Leach, F.S., Petersen, G.M., Hamilton, S.R., de la Chapelle, A., and Vogelstein, B. 1993. Genetic mapping of a locus predisposing to human colorectal cancer. *Science* 260:810–812.

Pennisi, E. 1996. Homing in on a prostate cancer gene. *Science* 274:1301.

Peraino, D.R., Fry, R.J., and Staffeldt, E. 1971. Reduction and enhancement by phenobarbital of hepatocarcinogenesis induced in the rat by 2-acetylaminofluorene. *Cancer Res* 31:1506–1512.

Perantoni, A.O., Rice, J.M., Reed, C.D., Watatani, M., and Wenk, M.L. 1987. Activated neu oncogene sequences in primary tumors of the peripheral nervous system induced in rats by transplacental exposure to ethylnitrosourea. *Proc Natl Acad Sci USA* 84:6317–6321.

Perantoni, A.O., Turusov, V.S., Buzard, G.S., and Rice, J.M. 1994. Infrequent transforming mutations in the transmembrane domain of the neu oncogene in spontaneous rat schwannomas. *Mol Carcinog* 9:230–235.

Perris, R., and Bronner-Fraser, M. 1989. Recent advances in defining the role of the extracellular matrix in neural crest development. *Comments Dev Neurobiol* 1:61–83.

Pesenti-Barili, B., Ferdandi, E., Mosti, M., and Degli-Innocenti, F. 1991. Survival of Agrobacterium radiobacter K84 on various carriers for crown gall control. *Appl Environ Microbiol* 57:2047–2051.

Peter, M., and Herskowitz, I. 1994. Joining the complex: Cyclin-dependent kinase inhibitory proteins and the cell cycle. *Cell* 79:181–184.

Peters, E.C. 1989. Disease processes in marine bivalve molluscs. *Am Fish Soc* 18:74–92.

Petruzzi, M.J., and Green, D.M. 1997. Wilms' tumor. *Pediatr Clin North Am* 44:939–952.

Phelps, W.C., Yee, C.L., Munger, K., and Howley, P.M. 1988. The human papillomavirus type 16 E7 gene encodes transactivation and transformation functions similar to adenovirus E1a. *Cell* 53:539–547.

Phillips, D.H., Carmichael, P.L., Hewer, A., Cole, K.J., Hardcastle, I.R., Poon, G.K., Keogh, A., and Strain, A.J. 1996. Activation of tamoxifen and its metabolite alpha-hyroxytamoxifen to DNA-binding products: Comparisons between human, rat and mouse hepatocytes. *Carcinogenesis* 17:89–94.

Pickle, L.W., Mason, T.J., Howard, N., Hoover, R., and Fraumeni, J.F. 1987. *Atlas of U.S. Cancer Mortality Among Whites.* Bethesda, MD: Public Health Service, National Institutes of Health.

Pickle, L.W., Mungiole, M., Jones, G.K., and White, A.A. 1996. *Atlas of United States Mortality.* Hyattsville, MD: U.S. Department of Health and Human Services.

Pierce, G.B. 1970. Differentiation of normal and malignant cells. *Fed Proc* 29:1248–1254.

Pierce, G.B. 1974. Neoplasms, differentiation, mutations. *Am J Pathol* 77:103–118.

Pierce, G.B. 1983. The cancer cell and its control by the embryo. *Am J Pathol* 113:117–124.

Pierce, G.B., and Dixon, F.J. 1959. The demonstration of teratogenesis by metamorphosis of multipotential cells. *Cancer* (Phila.) 12:573–589.

Pierce, G.B., Dixon, F.J., and Verney, E.L. 1960. Teratocarcinogenic and tissue forming potentials of the cell types comprising neoplastic embryoid bodies. *Lab Invest* 9:583–602.

Pierce, G.B., Gramzinski, R.A., and Parchment, R.E. 1990. Amine oxidases, programmed cell death, and tissue renewal. *Philos Trans R Soc Lond Biol* 327:67–74.

Pierce, G.B., Lewellyn, A.L., and Parchment, R.E. 1989. Mechanisms of programmed cell death in the blastocyst. *Proc Natl Acad Sci USA* 86:3654–3658.

Pierce, G.B., and Parchment, R. 1991. Progression and teratocarcinoma. In *Boundaries Between Promotion and Progression During Carcinogenesis,* O. Sudilovsky, H. Pitot, and L. Liotta, eds., pp. 71–81. Collection of Basic Life Science. New York: Plenum Press.

Pierce, G.B., Shikes, R.H., and Fink, L.M. 1978. *Cancer: A problem of developmental biology.* Englewood Cliffs, NJ: Prentice-Hall.

Pierce, G.B., and Speers, W.C. 1988. Tumors as caricatures of the process of tissue renewal: Prospects for therapy by directing differentiation. *Cancer Res* 48:1996–2004.

Pierce, G.B., Stevens, L.C., and Nakane, P.K. 1967. Ultrastructural analysis of the early development of teratocarcinomas. *J Natl Cancer Inst* 39:755–773.

Pierce, G.B., and Wallace, C. 1971. Differentiation of malignant to benign cells. *Cancer Res* 31:127–134.

Pines, J. 1994. The cell cycle kinases. *Semin Cancer Biol* 5:305–313.

Piper, K.R., Beck von Bodman, S., and Farrand, S.K. 1993. Conjugation factor of Agrobacterium tumefaciens regulates Ti plasmid transfer by autoinduction. *Nature* 362:448–450.

Pitot, H.C. 1989. Principles of carcinogenesis: Chemical. In *Cancer: Principles and Practice of Oncology*, 3rd ed., V.T. DeVita, S. Hellman, and S.A. Rosenberg, eds., pp. 116–135. Philadelphia: J.B. Lippincott.

Pittillo, R.F., Schabel, F.M., Jr., and Skipper, H.E. 1970. The "sensitivity" of resting and dividing cells. *Cancer Chemother Rep* 54:137–142.

Plasterk, R.H. 1992. Genetic switches: Mechanisms and function. *Trends Genet* 8:403–406.

Platanias, L.C., Pfeffer, L.M., Barton, K.P., Vardiman, J.W., Golomb, H.M., and Colamonici, O.R. 1992. Expression of the IFNa receptor in hairy cell leukaemia. *Br J Haematology* 82:541–546.

Podesta, A.H., Mullins, J., Pierce, G.B., and Wells, R.S. 1984. The neurula stage mouse embryo in control of neuroblastoma. *Proc Natl Acad Sci USA* 81:7608–7611.

Poland, A., Glover, E., and Kende, A.S. 1976. Stereospecific high affinity binding of 2,3,7,8-tetra-chlorodibenzo-p-dioxin by hepatic cytosols. Evidence that the binding species is a receptor for induction of aryl hydrocarbon hydroxylase. *J Biol Chem* 251:4936–4946.

Polednak, A.P. 1976. College athletics, body size, and cancer mortality. *Cancer* 38:382–387.

Polyak, K., Kato, J.Y., Solomon, M.J., and Sherr, C.J. 1994. p27^{Kip1}, a cyclin-Cdk inhibitor, links transforming growth factor-beta and contact inhibition to cell cycle arrest. *Genes Dev* 8:9–22.

Postel, E.H., Berberich, S.J., Flint, S.J., and Ferrone, C.A. 1993. Human c-*myc* transcription factor PuF identified as nm23-H2 nucleoside diphosphate kinase, a candidate suppressor to tumor metastasis. *Science* 261:478–480.

Post-White, J., Herzan, D., Drew, D., and Anglim, M.A. 1989. Nutrition and cancer: Educating the public through a health fair. *Oncol Nurs Forum* 16:115–118.

Pott, P. 1775. *Chirurgical observations relative to the cataract, the polypus of the nose, the cancer of the scrotum, the different kinds of ruptures and the mortification of the toes and feet.* London: Hawkes, Clarke, and Collins.

Potten, C.S., Booth, C., Chadwick, C.A., and Evans, G.S. 1994. A potent stimulator of small intestinal cell proliferation extracted by simple diffusion from intact irradiated intestine: In vitro studies. *Growth Factors* 10:53–61.

Potten, C.S., and Chadwick, C.A. 1994. Small intestinal regulatory factors extracted by simple diffusion from intact irradiated intestine and tested in vivo. *Growth Factors* 10:63–75.

Potten C.S., Watson R.J., Williams G.T., Tickle, S., Roberts, S.A., Harris, M., and Howell, A. 1988. The effect of age and menstrual cycle upon proliferative activity of normal human breast. *Br J Cancer* 58:163–170.

Powell, S.M., Zilz, N., Beazer-Barclay, Y., Bryan, T.M., Hamilton, S.R., Thibodeau, S.N., Vogelstein, B., and Kinzler, K.W. 1992. APC mutations occur early during colorectal tumorigenesis. *Nature* 359:235–237.

Pulciani, S., Santos, E., Lauver, A.V., Long, L.K., Robbins, K.C., and Barbacid, M. 1982. Oncogenes in human tumor cell lines: Molecular cloning of a transforming gene from human bladder cell carcinoma cells. *Proc Natl Acad Sci USA* 79:2845–2849.

Pulverer, B.J., Kyriakis, J.M., Avruch, J., Nikolakaki, E., and Woodgett, J.R. 1991. Phosphorylation of c-jun mediated by MAP kinases. *Nature* 353:670–674.

Purchio, A.F., Erikson, E., Brugge, J.S., and Erikson, R.L. 1978. Identification of a polypeptide encoded by the avian sarcoma virus *src* gene. *Proc Natl Acad Sci USA* 75:1567–1571.

Pye, G., Evans, D.F., Ledingham, S., and Hardcastle, J.D. 1990. Gastrointestinal intraluminal pH in normal subjects and those with colorectal adenoma or carcinoma. *Gut* 31:1355–1357.

Quintanilla, M., Brown, K., Ramsden, M., and Balmain, A. 1986. Carcinogen-specific mutation and amplification of Ha-*ras* during mouse skin carcinogenesis. *Nature* 322:78–80.

Rabinowitz, Y., and Wilhite, B.A. 1969. Thymidine salvage pathway in normal and leukemic leukocytes with effects of ATP on enzyme control. *Blood* 33:759–771.

Radonski, J.L. 1979. The primary aromatic amines, their biological properties and structure-activity relationships. *Annu Rev Pharmacol Toxicol* 19:129–157.

Recamier, J.C.A. 1829. *Recherches sur le traitement du cancer, par la compression methodique simple ou combinée et sur l'histoire générale de la même maladie,* vol. II. Paris: Gabon.

Recny, M.A., Scoble, H.A., and Kim, Y. 1987. Structural characterization of natural human urinary and recombinant DNA-derived erythropoietin. *J Biol Chem* 262:17156–17163.

Reddel, R.R., Alexander, I.E., Koga, M., Shine, J., and Sutherland, R.L. 1988. Genetic instability and the development of steroid hormone insensitivity in cultured T 47D human breast cancer cells. *Cancer Res* 48:4340–4347.

Redman, B.G., Kawachi, M., and Schwartz, D. 1996. Urothelial and kidney cancers. In *Cancer Management: A Multidisciplinary Approach,* R. Pazdur, L. R. Coia, W. J. Hoskins, and L. D. Wagman, eds., pp. 417–431. Huntington, NY: PRR.

Reed, E., and Kohn, K.W. 1990. Platinum analogues. In *Cancer Chemotherapy,* B.A. Chabner and J.M. Collins, eds., pp. 465–490. Philadelphia: J.B. Lippincott.

Regaud, C. 1930. Sur les principes radiophysiologiques de la radiotherapie des cancers. *Acta Radiol* 11:456–486.

Regaud, C., and Ferroux, R. 1927. Discordance des effets des rayons X, d'une part dans la peau, d'autre part dans le testicule, par le fractionnement de la dose; diminution de l'efficacité dans le peau, maintien de l'efficacité dans le testicule. *C R Soc Biol* 97:431–434.

Rehn, L. 1895. Bladder tumours in Fuchsine workers. *Arch Klin Chir* 50:588–600.

Reich, R., Stratford, B., Klein, K., Martin, G.R., Mueller, R.A., and Fuller, G.C. 1988. Inhibitors of collagenase IV and cell adhesion reduce the invasive activity of malignant tumour cells. *Ciba Found Symp* 141:193–210.

Reiss, M., Gamba-Vitalo, C., and Sartorelli, A.C. 1986. Induction of tumor differentiation as a therapeutic approach: Preclinical models for hematopoietic and solid neoplasms. *Cancer Treat Rep* 70:201–218.

Remmer, H. 1962. Drug tolerance. In *Ciba Foundations Symposium on Enzymes and Drug Action,* J.L. Mongar and A.V.S. De Reuck, eds., pp. 276–298. Boston: Little, Brown.

Rendina, G.M., Donadio, C., and Giovannini. M. 1982. Steroid receptors and progestinic therapy in ovarian endometrioid carcinoma. *Eur J Gynaecol Oncol* 3:241–246.

Repesh, L.A., Drake, S.R., Warner, M.C., Downing, S.W., Jyring, R., Seftor, E.A., Hedrix, M.J.C., and McCarthy, J.B. 1993. Adriamycin-induced inhibition of melanoma cell invasion is correlated with decreases in tumour cell motility and increases in focal contact formation. *Clin Exp Metastasis* 11:91–102.

Richardson, S., Gerber, M., and Cenee, S. 1991. The role of fat, animal protein and some vitamin consumption in breast cancer: A case control study in southern France. *Int J Cancer* 48:1–9.

Richter, C., Park, J.-W., and Ames, B.N. 1988. Normal oxidative damage to mitochondrial and nuclear DNA is extensive. *Proc Natl Acad Sci USA* 85:6465–6467.

Rickles, F.R., Hancock, W.W., Edwards, R.L., and Zacharski, L.R. 1988. Antimetastatic agents: I. Role of cellular procoagulants in the pathogenesis of fibrin deposition in cancer and the use of anticoagulants and/or antiplatelet drugs in cancer treatment. *Semin Thromb Hemostasis* 14:88–94.

Rigel, D.S. 1996. Malignant melanoma: Perspectives on incidence and its effects on awareness, diagnosis, and treatment. *CA Cancer J Clin* 46:195–198.

Rigel, D.S., Kopf, A.W., and Friedman, R.J. 1987. The rate of malignant melanoma in the United States: Are we making an impact? *J Am Acad Dermatol* 17:1050–1053.

Rijke, R.P., Plaisier, H.M., and Langendoen, N.J. 1979. Epithelial cell kinetics in the descending colon of the rat. *Virchows Arch B* 30:85–94.

Ringstad, J., Jacobsen, B.J., Tretli, S., and Thomassen, Y. 1988. Serum selenium concentration associated with risk of cancer. *J Clin Pathol* 41:454–457.

Riou, G., Favre, M., Jeannel, D., Bourhis, J., Le Doussal, V., and Orth, G. 1990. Association between poor prognosis in early-stage invasive cervical carcinomas and non-detection of HPV DNA. *Lancet* 335:1171–1174.

Rizzino, A., and Crowley, C. 1980. The growth and differentiation of the embryonal carcinoma cell line, F9, in defined media. *Proc Natl Acad Sci USA* 77:457–461.

Robertson, A.M., Bird, C.C., Waddell, A.W., and Currie, A.R. 1978. Morphological aspects of glucocorticoid-induced cell death in human lymphoblastoid cells. *J Pathol* 126:181–187.

Robinson, E.S., VandeBerg, J.L., Hubbert, G.B., and Dooley, T.P. 1994. Malignant melanoma in ultraviolet irradiated laboratory opossums: Initiation in suckling young, metastasis in adult, and xenograph behavior in nude mice. *Cancer Res* 54:5986–5991.

Roffo, A.H. 1935. Krebs und Sarkom durch Ultraviolett- und Sonnenstrahlen. *Z Krebsforsch* 41:448–467.

Rohan, T.E., McMichael, A.J., and Baghurst, P.A. 1988. A population-based case-control study of diet and breast cancer in Australia. *Am J Epidemiol* 128:478–489.

Roizman, B., Desrosiers, R.C., Fleckenstein, B., Lopez, C., Minson, A.C., and Studdert, M.J. 1995. Family Herpesviridae. In *Virus Taxonomy*, F.A. Murphy, C.M. Fauquet, D.H.L. Bishop, S.A. Ghabrial, A.W. Jarvis, G.P. Martelli, M.A. Mayo, and M.D. Summers, eds., pp. 114–127. New York: Springer-Verlag.

Rong-Fu, W., Beasley, J.N., Cao, W.W., Slavik, M.F., and Johnson, M.G. 1993. Development of PCR method specific for Marek's disease virus. *Mol Cell Probes* 7:127–131.

Rose, F.L., and Harshbarger, J.C. 1977. Neoplastic and possibly related skin lesions in neotenic tiger salamanders from a sewage lagoon. *Science* 298:270–279.

Rosen, M., Nystrom, L., and Wall, S. 1988. Diet and cancer mortality in the counties of Sweden. *Am J Epidemiol* 127:42–49.

Rosenberg, B., VanCamp, L., Trosko, J.E., and Mansour, V.H. 1965. Inhibition of cell division in *Escherichia coli* by electrolysis products from a platinum electrode. *Nature* 205:698–699.

Rosenberg, B., VanCamp, L., Trosko, J.E., and Mansour, V.H. 1969. Platinum compounds: A new class of potent anti-tumour agents. *Nature* 222:385–386.

Rosenberg, S.A. 1988. The development of new immunotherapies for the treatment of cancer using interleukin-2. A review. *Ann Surg* 208:121–135.

Rosenberg, S.A. 1992. Karnofsky Memorial Lecture. The immunotherapy and gene therapy of cancer. *J Clin Oncol* 10:180–199.

Rostami, M., Tateyama, S., Uchida, K., Naitou, H., Yamaguchi, R., and Ostuka, H. 1994. Tumors in domestic animals examined during a ten-year period 1980 to 1990 at Miyazaki University. *J Vet Med Sci* 56:403–405.

Rous, P. 1911a. A sarcoma of the fowl transmissible by an agent separable from the tumor cells. *J Exp Med* 13:397–411.

Rous, P. 1911b. Transmission of a malignant new growth by means of a cell-free filtrate. *J Am Med Assoc* 56:198–208.

Rowley, J.D. 1973. A new consistent chromosomal abnormality in chronic myelogenous leukaemia identified by quinacrine fluorescence and Giemsa staining. *Nature* 243:290–293.

Rubin, P., and Cooper, R.A. 1993. Statement of the clinical oncologic problem. In *Clinical Oncology: A Multidisciplinary Approach for Physicians and Students,* 7th ed., P. Rubin, ed., pp. 8–11. Philadelphia: W.B. Saunders.

Ruddon, R.W. 1995. *Cancer Biology,* 3rd ed. New York: Oxford University.

Rudkin, G.T., Hungerford, D.A., and Nowell, P.C. 1964. DNA contents of chromosome Ph[1] and chromosome 21 in human granulocytic leukemia. *Science* 144:1229–1232.

Ruggeri, B., Zhang, S.Y., Caamano, J., DiRado, M., Flynn, S.D., and Klein-Szanto, A.J.P. 1992. Human pancreatic carcinomas and cell lines reveal frequent and multiple alterations in the p53 and Rb-1 tumor-suppressor genes. *Oncogene* 7:1503–1511.

Ruley, H.E. 1990. Transforming collaborations between ras and nuclear oncogenes. *Cancer Cells* 2:258–268.

Rusch, E.P., and Kline, B.E. 1944. The effect of exercise on the growth of a mouse tumor. *Cancer Res* 4:116–118.

Russell, S.W., Doe, W.F., Hoskins, R.G., and Cochrane, C.G. 1976. Inflammatory cells in solid murine neoplasms: I. Tumor disaggregation and identification of constituent inflammatory cells. *Int J Cancer* 18:322–330.

Ryan, J.J., Ketcham, A.S., and Wexler, H. 1968. Warfarin treatment of mice bearing autochthonous tumors: Effect on spontaneous metastases. *Science* 162:1493–1494.

Sachs, L. 1987. Cell differentiation and bypassing genetic defects in the suppression of malignancy. *Cancer Res* 47:1981–1986.

Sachs, L. 1993. Regulation of normal development and tumor suppression. *Int J Dev Biol* 37:51–59.

Safai, B. 1997. Management of skin cancer. In *Cancer: Principles and Practice of Oncology,* 5th ed., V.T. DeVita, S. Hellman, and S.A. Rosenberg, eds., pp. 1883–1933. Philadelphia: Lippincott-Raven.

Safai, B., Diaz, B., and Schwartz, J. 1992. Malignant neoplasms associated with human immunodeficiency virus infection. *CA Cancer J Clin* 42:74–95.

Salonen, J.T. 1986. Selenium and human cancer. *Ann Clin Res* 18:18–21.

Salsbury, A.J. 1975. The significance of the circulating cancer cell. *Cancer Treat Rev* 2:55–72.

Samet, J.M., Kutvirt, D.M., Waxweiler, R.J., and Key, C.R. 1984. Uranium mining and lung cancer in Navajo men. *N Engl J Med* 310:1481–1484.

Samuels, M.L., Weber, M.J., Bishop, J.M., and McMahon, M. 1993. Conditional transformation of cells and rapid activation of the mitogen-activated protein kinase cascade by an estradiol-dependent human *raf*-1 protein kinase. *Mol Cell Biol* 13:6241–6252.

Sancar, G.B., Siede, W., and van Zeeland, A.A. 1996. Repair and processing of DNA damage: A summary of recent progress. *Mutat Res* 362:127–146.

Sandford, N.L., Searle, J.W., and Kerr, J.F. 1984. Successive waves of apoptosis in the rat prostate after repeated withdrawal of testosterone stimulation. *Pathology* 164:406–410.

Sandow, B.A., West, N.B., Norman, R.L., and Brenner, R.M. 1979. Hormonal control of apoptosis in hamster uterine luminal epithelium. *Am J Anat* 156:15–35.

Sano, T., Fukuda, H., and Furukawa, M. 1985. *Herpesvirus cyprini:* Biological and oncogenic properties. *Fish Pathol* 20:381–388.

Santen, R.J., Manni, A., and Harvey, H. 1986. Gonadotropin releasing hormone (GnRH) analogs for the treatment of breast and prostatic carcinoma. *Breast Cancer Res Treat* 7:129–145.

Santos, E., and Nebreda, A.R. 1989. Structural and functional properties of *ras* proteins. *Fed Am Soc Exp Biol J* 3:2151–2163.

Saracci, R. 1987. The interactions of tobacco smoking and other agents in cancer etiology. *Epidemiol Rev* 9:175–193.

Sasaki, M. 1982. Current status of cytogenetic studies in animal tumors with special reference to nonrandom chromosome changes. *Cancer Genet Cytogenet* 5:153–172.

Sauerbier, W., Rollins-Smith, L.A., Carlson, D.L., Williams, C.S., Williams, J.W., III, and McKinnell, R.G. 1995. Sizing of the Lucké tumor herpesvirus genome by field inversion gel electrophoresis and restriction analysis. *Herpetopathologia* 2:137–143.

Scarpelli, D.G. 1975. Neoplasia in poikilotherms. In *Cancer 4, Biology of Tumors,* F.F. Becker, ed., pp. 375–410. New York: Plenum.

Schaaper, R.M., and Dunn, R.L. 1987. Spectra of spontaneous mutations in Escherichia coli strains defective in mismatch repair correction: The nature of in vivo DNA replication errors. *Proc Natl Acad Sci USA* 84:6220–6224.

Schabel, F.M., Jr. 1969. The use of tumor growth kinetics in planning "curative" chemotherapy of advanced solid tumors. *Cancer Res* 29:2384–2389.

Schabel, F.M., Jr. 1975. Concepts for systemic treatment of micrometastases. *Cancer* 35:15–24.

Schabel, F.M., Jr., Skipper, H.E., Trader, M.W., Laster, W.R., Jr., Corbett, T.H., and Griswold, D.P., Jr. 1980. Concepts for controlling drug-resistant tumor cells. *Eur J Cancer* Suppl. 1:199–211.

Schabel, F.M., Jr., Skipper, H.E., Trader, M.W., and Wilcox, W.S. 1965. Experimental evaluation of potential anticancer agents: XIX. Sensitivity of nondividing leukemic cell populations to certain classes of drugs in vivo. *Cancer Chemother Rep* 48:17–30.

Scharrer, B., and Lochhead, M.S. 1950. Tumors in the invertebrates: A review. *Cancer Res* 10:403–419.

Schartl, A., and Schartl, M. 1996. Tumor induction and tumor regression in Xiphorous. *In Vivo* 10:179–184.

Scheffner, M., Werness, B.A., Huibregtse, J.M., Levine, A.J., and Howley, P.M. 1990. The E6 oncoprotein encoded by human papillomavirus types 16 and 18 promotes the degradation of p53. *Cell* 63:1129–1136.

Scheumman, G.F.W., Hoang-Vu, C., Cetin, Y., Gimm, O., Behrends, J., Von Wasielewski, R., Georgii, A., Birchmeier, W., Von zur Mühlen, A., Dralle, H., and Brabant, G. 1995. Clinical significance of E-cadherin as a prognostic marker in thyroid carcinomas. *Clin Endocrinol Metab* 80:2168–2172.

Schiffer, C.A. 1993. Acute lymphocytic leukemia in adults. In *Cancer Medicine,* vol. 2, J.F. Holland, E. Frei III, R.C. Bast, Jr., D.W. Kufe, D.L. Morton, and R.R. Weichselbaum, eds., pp. 1946–1955. Philadelphia: Lea and Febiger.

Schiffmann, E. 1990. Motility as a principal requirement for metastasis. *Cancer Invest* 8:673–674.

Schlegel, R. 1990. Papilloma viruses and human cancer. *Semin Virol* 1:297–306.

Schlumberger, H.G., and Lucké, B. 1948. Tumors of fishes, amphibians, and reptiles. *Cancer Res* 8:657–754.

Schmeer, A.C. 1969. Mercenene: An antineoplastic agent extracted from the marine clam, *Mercenaria mercenaria. Natl Cancer Inst Monogr* 31:581–591.

Schröck, E., du Manoir, S., Veldman, T., Schoell, B., Wienberg, J., Ferguson-Smith, M.A., Ning, Y., Ledbetter, D.H., Bar-Am, I., Soenksen, D., Garini, Y., and Ried, T. 1996. Multicolor spectral karyotyping of human chromosomes. *Science* 273:494–497.

Schuchardt, A., D'Agati, V., Larsson-Blomberg, L., Costantini, F., and Pachnis, V. 1994. Defects in the kidney and enteric nervous system of mice lacking the tyrosine kinase receptor Ret. *Nature* 367:380–383.

Schuster, S.J., Wilson, J.H., Erslev, A.J., and Caro, J. 1987. Physiologic regulation and tissue localization of renal erythropoietin messenger RNA. *Blood* 70:316–318.

Schwartz, M.K. 1993. Cancer markers. In *Cancer: Principles and Practice of Oncology,* 4th ed., V.T. DeVita, Jr., S. Hellman, and S.A. Rosenberg, eds., pp. 531–542. Philadelphia: J.B. Lippincott.

Seger, R., Ahn, N.G., Posada, J., Manur, E.S., Jensen, A.M., Cooper, J.A., Cobb, M.H., and Krebs, E.G. 1992. Purification and characterization of mitogen-activated protein kinase activator(s) from epidermal growth factor-stimulated A431 cells. *J Biol Chem* 267:14373–14381.

Selikoff, I.J., Seidman, H., and Hammond, E.C. 1980. Mortality effects of cigarette smoking among amosite asbestos factory workers. *J Natl Cancer Inst* 65:507–513.

Sell, S., ed. 1980. *Cancer Markers. Diagnostic and Developmental Significance.* Clifton, NY: Humana Press.

Sell, S., and Leffert, H.L. 1982. An evaluation of cellular lineages in the pathogenesis of hepatocellular carcinomas. *Hepatology* 2:77–86.

Sell, S., and Pierce, G.B. 1994. Maturation arrest of stem cell differentiation is a common pathway for the cellular origin of teratocarcinomas and epithelial cancers. *Lab Invest* 70:6–22.

Seppanen, E.D., McKinnell, R.G., Rollins-Smith, L.A., and Hanson, W. 1984. Temperature-dependent dissociation of Lucké renal adenocarcinoma cells. *Differentiation* 26:227–230.

Serrano, M., Hannon, G.J., and Beach, D. 1993. A new regulatory motif in cell-cycle control causing specific inhibition of cyclin D/CDK4. *Nature* 366:704–707.

Serrano, M., Lee, H-W., Chin, L., Cordon-Cardo, C., Beach, D., and DePinho, R.A. 1996. Role of the INK4a locus in tumor suppression and cell mortality. *Cell* 85:27–37.

Setlow, R.B. 1974. The wavelength of sunlight effective in producing skin cancer: A theoretical analysis. *Proc Natl Acad Sci USA* 71:3363–3366.

Shackney, S.E., McCormack, G.W., and Cuchural, G.J., Jr. 1978. Growth rate patterns of solid tumors and their relation to responsiveness to therapy: An analytical review. *Ann Int Med* 89.107–121.

Shankar, S., and Lanza, E. 1991. Dietary fiber and cancer prevention. *Hematol Oncol Clin North Am* 5:25–41.

Shapiro, D.M., and Fugmann, R.A. 1957. A role for chemotherapy as an adjunct to surgery. *Cancer Res* 17:1098–1101.

Shaw, P.E., Schroter, H., and Nordheim, A. 1989. The ability of a ternary complex to form over the serum response element correlates with serum inducibility of the human c-*fos* promoter. *Cell* 56:563–572.

Shelby, M.D., and Zeiger, E. 1990. Activity of human carcinogens in the Salmonella and rodent bone-marrow cytogenetics tests. *Mutat Res* 34:257–261.

Sherr, C.J. 1996. Cancer cell cycles. *Science* 274:1672–1677.

Shibutani, S., Takeshita, M., and Grollman, A.P. 1991. Insertion of specific bases during DNA synthesis past the oxidation-damaged base 8-oxodG. *Nature* 349:431–434.

Shields, S.E., Ogilvie, D.J., McKinnell, R.G., and Tarin, D. 1984. Degradation of basement membrane collagens by metalloproteases released by human, murine, and amphibian tumours. *J Pathol* 143:193–197.

Shih, C., Padhy, L.C., Murray, M., and Weinberg, R.A. 1981. Transforming genes of carcinomas and neuroblastomas introduced into mouse fibroblasts. *Nature* 290:261–264.

Shih, C., and Weinberg, R.A. 1982. Isolation of a transforming sequence from a human bladder carcinoma cell line. *Cell* 29:161–169.

Shimizu, Y., Kato, H., and Schull, W.J. 1990. Studies of the mortality of A-bomb survivors; mortality, 1950–1985: Part 2. Cancer mortality based on the recently revised doses DS86. *Radiation Res* 121:120–141.

Shimizu, Y., Kato, H., Schull, W.J., Preston, D.L., Fujita, S., and Pierce, D.A. 1989. Studies of the mortality of A-bomb survivors; mortality, 1950–1985: Part 1. Comparison of risk coefficients for site-specific cancer mortality based on the DS86 and T65DR shielded kerma and organ doses. *Radiation Res* 118:502–524.

Shimkin, M.B. 1977. *Contrary to Nature.* Pub NIH 76–720. Washington, DC: U.S. Dept. Health, Education and Welfare.

Sikic, B.I., Fisher, G.A., Lum, B.L., Brophy, N.A., Yahanda, A.M., Alder, K.M., and Halsey, J. 1994. Clinical reversal of multidrug resistance. *Cancer Treat Res* 73:167–200.

Silvennoinen, O., Ihle, J.N., Schlessinger, J., and Levy, D.E. 1993. Interferon-induced nuclear signalling by Jak protein tyrosine kinases. *Nature* 366:583–585.

Simard, J., Tonin, P., Durocher, F., Morgen, K., Rommens, J., Gingras, S., Samson, C., Leblanc, J.-F., Bélanger, C., Dion, F., Liu, Q., Skolnick, M., Goldgar, D., Shattuck-Eidens, D., Labrie, F., and Narod, S.A. 1994. Common origins of *BRCA1* mutations in Canadian breast and ovarian cancer families. *Nat Genet* 8:392–398.

Simpson-Herren, L., and Noker, P.E. 1990. Effects of the initial chemotherapy on subsequent therapy. *Prog Clin Biol Res* 354A:21–29.

Sincock, A.M., Delhanty, J.D., and Casey, G.A. 1982. A comparison of the cytogenetic response to asbestos and glass fibers in CHO cell lines. *Mutat Res* 101:257–268.

Sinha, A.A., ed. 1995. Biology of normal, abnormal and aging prostate. *Micro Res Tech* 30:269–350.

Sinha, A.A., Gleason, D.F., Deleon, O.F., Wilson, M.J., Limas, C., Reddy, P.K., and Furcht, L.T. 1991. Localization of Type IV collagen in the basement membranes of human prostate and lymph nodes by immunoperoxidase and immunoalkaline phosphatase. *Prostate* 18:93–104.

Sinha, A.A., Gleason, D.F., Deleon, O.F., Wilson, M.J., and Sloane, B.F. 1993. Localization of a biotinylated cathepsin B oligonucleotide probe in human prostate including invasive cells and invasive edges by in situ hybridization. *Anat Rec* 235:233–240.

Sinha, A.A., Gleason, D.F., Limas, C., Reddy, P.K., Wick, M.R., Hagen, K.A., and Wilson, M.J. 1989. Localization of cathepsin B in normal and hyperplastic human prostate by immunoperoxidase and protein A-gold techniques. *Anat Rec* 223:266–275.

Sinha, A.A., Gleason, D.F., Staley, N.A., Wilson, M.J., Sameni, M., and Sloane, B.F. 1995. Cathepsin B in angiogenesis of human prostate: An immunohistochemical and immunoelectron microscope analysis. *Anat Rec* 241:353–362.

Sisskin, E.E., Gray, T., and Barrett, J.C. 1982. Correlation between sensitivity to tumor promotion and sustained epidermal hyperplasia of mice and rats treated with 12-O-tetradecanoylphorbol-13-acetate. *Carcinogenesis* 3:403–407.

Skipper, H.E., and Perry, S. 1970. Kinetics of normal and leukemic leukocyte populations and relevance to chemotherapy. *Cancer Res* 30:1883–1897.

Skipper, H.E., Schabel, F.M., Jr., and Lloyd, H.H. 1978. Experimental therapeutics and kinetics: Selection and overgrowth of specifically and permanently drug-resistant tumor cells. *Semin Hematol* 15:207–219.

Skipper, H.E., Schabel, F.M., Jr., Mellett, L.B., Montgomery, J.A., Wilkoff, L.J., Lloyd, H.H., and Brockman, R.W. 1970. Implications of biochemical, cytokinetic, pharmacologic, and toxicologic relationships in the design of optimal therapeutic schedules. *Cancer Chemother Rep* 54:431–450.

Skipper, H.E., Schabel, F.M., Jr., and Wilcox, W.S. 1967. Experimental evaluation of potential anticancer agents: XXI. Scheduling of arabinosylcytosine to take advantage of its S-phase specificity against leukemia cells. *Cancer Chemother Rep* 51:125–165.

Slack, J.M.W. 1983. *From Egg to Embryo: Determinative Events in Early Development,* pp. 136–161. Cambridge, London: Cambridge University Press.

Slaga, T., Bowden, G.T., and Boutwell, R.K. 1975. Acetic acid, a potent stimulator of mouse epidermal macromolecular synthesis and hyperplasia but weak promoting activity. *J Natl Cancer Inst* 55:983–987.

Slamon, D.J., Clark, G.M., Wong, S.G., Levin, W.J., Ullrich, A., and McGuire, W.L. 1987. Human breast cancer: Correlation of relapse and survival with amplification of the HER2/*neu* oncogene. *Science* 235:177–182.

Slamon, D.J., and Cline, M.J. 1984. Expression of cellular oncogenes during embryonic and fetal development of the mouse. *Proc Natl Acad Sci USA* 81:7141–7145.

Slattery, M.L., Schuman, K.L., West, D.W., French, T.K., and Robison, L.M. 1989. Nutrient intake and ovarian cancer. *Am J Epidemiol* 130:497–502.

Slavin, J.L. 1987. Dietary fiber: Classification, chemical analyses, and food sources. *J Am Diet Assoc* 87:1164–1171.

Slichenmyer, W.J., Rowinsky, E.K., Grochow, L.B., Kaufmann, S.H., and Donehower, R.C. 1994. Camptothecin analogues: Studies from the Johns Hopkins Oncology Center. *Cancer Chemother Pharmacol* 34 Suppl:S53–57.

Sloan, D.A., Schwartz, R.W., McGrath, P.C., and Kenady, D.E. 1996. Diagnosis and management of adrenal tumors. *Curr Opin Oncol* 8:30–36.

Sloane, B.F., and Honn, K.V. 1984. Cysteine proteinases and metastasis. *Cancer Metastasis Rev* 3:249–263.

Smeyne, R.J., Klein, R., Schnapp, A., Long, L.K., Bryant, S., Lewin, A., Lira, S.A., and Barbacid, M. 1994. Severe sensory and sympathetic neuropathies in mice carrying a disrupted Trk/NGF receptor gene. *Nature* 368:246–249.

Smith, E.F., and Townsend, C.O. 1907. A plant-tumor of bacterial origin. *Science* 25:671–673.

Smith, J.R., Freije, D., Carpten, J.D., Grönberg, H., Xu, J., Isaacs, S.D., Brownstein, M.J., Bova, G.S., Guo, H., Bujnovszky, P., Nusskern, D.R., Damber, J-E, Bergh, A., Emanuelsson, M., Kallioniemi, O.P., Walker-Daniels, J., Bailey-Wilson, J.E., Beaty, T.H., Meyers, D.A., Walsh, P.C., Collins, F.S., Trent, J.M., and Isaacs, W.B. 1996. Major susceptibility locus for prostate cancer on chromosome 1 suggested by a genome-wide search. *Science* 274:1371–1374.

Smith, K.J., Johnson, K.A., Bryan, T.M., Hill, D.E., Markowitz, S., Willson, J.K.V., Paraskeva, C., Petersen, G.M., Hamilton, S.R., Vogelstein, B., and Kinzler, K.W. 1993. The *APC* gene product in normal and tumor cells. *Proc Natl Acad Sci USA* 90:2846–2850.

Smith, M.W., and Jarvis, L.G. 1980. Use of differential interference contrast microscopy to determine cell renewal times in mouse intestine. *J Micros* 118:153–159.

Smith, R.B., and Haskell, C.M. 1990. Testis. In *Cancer Treatment*, 3rd ed., C.M. Haskell, ed., pp. 779–797. Philadelphia: W.B. Saunders.

Smolev, J.K., Coffey, D.S., and Scott, W.W. 1977. Experimental models for the study of prostatic adenocarcinoma. *J Urol* 118:216–220.

Smolev, J.K., Heston, W.D., Scott, W.W., and Coffey, D.S. 1977. Characterization of the Dunning R3327H prostatic adenocarcinoma: An appropriate animal model for prostatic cancer. *Cancer Treat Rep* 61:273–287.

Smulson, M.E., Schein, P., Mullins, D.W., Jr., and Sudhakar, S.A. 1977. A putative role for nicotinamide adenine dinucleotide-promoted nuclear protein modification in the antitumor activity of N-methyl-N-nitrosourea. *Cancer Res* 37:3006–3012.

Snell, R.S. 1978. *Clinical and Functional Histology for Medical Students*. Boston: Little, Brown.

Snellwood, R.A., Kuper, S.W.A., Payne, P.M., and Burn, J.I. 1969. Factors affecting the finding of cancer cells in the blood. *Br J Surg* 56:649–652.

Snodgrass, M.J., and Burke, J.D. 1976. Inhibitory effect of shark serum on the Lewis lung carcinoma. *J Natl Cancer Inst* 55:981–984.

Song, M.J., Reilly, A.A., Parsons, D.F., and Hussain, M. 1986. Patterns of blood-vessel invasion by mammary tumor cells. *Tissue Cell* 18:817–825.

Sonnenberg, E., Godecke, A., Walter, B., Bladt, F., and Birchmeier, C. 1991. Transient and locally restricted expression of the ros1 protooncogene during mouse development. *Eur Mol Biol Organ J* 10:3693–3702.

Sonnenberg, E., Meyer, D., Weidner, M., and Birchmeier, C. 1993. Scatter factor/hepatocyte growth factor and its receptor, the c-met tyrosine kinase, can mediate a signal exchange between mesenchyme and epithelia during mouse development. *J Cell Biol* 123:223–235.

Sora, M., Einisto, P., Husgafvel-Pdursiainen, K., Jarventaus, H., Kivisto, H., Peltonen, Y., Tuomi, T., and Valkonen, S. 1985. Passive and active exposure to cigarette smoke in a smoking experiment. *J Toxicol Environ Health* 16:523–534.

Sparkes, R.S., Murphree, A.L., Lingua, R.W., Sparkes, M.C., Field, L.L., Funderburk, S.J., and Benedict, W.F. 1983. Gene for hereditary retinoblastoma assigned to human chromosome 13 by linkage to esterase D. *Science* 219:971–973.

Sparnins, V.L., Barany, G., and Wattenberg, L.W. 1988. Effects of organo-sulfur compounds from garlic and onions on benzo(a)pyrene-induced neoplasia and glutathione S-transferase activity. *Carcinogenesis* 9:131–134.

Spinelli, G., Bardazzi, N., Citernesi, A., Fontanarosa, M., and Curiel, P. 1991. Endometrial carcinoma in tamoxifen-treated breast cancer patients. *J Chemother* 3:267–270.

Spratt, N.T. 1946. Formation of the primitive streak in the explanted chick blastoderm marked with carbon particles. *J Exp Zool* 103:259–304.

Squier, C.A. 1988. The nature of smokeless tobacco and pattern of its use. *CA Cancer J Clin* 38:226–229.

Squires, D.F. 1965. Neoplasia in a coral? *Science* 148:503–505.

Stahelin, H.B., Gey, K.F., Eichholzer, M., Ludin, E., Bernasconi, F., Thurneysen, J., and Brubacher, G. 1991. Plasma antioxidant vitamins and subsequent cancer mortality in the 12-year follow-up of the prospective Basel study. *Am J Epidemiol* 133:766–775.

Stanton, B.R., Perkins, A.S., Tessarollo, L., Sassoon, D.A., and Parada, L.F. 1992. Loss of N-*myc* function results in embryonic lethality and failure of the epithelial component of the embryo to develop. *Genes Dev* 6:2235–2247.

Stanwell, C., Gescher, A., Bradshaw, T.D., and Pettit, G.R. 1994. The role of protein kinase C isoenzymes in the growth inhibition caused by bryostatin 1 in human A549 lung and MCF-7 breast carcinoma cells. *Int J Cancer* 56:585–592.

Stayner, L.T., Dankovic, D.A., and Lemen, R.A. 1996. Occupational exposure to chrysotile asbestos and cancer risk: A review of the amphibole hypothesis. *Am J Public Health* 86:179–186.

Steeg, P.S., Bevilacqua, G., Kopper, L., Thorgeirsson, U.P., Talmadge, J.E., Liotta, L.A., and Sobel, M.E. 1988. Evidence for a novel gene associated with low tumor metastatic potential. *J Natl Cancer Inst* 80:200–204.

Steer, H.W. 1971. Implantation of the rabbit blastocyst: The invasive phase. *J Anat London* 110:445–462.

Stehelin, D., Varmus, H.E., Bishop, J.M., and Vogt, P.K. 1976. DNA related to the transforming gene(s) of avian sarcoma virus is present in normal avian DNA. *Nature* 260:170–173.

Steinmetz, K.A., and Potter, J.D. 1991. Vegetables, fruit, and cancer: II. Mechanisms. *Cancer Causes, Control* 2:427–442.

Stellman, J.M., and Stellman, S.D. 1996. Cancer and the workplace. *CA Cancer J Clin* 46:70–92.

Stevens, L.C. 1967. Origin of testicular teratomas from primordial germ cells. *J Natl Cancer Inst* 38:549–552.

Stevens, L.C. 1970. The development of transplantable teratocarcinomas from intertesticular grafts of pre- and post-implantation mouse embryos. *Dev Biol* 21:364–382.

Stevens, L.C. 1973. A new inbred subline of mice 129/terSv with a high incidence of spontaneous testicular teratoma. *J Natl Cancer Inst* 50:235–242.

Stevens, L.C. 1981. Genetic influences on the development of gonadal tumors in mice with emphasis on teratomas. In *Neoplasms – Comparative Pathology of Growth in Animals, Plants, and Man*, H.E. Kaiser, ed., pp. 467–474. Baltimore: Williams and Wilkins.

Stoker, A.W., Hatier, C., and Bissell, M.J. 1990. The embryonic environment strongly attentuates v-*src* oncogenesis in mesenchymal and epithelial tissues but not in endothelia. *J Cell Biol* 111:217–228.

Stone, R.M. 1997. Bryostatin 1: Differentiating agent from the depths. *Leuk Res* 21:399–401.

Stracke, M.L., and Liotta, L.A. 1992. Multi-step cascade of tumor cell metastasis. *In Vivo* 6:309–316.

Strasser, A., Harris, A.W., Bath, M.L., and Cory, S. 1990. Novel primitive lymphoid tumours induced in transgenic mice by cooperation between *myc* and *bcl*-2. *Nature* 348:331–333.

Strauli, P., and Haemmerli, G. 1984. Cancer cell locomotion: Its occurrence during invasion. In *Invasion: Experimental and Clinical Implications*, M.M. Mareel and K.C. Calman, eds., pp. 252–274. Oxford: Oxford University Press.

Strickland, S., Reich, E., and Sherman, M.I. 1976. Plasminogen activator in early embryogenesis: Enzyme production by trophoblast and parietal endoderm. *Cell* 9:231–240.

Strickland, S., Smith, K.K., and Marotti, K.R. 1980. Hormonal induction of differentriation in teratocarcinoma stem cells. Generation of endoderm with retinoic acid and dibuteryl cAMP. *Cell* 21:347–355.

Strong, L.C., Stine, M., and Norsted, T.L. 1987. Cancer in survivors of childhood soft-tissue sarcoma and their relatives. *J Natl Cancer Inst* 79:1213–1220.

Studzinski, G.P., and Moore, D.C. 1995. Sunlight – Can it prevent as well as cause cancer? *Cancer Res* 55:4014–4022.

Sugerbaker, E.V. 1981. Patterns of metastasis in human malignancies. *Cancer Biol Rev* 2:235–278.

Sugio, K., Kishimoto, Y., Virmani, A.K., Hung, J.Y., and Gazdar, A.F. 1994. K-*ras* mutations are a relatively late event in the pathogenesis of lung carcinomas. *Cancer Res* 54:5811–5815.

Sutherland, B.M., Delihas, N.C., Oliver, R.P., and Sutherland, J.C. 1981. Action spectra for ultraviolet light-induced transformation of human cells to anchorage-independent growth. *Cancer Res* 41:2211–2214.

Sweeney, D.C., and Johnston, G.S. 1995. Radioiodine therapy for thyroid cancer. *Endocrinol Metabol Clin North Am* 24:803–839.

Swift, C.H. 1914. Origin and early history of the primordial germ cells in the chick. *Am J Anat* 15:483–516.

Symonds, G., and Sachs, L. 1982. Autoinduction of differentiation in myeloid leukemic cells: Restoration of normal coupling between growth and differentiation in leukemic cells that constitutively produce their own growth-inducing protein. *Eur Mol Biol Organ J* 1:1343–1346.

Symonds, H., Krall, L., Remington, L., Saenz-Robles, M., Lowe, S., Jacks, T., and Van Dyke, T. 1994. p53-dependent apoptosis suppresses tumor growth and progression in vivo. *Cell* 78:614–617.

Szekely, L., Salivanova, G., Magnusson, K.P., Klein, G., and Wiman, K.G. 1993. EBNA-5, an Epstein-Barr virus-encoded nuclear antigen, binds to the Rb and p53 proteins. *Proc Natl Acad Sci USA* 90:5455–5459.

Takahashi, S., Maecker, H.T., and Levy, R. 1989. DNA fragmentation and cell death mediated by T cell antigen receptor/CD3 complex on a leukemia T cell line. *Eur J Immunol* 19:1911–1919.

Takamiya, Y., Short, M.P., Ezzeddine, Z.D., Moolten, F.L., Breakefield, X.O., and Martuza, R.L. 1992. Gene therapy of malignant brain tumors: A rat glioma line bearing the herpes simplex virus type 1-thymidine kinase gene and wild type retrovirus kills other tumor cells. *J Neurosci Res* 33:493–503.

Takeichi, M. 1991. Cadherin cell adhesion receptors as a morphogenetic regulator. *Science* 251:1451–1455.

Tallman, M.S. 1994. All-trans-retinoic acid in acute promyelocytic leukemia and its potential in other hematologic malignancies. *Semin Hematol* 31 (Suppl 5):38–48.

Tanaka, K., Oshimura, M., Kikuchi, R., Seki, M., Hayashi, T., and Miyaki, M. 1991. Suppression of tumorigenicity in human colon carcinoma cells by introduction of normal chromosome 5 or 18. *Nature* 349:340–342.

Tannenbaum, A. 1942. The genesis and growth of tumors: III. Effects of a high fat diet. *Cancer Res* 2:468–475.

Tannenbaum, A., and Silverstone, H. 1953. Nutrition in relation to cancer. *Adv Cancer Res* 1:451–501.

Tarin, D. 1992. Tumour metastasis. In *Oxford Textbook of Pathology,* J.O'D. McGee, P.G. Isaacson, and N.A. Wright, eds., pp. 607–633. Oxford: Oxford University Press.

Tarin, D. 1996. Prognostic markers and mechanisms of metastasis. In *Recent Advances in Histopathology 17,* P.P. Anthony, R.N.M. MacSween, and D. Lowe, eds. Edinburgh: Churchill Livingstone, in press.

Tarin, D., Hoyt, B.J., and Evans, D.J. 1982. Correlation of collagenase secretion with metastatic-colonization potential in naturally occurring murine mammary tumors. *Br J Cancer* 46:266–278.

Tarin, D., and Price, J.E. 1979. Metastatic colonization potential of primary tumour cells in mice. *Br J Cancer* 39:740–754.

Tarin, D., Price, J.E., Kettlewell, M.G.W., Souter, R.G., Vass, A.C.R., and Crossley, B. 1984a. Mechanisms of human tumor metastasis studied in patients with peritoneovenous shunts. *Cancer Res* 44:3584–3589.

Tarin, D., Price, J.E., Kettlewell, M.G.W., Souter, R.G., Vass, A.C.R., and Crossley, B. 1984b. Clinicopathological observations on metastasis in man studied in patients with peritoneovenous shunts. *Br Med J* 288:749–751.

Tarkowski, A.K., and Wroblewska, J. 1967. Development of blastomeres of mouse eggs isolated at the 4–8 cell stage. *J Embryol Exp Morph* 18:155–180.

Tartour, E., Mathiot, C., and Fridman, W.H. 1992. Current status of interleukin-2 therapy in cancer. *Biomed Pharmacother* 46:473–484.

Tavassoli, M., and Hardy, C.L. 1990. Molecular basis of homing of intravenously transplanted stem cells to the marrow. *Blood* 76:1059–1070.

Tavassoli, M., and Yoffey, Y. 1983. *Bone Marrow: Structure and Function.* New York: Liss.

Tefre, T., Ryberg, D., Haugen, A., and Nebert, D.W. 1991. Human CYP1A1 cytochrome P(1)450 gene: Lack of association between the Msp I restriction fragment length polymorphism and incidence of lung cancer in a Norwegian population. *Pharmacogenetics* 1:20–25.

Terranova, V.P., and Maslow, D.E. 1991. Interactions of tumor cells with basement membrane. In

Microcirculation in Cancer Metastasis, F.W. Orr, M.R. Buchanan, and L. Weiss, eds., pp. 23–44. Boca Raton: CRC Press.

Thompson, H.J., Ronon, A.M., Ritacco, K.A., and Tagliaferro, A.R. 1989. Effect of type and amount of fat on the enhancement of rat mammary tumorigenesis by exercise. *Cancer Res* 49:1904–1908.

Thompson, J.A., Wiesner, G.L., Sellers, T.A., Vachon, C., Ahrens, M., Potter, J.D., Sumpmann, M., and Kersey, J. 1995. Genetic services for familial cancer patients: A survey of National Cancer Institute cancer centers. *J Natl Cancer Inst* 87:1446–1455.

Thompson, J.S., and Kostiala, A.A.I. 1990. Immunological and ultrastructural characterization of true histiocytic lymphoma in the northern pike, *Esox lucius* L. *Cancer Res* 50 Suppl 5668s–5670s.

Thompson, T.C., Southgate, J., Kitchener, G., and Land, H. 1990. Multistage carcinogenesis induced by *ras* and *myc* oncogenes in a reconstituted organ. *Cell* 56:917–930.

Thorgeirsson, S.S., Gant, T.W., and Silverman, J.A. 1994. Transcriptional regulation of multidrug resistance gene expression. *Cancer Treat Res* 73:57–68.

Thornton, J.R. 1981. High colonic pH promotes colorectal cancer. *Lancet* 1:1081–1083.

Thrush, G.R., Lark, L.R., Clinchy, B.C., and Vitetta, E.S. 1996. Immunotoxins: An update. *Ann Rev Immunol* 14:49–71.

Tienari, J., Alanko, T., Saksela, O., Vesterinen, M., and Lehtonen, E. 1995. Fibroblast growth factor-mediated stimulation of differentiating teratocarcinoma cells: Evidence for paracine growth regulation. *Differentiation* 59:193–199.

Tirmarche, M., Raphalen, A., Allin, F., Chameaud, J., and Bredon, P. 1993. Mortality of a cohort of French uranium miners exposed to relatively low radon concentrations. *Br J Cancer* 67:1090–1097.

Tjio, J.H., and Levan, A. 1956. The chromosome number of man. *Hereditas* 42:1–6.

Todaro, G.J., DeLarco, J.E., and Sporn, M.B. 1978. Retinoids block phenotypic cell transformation produced by sarcoma growth factor. *Nature* 276:272–274.

Tomatis, L. 1988. The contribution of the IARC monographs program to the identification of cancer risk factors. *Ann NY Acad Sci* 534:31–38.

Tonks, N.K., and Neel, B.G. 1996. From form to function: Signaling by protein tyrosine phosphatases. *Cell* 87:365–368.

Trayner, I.D., and Farzaneh, F. 1993. Retinoid receptors and acute promyelocytic leukaemia. *Eur J Cancer* 29A:2046–2054.

Treisman, R. 1990. The SRE: A growth factor responsive transcription regulator. *Semin Cancer Biol* 1:47–58.

Treisman, R. 1994. Ternary complex factors: Growth factor regulated transcriptional activators. *Curr Opin Genet Dev* 4:96–101.

Trichopoulos, D., Petridou, E., Lipworth, L., and Adami, H-O. 1997. Epidemiology of cancer. In *Cancer: Principles and Practice of Oncology*, 5th ed., V.T. DeVita, Jr., S. Hellman, and S.A. Rosenberg, eds., pp. 231–257. Philadelphia: Lippincott-Raven.

Trigg, M.E. 1993. Acute lymphoblastic leukemia in children. In *Cancer Medicine*, vol. 2, J.F. Holland, E. Frei III, R.C. Bast, Jr., D.W. Kufe, D.L. Morton, and R.R. Weichselbaum, eds., pp. 2153–2166. Philadelphia: Lea and Febiger.

Trock, B.J., Lanza, E., and Greenwald, P. 1990. High fiber diet and colon cancer: A critical review. *Prog Clin Biol Res* 346:145–157.

Trosko, J.E., Yotti, L.P., Warren, S.T., Tsushimoto, G., and Chang, C. 1982. Inhibition of cell-cell communication by tumor promoters. In *Carcinogenesis: Cocarcinogenesis and Biological Effects of*

Tumor Promoters, E. Hecker, N.E. Fusenig, W. Kunz, F. Marks, and H.W. Thielmann, eds., pp. 565–585. New York: Raven Press.

Tsang, K.R., and Brooks, M.A. 1981. Malignant transformation of insect cells *in vitro.* In *Phyletic Approaches to Cancer,* C.J. Dawe, J.C. Harshbarger, S. Kondo, T. Sugimura, and S. Takayama, eds., pp. 267–274. Tokyo: Japan Scientific Societies Press.

Tsurumi, H., Tojo, A., Takahashi, T., Ozawa, K., Moriwaki, H., Asano, S., and Muto, Y. 1993. Differentiation induction of human promyelocytic leukemia cells by acyclic retinoid polyprenoic acid. *Int J Hematol* 59:9–15.

Turpeenniemik-Hujanen, T., Thorgeirsson, U.P., Hart, I.R., Grant, S.S., and Liotta, L.A. 1985. Expression of collagenase IV basement membrane collagenase activity in murine tumor cell hybrids that differ in metastatic potential. *J Natl Cancer Inst* 75:99–103.

Turusov, V.S., and Mohr, U., eds. 1994. *Pathology of Tumours in Laboratory Animals,* II: *Tumours of the Mouse.* Lyon, France: Int. Agency Res. Cancer.

Tuyns, A.J. 1990. Alcohol and cancer. *Proc Nutr Soc* 49:145–151.

Tylecote, F.E. 1927. Cancer of the lung. *Lancet* 2:256–257.

Ueda, K., Cornwell, M.M., Gottesman, M.M., Pastan, I., Robinson, I.B., Ling, V., and Riordan, J.R. 1986. The *mdr1* gene, responsible for multidrug-resistance, codes for P-glycoprotein. *Biochem Biophys Res Commun* 141:956–962.

Upton, A.H. 1982. The biological effect of low-level ionizing radiation. *Sci Am* 246:41–49.

Urbach, F. 1993. Environmental risk factors for skin cancer. *Recent Res Cancer Res* 128:243–262.

Vadhan-Raj, S. 1996. Appropriate use of hematopoietic growth factors. In *Cancer Management: A Multidisciplinary Approach,* R. Pazdur, L. R. Coia, W. J. Hoskins, and L. D. Wagman, eds., pp. 619–628. Huntington, NY: PRR.

Vakaet, L. 1984. The initiation of gastrular ingression in the chick blastoderm. *Am Zool* 24:555–562.

Vakaet, L., VanRoelen, C., and Andries, L. 1980. An embryological model of non-malignant invasion or ingression. In *Cell Movement and Neoplasia,* M. DeBrabander, M. Mareel, and L. DeRider, eds., pp. 65–75. Oxford: Pergamon.

Van Beneden, R.J., Henderson, K.W., Blair, D.G., Papas, T.S., and Gardner, H.S. 1990. Oncogenes in hematopoietic and hepatic fish neoplasms. *Cancer Res* 50 Suppl:5671s–5674s.

van der Bliek, A.M., and Borst, P. 1989. Multidrug resistance. *Adv Cancer Res* 52:165–203.

van Steenbrugge, G.J., Groen, M., Romijn, J.C., and Schroder, F.H. 1984. Biological effects of hormonal treatment regimens on a transplantable human prostatic tumor line (PC-82). *J Urol* 131:812–817.

Vaughan, W.P., Karp, J.E., and Burke, P.J. 1984. Two-cycle-timed sequential chemotherapy for adult acute nonlymphocytic leukemia. *Blood* 64:975–980.

Vazquez, A., Auffredou, M.T., Chaouchi, N., Taieb, J., Sharma, S., Galanaud, P., and Leca, G. 1991. Differential inhibition of interleukin 2- and interleukin 4-mediated human B cell proliferation by ionomycin: A possible regulatory role for apoptosis. *Eur J Immunol* 21:2311–2316.

Vedantham, S., Gamliel, H., and Golomb, H.M. 1992. Mechanism of interferon action in hairy cell leukemia: A model of effective cancer biotherapy. *Cancer Res* 52:1056–1066.

Vena, J.E., Graham, S., Zielezny, M., Brasure, J., and Swanson, M.K. 1987. Occupational exercise and risk of cancer. *Am J Clin Nutr* 45:318–327.

Verma, R.S. 1990. *The Genome.* New York: VCH.

Vermeulen, S.J., Bruyneel, E.A., Bracke, M.E., De Bruyne, G.K., Vennekens, K.M., Vleminckx, K.L., Berx, G.J., van Roy, F.M., and Mareel, M.M. 1995. Transition from the noninvasive to the invasive phenotype and loss of α-catenin in human colon cancer cells. *Cancer Res* 55:4722–4728.

Viadana, E., Bross, I.D.J., and Pickren, J.W. 1978a. Cascade spread of blood-borne metastases in solid and nonsolid cancers of humans. In *Pulmonary Metastasis,* L. Weiss and H.A. Gilbert, eds., pp. 142–167. Boston: G.K. Hall.

Viadana, E., Bross, I.D.J., and Pickren, J.W. 1978b. The metastatic spread of cancers of the digestive system in man. *Oncology* 35:114–126.

Virchow, R. 1863. Über bewegliche thierische Zellen. *Arch Pathol Anat* 28:237–240.

Vitaliano, P.P. 1978. The use of logistic regression for modeling risk factors: With application to non-melanoma skin cancer. *Am J Epidemiol* 108:402–414.

Vogel, C.L. 1996. Hormonal approaches to breast cancer treatment and prevention: An overview. *Semin Oncol* 23 (4 Suppl 9):2–9.

Vogel, V.G., and McPherson, R.S. 1989. Dietary epidemiology of colon cancer. *Hematol Oncol Clin North Am* 3:35–63.

Vogelbein, W.K., Fournie, J.W., Van Veld, P.A., and Huggett, R.J. 1990. Hepatic neoplasms in the mummichog *Fundulus heteroclitus* from a creosote-contaminated site. *Cancer Res* 50:5978–5986.

Vogelstein, B., Fearon, E.R., Hamilton, S.R., and Kern, S.E. 1988. Genetic alteration during colorectal-tumor development. *N Engl J Med* 319:525–532.

Vogelstein, B., and Kinzler, K.W. 1993. The multistep nature of cancer. *Trends Genet* 9:138–142.

von Rohr, A., and Thatcher, N. 1992. Clinical applications of interleukin-2. *Prog Growth Factor Res* 4:229–246.

Waga, S., Hannon, G.J., Beach, D., and Stillman, B. 1994. The p21 inhibitor of cyclin-dependent kinases controls DNA replication by interaction with PCNA. *Nature* 369:574–578.

Wagner, J.C., Newhouse, M.L., Corrin, B., Rossiter, C.E.R., and Griffiths, D.M. 1988. Correlation between fibre content of the lung and disease in east London asbestos factory workers. *Br J Ind Med* 45:305–308.

Wagner, J.C., Sleggs, C.A., and Marchand, P. 1960. Diffuse pleural mesothelioma and asbestos exposure in the North Western Cape Province. *Br J Ind Med* 17:260–271.

Waldron, H.A., Waterhouse, J.A., and Tessema, N. 1984. Scrotal cancer in the West Midlands 1936–1976. *Br J Ind Med* 41:473–474.

Wallet, V., Mutzel, R., Troll, H., Barzu, O., Wurster, B., Veron, M., and Lacombe, M. 1990. Dictyostelium nucleoside diphosphate kinase highly homologous to *nm23* and awd proteins involved in mammalian tumor metastasis and Drosophila development. *J Natl Cancer Inst* 82:1199–1202.

Walsh, P.C., Lepor, H., and Eggleston, J.C. 1983. Radical prostatectomy with preservation of sexual function: Anatomical and pathological considerations. *Prostate* 4:473–485.

Walsh, P.C., Quinlan, D.M., Morton, R.A., and Steiner, R.A. 1990. Radical retropubic prostatectomy. Improved anastomosis and urinary continence. *Urol Clin North Am* 17:679–684.

Walter, J.B. 1982. *An Introduction to the Principles of Disease,* p. 245. Philadelphia: W.B. Saunders.

Wang, J.Y.J. 1994. Nuclear protein tyrosine kinases. *Trends Biochem Sci* 19:373–376.

Wang, L., Patel, U., Lagnajita, G., and Banerjee, S. 1992. DNA polymerase b mutations in human colorectal cancer. *Cancer Res* 52:4824–4827.

Wang, R.F., and Rosenberg, S.A. 1996. Human tumor antigens recognized by T lymphocytes: Implications for cancer therapy. *J Leukocyte Biol* 60:296–309.

Wang, Y., Dang, J. Xiaoming, L., and Doe, W. 1995. Amiloride modulated urokinase gene expression at both transcription and post-transcription levels in human colon cancer cells. *Clin Exp Metastasis* 13:196–202.

Wani, M.C., Taylor, H.L., and Wall, M.E. 1971. Plant antitumor agents: VI. The isolation and structure of Taxol, a novel antileukemic and antitumor agent from *Taxus brevifolia. J Am Chem Soc* 93:2325–2327.

Ward, M., Richardson, C., Pioli, P., Smith, L., Podda, S., Goff, S., Hesdorffer, C., and Bank, A. 1994. Transfer and expression of the human multiple drug resistance gene in human CD34+ cells. *Blood* 84:1408–1414.

Warne, P.H., Viciana, P.R., and Downward, J. 1993. Direct interaction of Ras and the amino-terminal region of Raf-1 *in vitro. Nature* 364:353–355.

Warren, J.R., Scarpelli, D.G., Reddy, J.K., and Kanwar, Y.S. 1987. *Essentials of General Pathology.* New York: Macmillan.

Wasson, T. 1987. *Nobel Prize Winners,* pp. 486–488. New York: H.W. Wilson.

Watson, R.R. 1984. Regulation of immunological resistance to cancer by beta-carotene and retinoids. In *Nutrition, Disease Resistance, and Immune Function,* R.R. Watson, ed., pp. 345–355. New York: Marcel Decker.

Watson, R.R., and Mufti, S.I., eds. 1996. *Nutrition and Cancer Prevention.* Boca Raton: CRC Press.

Wattenberg, L.W. 1985. Chemoprevention of cancer. *Cancer Res* 45:1–8.

Wattenberg, L.W. 1990. Inhibition of carcinogenesis by naturally occurring and synthetic compounds. *Basic Life Sci* 52:155–166.

Wattenberg, L.W. 1992. Inhibition of carcinogenesis by minor dietary constituents. *Cancer Res* Suppl 52:2085s–2091s.

Wattenberg, L.W., and Bueding, E. 1986. Inhibitory effects of 5-(2-pyrazinyl)-4-methyl-1,2-dithiol-3-thione (Oltipraz) on carcinogenesis induced by benz(a)pyrene, diethylnitrosamine and uracil mustard. *Carcinogenesis* 7:1379–1381.

Wattenberg, L.W., and Coccia, J.B. 1991. Inhibition of 4-(methylnitrosamino)-1-(3-pyridyl)-1-butanone carcinogenesis in mice by D-limonene and citrus fruit oils. *Carcinogenesis* 12:115–117.

Wattenberg, L.W., Sparnins, V.L., and Barany, G. 1989. Inhibition of N-nitrosodiethylamine carcinogenesis by naturally occurring organosulfur compounds and monoterpenes. *Cancer Res* 49:2689–2692.

Weichselbaum, R.R., Hallahan, D.E., and Chen, G.T.Y. 1993. Biological and physical basis to radiation oncology. In *Cancer Medicine,* vol. 2, J.F. Holland, E. Frei III, R.C. Bast, Jr., D.W. Kufe, D.L. Morton, and R.R. Weichselbaum, eds., pp. 539–566. Philadelphia: Lea and Febiger.

Weinberg, R.A. 1995. The retinoblastoma protein and cell cycle control. *Cell* 81:323–330.

Weinhouse, S. 1986. The role of diet and nutrition in cancer. *Cancer* 58 (Suppl 8):1791–1794.

Weintraub, S.J., Prater, C.A., and Dean, D.C. 1992. Retinoblastoma protein switches the E2F site from positive to negative element. *Nature* 358:259–261.

Weisburger, J.H., and Wynder, E.L. 1987. Etiology of colorectal cancer with emphasis on mechanisms of action and prevention. *Important Adv Oncol* 1987:197–221.

Weiss, L. 1985. *Principles of Metastasis.* Orlando: Academic Press.

Welch, D.R., Bisi, J.E., Miller, B.E., Conaway, D., Jeftor, E.A., Yohem, K.H., Gilmore, L.B., Seftor, R.E., Nakajima, M., and Hendrix, M.J. 1991. Characterization of a highly invasive and spontaneously metastatic human malignant melanoma cell line. *Int J Cancer* 47:227–237.

Wendling, F., Maraskovsky, E., Debili, N., Florindo, C., Teepe, M., Titeux, M., Methia, N., Breton-Gorius, J., Cosman, D., and Vainchenker, W. 1994. cMpl ligand is a humoral regulator of megakaryocytopoiesis. *Nature* 369:571–574.

Weng, Z., Taylor, J.A., Turner, C.E., Brugge, J.S., and Seidel-Dugan, C. 1993. Detection of Src homology 3-binding proteins, including paxillin, in normal and v-src-transformed Balb/c 3T3 cells. *J Biol Chem* 268:14956–14963.

Werness, B.A., Munger, K., and Howley, P.M. 1991. The role of the human papillomavirus oncoproteins in transformation and carcinogenic progression. In *Important Advances in Oncology,* V.T. DeVita, Jr., S. Hellman, and S.A. Rosenberg, eds., pp. 2–18. Philadelphia: J.B. Lippincott.

Whang-Peng, J., Lee, E.C., and Knutsen, T.A. 1974. Genesis of the Ph chromosome. *J Natl Cancer Inst* 52:1035–1036.

White, E. 1996. Life, death and the pursuit of apoptosis. *Genes Dev* 10:1–15.

White, P.R. 1965. Abnormal corallites. *Science* 150:677–678.

White, P.R., and Braun, A.C. 1942. Cancerous neoplasm of plants. Autonomous bacteria-free crown gall tissue. *Cancer Res* 2:597–617.

Whitehead, V.M., Rosenblatt, D.S., Vuchich, M.J., and Beaulieu, D. 1987. Methotrexate polygluta-mate synthesis in lymphoblasts from children with acute lymphoblastic leukemia. *Dev Pharmacol Ther* 10:443–448.

Whiteside, T.L., Jost, L.M., and Herberman, R.B. 1992. Tumor-infiltrating lymphocytes. Potential and limitations to their use for cancer therapy. *Crit Rev Oncol Hematol* 12:25–47.

Whittlake, E.B. 1981. Fossil plant galls. In *Neoplasms – Comparative Pathology of Growth in Animals, Plants, and Man,* H.E. Kaiser, ed., pp. 729–731. Baltimore: Williams and Wilkins.

Whyte, P., Williamson, N.M., and Harlow, E. 1989. Cellular targets for transformation by the ade-novirus E1A proteins. *Cell* 56:67–75.

Wieland, H., and Dane, E. 1933. Untersuchungen über die Konstitution der Gallensauren. LII. Mitteilung über die Haftstelle der Seitenkette. *Hoppe Seylers Z Physiol Chem* 219:240–245.

Wilbourn, J., Haroun, L., Heseltine, E., Kaldor, J., Partensky, C., and Vainio, H. 1986. Response of experimental animals to human carcinogens: An analysis based upon the IARC monographs pro-gramme. *Carcinogenesis* 7:1853–1863.

Wilcox, W.S. 1966. The last surviving cancer cell: The chances of killing it. *Cancer Chemother Rep* 50:541–542.

Wilcox, W.S., Griswold, D.P., Laster, W.R., Jr., Schabel, F.M., Jr., and Skipper, H.E. 1965. Experi-mental evaluation of potential anticancer agents: XVII. Kinetics of growth and regression after treatment of certain solid tumors. *Cancer Chemother Rep* 47:27–39.

Wilder, R.J. 1956. Historical development of concept of metastasis. *J Mt Sinai Hosp (NY)* 23:728–734.

Wilhelmsson, A., Cuthill, S., Denis, M., Wikström, A.-C., Gustaffson, J.-A., and Poellinger, L. 1990. *Eur Mol Biol Organ J* 9:69–76.

Willett, W.C., Stampfer, M.J., Colditz, G.A., Rosner, B.A., Hennekens, C.H., and Speizer, F.E. 1987. Dietary fat and the risk of breast cancer. *N Engl J Med* 316:22–28.

Williams, B.O., Remington, L., Albert, D.M., Mukai, S., Bronson, R.T., and Jacks, T. 1994. Coop-erative tumorigenic effects of germline mutations in Rb and p53. *Nat Genet* 7:480–484.

Williams, J.W. III, Carlson, D.L., Gadson, R.G., Rollins-Smith, L., Williams, C.S., and McKinnell, R.G. 1993. Cytogenetic analysis of triploid renal carcinoma in *Rana pipiens. Cytogenet Cell Genet* 64:18–22.

Williams, J.W. III, Tweedell, K.S., Sterling, D., Marshall, N., Christ, G.C., Carlson, D.L., and McKinnell, R.G. 1996. Oncogenic herpesvirus DNA absence in kidney cell lines established from the northern leopard frog, *Rana pipiens. Dis Aquat Organ* 27:1–4.

Willis, R.A. 1967. *Pathology of Tumors,* 4th ed. London: Butterworth.

Willis, R.A. 1973. *The Spread of Tumours in the Human Body,* 3rd ed. London: Butterworth.

Wingo, P.A., Landis, S., and Ries, L.A.G. 1997. An adjustment to the 1997 estimate for new prostate cancer cases. *CA Cancer J Clin* 47:239–242.

Winn, D.M. 1988. Smokeless tobacco and cancer: The epidemiological evidence. *CA Cancer J Clin* 33:236–243.

Winton, D.J., and Ponder, B.A. 1990. Stem-cell organization in mouse small intestine. *Proc R Soc Lond,* Part B 241:13–18.

Witkop, C.J., Quevedo, W.C., Fitzpatrick, T.B., and King, R.A. 1989. Albinism. In *The Metabolic Basis of Inherited Disease,* 6th ed., R. Scriver, A.L. Beaudef, W.S. Sly, and D. Valle, eds., pp. 2905–2949. New York: McGraw-Hill.

Witschi, E. 1948. Migrations of germ cells of human embryos from the yolk sac to the primitive gonadal folds. *Contr Embryol 209, Carnegie* 32:67–80.

Wittes, R.E. 1986. Adjuvant chemotherapy – Clinical trials and laboratory models. *Cancer Treat Rep* 70:87–103.

Wogan, G.N. 1986. Diet and nutrition as risk factors for cancer. *Int Symp Princess Takamatsu Cancer Res Fund* 16:3–10.

Wojtowicz-Praga, S.M., Dickson, R.B., and Hawkins, M.J. 1997. Matrix metalloproteinase inhibitors. *Invest New Drugs* 15:61–75.

Won, K.A., and Reed, S.I. 1996. Activation of cyclin E/CDK2 is coupled to site-specific autophosphorylation and ubiquitin-dependent degradation of cyclin E. *Eur Mol Biol Organ J* 15:4182–4193.

Wood, S. 1958. Pathogenesis of metastasis formation observed in vivo in the rabbit ear chamber. *Arch Pathol* 66:550–568.

Wood, S., Baker, R.R., and Marzocchi, B. 1968. In vivo studies of tumor behavior: Locomotion of and interrelationships between normal cells and cancer cells. In *The Proliferation and Spread of Neoplastic Cells,* pp. 495–509. Baltimore: Williams and Wilkins.

Wooster, R., Neuhausen, S.L., Mangion, J., Quirk, Y., Ford, D., Collins, N., Nguyen, K., Seal, S., Tran, T., Averill, D., Fields, P., Marshall, G., Narod, S., Lenoir, G.M., Lynch, H., Feunteun, J., Devilee, P., Cornelisse, C.J., Menko, F.H., Daly, P.A., Ormiston, W., McManus, R., Pye, C., Lewis, C.M., Cannon-Albright, L. S., Peto, J., Ponder, B.A.J., Skolnick, M.H., Easton, D.F., Goldgar, D.E., and Stratton, M.R. 1994. Localization of a breast cancer susceptibility gene, *BRCA2,* to chromosome 13q12–13. *Science* 265:2088–2090.

Wright, N., Watson, A., Morley, A., Appleton, D., Marks, J., and Douglas, A. 1973. The cell cycle time in the flat (avillous) mucosa of the human small intestine. *Gut* 14:603–606.

Wu, C.C., Fang, T.H., Yang, M.D., Wu, T.C., and Liu, T.J. 1992. Gastrectomy for advanced gastric carcinoma with invasion to the serosa. *Int Surg* 77:144–148.

Wu, J., Dent, P., Jelinek, T., Wolfman, A., Weber, M.J., and Sturgill, T.W. 1993. Inhibition of the EGF-activated MAP kinase signaling pathway by adenosine 3′,5′-monophosphate. *Science* 262:1065–1069.

Wu, X., Wang, X., Qien, X., Liu, H., Ying, J., Yang, Z., and Yao, H. 1993. Four years' experience with the treatment of all-trans retinoic acid in acute promyelocytic leukemia. *Am J Hematol* 43:183–189.

Wylie, C.V., Nakane, P.K., and Pierce, G.B. 1973. Degrees of differentiation in nonproliferating cells of mammary carcinoma. *Differentiation* 1:11–20.

Wyllie, A.H. 1995. The genetic regulation of apoptosis. *Curr Opin Genet Dev* 5:97–104.

Wyllie, A.H., Kerr, J.F., and Currie, A.R. 1980. Cell death: The significance of apoptosis. *Int Rev Cytol* 68:251–306.

Wyllie, A.H., and Morris, R.G. 1982. Hormone-induced cell death. Purification and properties of thymocytes undergoing apoptosis after glucocorticoid treatment. *Am J Pathol* 109:78–87.

Wynder, E.L. 1969. Identification of women at high risk for breast cancer. *Cancer* 24:1235–1240.

Wynder, E.L., and Graham, E.A. 1950. Tobacco smoking as a possible etiological factor in bronchiogenic carcinoma. A study of six hundred and eighty-four proved cases. *J Am Med Assoc* 143:329–336.

Xiong, Y., Hannon, G.J., Zhang, H., Casso, D., Kobayashi, R., and Beach, D. 1993. p21 is a universal inhibitor of cyclin kinases. *Nature* 366:701–704.

Yagel, S., Kerbel, R., Lala, P., Eldar-Gera T., and Dennis, J.W. 1990. Basement membrane invasion by first trimester human trophoblast: Requirement for branched complex-type Asn-linked. *Clin Exp Metastasis* 8:305–317.

Yamada, T., and McDevitt, D.S. 1974. Direct evidence for transformation of differentiated iris epithelial cells into lens cells. *Dev Biol* 38:104–118.

Yamagawa, K., and Ichikawa, K. 1918. Experimental study of the pathogenesis of carcinoma. *J Cancer Res* 3:1–29.

Yamasaki, E., and Ames, B.N. 1977. Concentration of mutagens from urine by adsorption with the nonpolar resin XAD-2: Cigarette smokers have mutagenic urine. *Proc Natl Acad Sci USA* 74:3555–3559.

Yandell, D.W., Campbell, T.A., Dayton, S.H., Petersen, R., Walton, D., Little, J.B., McConkie-Rosell, A., Buckley, E.G., and Dryja, T. P. 1989. Oncogenic point mutations in the human retinoblastoma gene: Their application to genetic counseling. *N Engl J Med* 321:1689–1695.

Yarden, Y., and Schlessinger, J. 1987. Epidermal growth factor induces rapid, reversible aggregation of the purified epidermal growth factor receptor. *Biochemistry* 26:1443–1451.

Yasumura, S., Lin, W-C., Weidmann, E., Hebda, P., and Whiteside, T.L. 1994. Expression of interleukin 2 receptors on human carcinoma cell lines and tumor growth inhibition by interleukin 2. *Int J Cancer* 59:225–234.

Yoshino, T., Kondo, E., Cao, L., Takahashi, K., Hayashi, K., Nomura, S., and Akagi, T. 1994. Inverse expression of bcl-2 protein and Fas antigen in lymphoblasts in peripheral lymph nodes and activated peripheral blood T and B lymphocytes. *Blood* 83:1856–1861.

Young, R.C. 1994. Management of early ovarian epithelial cancer. *NIH Consensus Development Conference on Ovarian Cancer: Screening, Treatment, and Followup.* April 5–7, Bethesda, MD: NIH.

Yunis, J.J. 1976. High resolution of human chromosomes. *Science* 191:1268–1270.

Yuspa, S.H. 1994. The pathogenesis of squamous cell cancer: Lessons learned from studies of skin carcinogenesis. *Cancer Res* 54:1178–1189.

Zacharski, L.R. 1984. The coagulation hypothesis of cancer dissemination. In *Treatment of Metastasis: Problems and Prospects,* K. Hellmann and S.A. Eccles, eds., pp. 77–80. London: Taylor and Francis.

Zacharski, L.R., Henderson, W.G., Rickles, F.R., Forman, W.B., Cornell, C.J., Forcier, R.J., Edwards, R., Headley, E., Kim, S.H., O'Donnell, J.R., O'Dell, R., Tornyos, K., and Kwann, H.C. 1981. Effect of warfarin on survival in small cell carcinoma of the lung. *J Am Med Assoc* 245:831–835.

Zarbl, H., Sukumar, S., Arthur, A.V., Martin-Zanca, D., and Barbacid, M. 1985. Direct mutagenesis of Ha-*ras*-1 oncogenes by N-nitroso-N-methylurea during initiation of mammary carcinogenesis in rats. *Nature* 315:382–385.

Zavanella, T. 1974. Il melanoma del tritone crestato: stato attuale delle ricerche. *Atti Acad Naz Lincei,* Ser. 8, 56:1031–1042.

Zech, L., Haglund, U., Nilsson, K., and Klein, G. 1976. Characteristic chromosomal abnormalities in biopsies and lymphoid-cell lines from patients with Burkitt and non-Burkitt lymphomas. *Int J Cancer* 17:47–56.

Zetter, B.R. 1990. Cell motility in angiogenesis and tumor metastasis. *Cancer Invest* 8:669–671.

Zhang, X., Settleman, J., Kyriakis, J.M., Takeuchi-Suzuki, E., Elledge, S.J., Marshall, M.S., Bruder, J.T., Rapp, U.R., and Avruch, J. 1993. Normal and oncogenic p21^{ras} proteins bind to the amino-terminal regulatory domain of c-Raf-1. *Nature* 364:308–313.

Zimmerman, K.A., Yancopoulos, G.D., Collum, R.G., Smith, R.K., Kohl, N.E., Denis, K.A., Nau, M.M., Witte, O.N., Toran-Allerand, D., Gee, C.E., Minna, J.D., and Alt, F.W. 1986. Differential expression of *myc* family genes during murine development. *Nature* 319:780–783.

zur Hausen, H. 1987. The role of papillomaviruses in human anogenital cancer. In *The Papovaviri-dae*, vol. 2, N.P. Salzman and P.M. Howley, eds., pp. 245–263. New York: Plenum Press.

Zwart, P., and Harshbarger, J.C. 1991. A contribution to tumors in reptiles. Description of new cases. In *4. Internationales Colloquium für Pathologie und Therapie der Reptilien und Amphibien*, K. Gabrisch, B. Schildger, and P. Zwart, eds., pp. 219–224. Bad Nauheim: Deutsche Veter-inärmedizinische Gesellschaft e.V.

Zwickey, R.E., and Davis, K.J. 1959. Carcinogenicity screening. In *Appraisal of the Safety of Chemi-cals in Foods, Drugs and Cosmetics*. Baltimore, MD: Association of Food and Drug Officials of the United States.

Index